Scheveningen Conference on Differential Equations, 1975.
New developments in differential equations : proceedings of the second Scheveningen Conference on Differential Equations, the Netherlands, August 25-29, 1975 / editor, Wiktor Eckhaus. Amsterdam : North Holland Pub. Co. ; New York : sole distributors for the U.S.A. and Canada, American Elsevier Pub. Co., c1976.
viii, 248 p. ; 24 cm. (North-Holland mathematics studies ; 21)

NORTH-HOLLAND
MATHEMATICS STUDIES 21

New Developments in Differential Equations

Proceedings of the second Scheveningen conference on differential equations, The Netherlands, August 25-29, 1975

Editor

WIKTOR ECKHAUS

University of Utrecht

1976

NORTH-HOLLAND PUBLISHING COMPANY
AMSTERDAM – NEW YORK – OXFORD

North-Holland ISBN: 0 7204 0466 5
American Elsevier ISBN: 0 444 11107 7

PUBLISHERS:
NORTH-HOLLAND PUBLISHING COMPANY
AMSTERDAM, NEW YORK, OXFORD

SOLE DISTRIBUTORS FOR THE U.S.A. AND CANADA:
AMERICAN ELSEVIER PUBLISHING COMPANY, INC.
52 VANDERBILT AVENUE, NEW YORK, N.Y. 10017

Library of Congress Cataloging in Publication Data

Scheveningen Conference on Differential Equations, 1975.
New developments in differential equations.

(North-Holland mathematics studies ; 21)
1. Differential equations--Congresses. 2. Nonlinear theories--Congresses. I. Eckhaus, Wiktor. II. Title.
QA371.S38 1975 515'.35 76-9754
ISBN 0-444-11107-7 (American Elsevier)

PRINTED IN THE NETHERLANDS

P R E F A C E

The field of differential equations is an ever flourishing branch of matnematics, attracting research workers who, in the traditional terminology, range from very "pure" to very "applied". New problems continue to arise from applications or just from the mathematicians' curiosity, old problems give rise to new developments through application of new methods of analysis.

In 1973 a group of Dutch mathematicians consisting of B.L.J.Braaksma, W.Eckhaus, E.M.de Jager and H.Lemei, organized a conference on differential equations with the aim of promoting contacts and stimulating the exchange of ideas among a purposely limited, relatively small, number of invited participants. The success of that meeting, the favourable reactions afterwards and the favourable reception of the proceedings (published as North Holland Mathematics Studies Vol.13), convinced the organizing committee of the usefulness of such conferences.

This volume is an account of the lectures delivered at the Second Scheveningen Conference on Differential Equations, held on Aug. 25-29, 1975. The organization was in the hands of the same committee and the conference was again made possible through the generous financial support of the Minister of Education and Sciences of the Netherlands. There were 51 participants from 9 countries. (A list of participants can be found on page 249 of these proceedings.)

The emphasis in this second conference was on nonlinear analysis. This is reflected by the fact that approximately half of the volume is devoted to nonlinear problems. However, linear problems in differential equations are still challenging and subject to new developments, as witnessed by the other half of the contributions.

It is a pleasure to acknowledge the gratitude to all authors who have contributed such stimulating accounts of their research. Particular thanks are due to Professor J.L.Lions, who has accepted the invitation to act as principal speaker and delivered a series of four lectures. His contribution opens this volume.

Wiktor Eckhaus, Editor
Utrecht. February 1976.

CONTENTS

LIST OF PARTICIPANTS

Invited speakers

H.Amann	Ruhr-Universität, Bochum, Germany
H.Brezis	Université P.et M.Curie, Paris, France
W.N.Everitt	The University, Dundee, Scotland
W.A.Harris Jr.	University of Southern California, USA
G.Iooss	Université de Nice, France
J.L.Lions	Collège de France, Paris, France
D.A.Lutz	University of Wisconsin, USA
H.-D.Niessen	University of Essen, Germany
A.Pleijel	Uppsala University, Sweden
A.Schneider	Gesamthochschule Wuppertal, Germany
G.Stampacchia	Scuole Normale Superiore, Pisa, Italy

Participants from the Netherlands

B.L.J.Braaksma	Rijksuniversiteit Groningen
T.M.T.Coolen	Mathematisch Centrum Amsterdam
C.Cuvelier	Technische Hogeschool Delft
B.R.Damsté	Landbouwhogeschool Wageningen
O.Diekmann	Mathematisch Centrum Amsterdam
A.Dijksma	Rijksuniversiteit Groningen
W.Eckhaus	Rijksuniversiteit Utrecht
J.A.v.Gelderen	Technische Hogeschool Delft
B.Gilding	Technische Hogeschool Delft
J.de Graaf	Rijksuniversiteit Groningen
J.Grasman	Mathematisch Centrum Amsterdam
P.P.N.de Groen	Vrije Universiteit Amsterdam
P.Habets	Institut Mathématique, Louvain-le-Neuve, Belgium
A.v.Harten	Rijksuniversiteit Utrecht
M.H.Hendriks	Landbouwhogeschool Wageningen
A.J.Hermans	Technische Hogeschool Delft
H.W.Hoogstraten	Rijksuniversiteit Groningen
F.J.Jacobs	Technische Hogeschool Eindhoven
E.M.de Jager	Universiteit van Amsterdam
M.Jansen	Vrije Universiteit Amsterdam
B.Kaper	Rijksuniversiteit Groningen
E.W.M.Koper	Universiteit van Amsterdam
H.A.Lauwerier	Universiteit van Amsterdam
C.G.Lekkerkerker	Universiteit van Amsterdam
H.Lemei	Technische Hogeschool Delft
R.Martini	Technische Hogeschool Delft
H.G.Meijer	Technische Hogeschool Delft
G.Y.Nieuwland	Vrije Universiteit Amsterdam
L.A.Peletier	Technische Hogeschool Delft
H.Pijls	Universiteit van Amsterdam
J.W.Reijn	Technische Hogeschool Delft
J.W.de Roever	Mathematisch Centrum Amsterdam
J.D.Siersma	Rijksuniversiteit Groningen
M.Sluijter	Mathematisch Centrum Amsterdam
I.G.Sprinkhuizen-Kuyper	Mathematisch Centrum Amsterdam
M.N.Spijker	Rijksuniversiteit Leiden
J.Sijbrand	Rijksuniversiteit Utrecht
N.M.Temme	Mathematisch Centrum Amsterdam
J.v.Tiel	Rijksuniversiteit Utrecht
P.J.Zandbergen	Technische Hogeschool Twente

W. Eckhaus (ed.), New Developments in Differential Equations

SOME TOPICS ON VARIATIONAL INEQUALITIES AND APPLICATIONS

J.L. LIONS
Collège de France, Paris
and IRIA-LABORIA, Le Chesnay, France.

INTRODUCTION.

We present here some results on Variational Inequalities of stationary (elliptic) type (Chapters I and II) and of evolution type (Chapter III).

Each Chapter gives a direct approach to some of the results (without any attempt to be exhaustive) and it also presents some open problems.

The plan is as follows :

(1) A programme somewhat similar to what is done here could be completed by replacing "optimal stopping time" by "optimal impulse control" in the sense of [1]: this would lead to Quasi-Variational Inequalities instead of V.I.. We refer to [1] [2] ; we do not study these questions here.

[1] A. Bensoussan and J.L. Lions, C.R.A.S., 276 (1973), pp. 1189-1192, pp. 1333-1338 ; 278(1974), pp. 675-679, pp. 747-751 ; 280 (1975), pp. 1049-1053.

A. Bensoussan, M. Goursat and J.L. LIONS. C.R.A.S. 276 (1973), pp. 1279-1284.

[2] A. Bensoussan and J.L. Lions. Book to appear. Hermann ed.

I. INTRODUCTION TO SOME VARIATIONAL INEQUALITIES

ORIENTATION

We give an elementary introduction to some of the methods of V.I. We prove the existence of a solution by the use of penalty methods. We give some error estimates which lead to a general and apparently open problem (cf. (3.13)). We present in Section 4 some remarks related to the question of the variation of the solution of a V.I. with respect to the geometrical domain.

1. AN EXAMPLE OF A V.I.

Let Ω be a bounded open set in $\mathbb{R}^n$ with a smooth boundary Γ. In Ω we consider the elliptic operator A defined by

(1.1) $$Av = -\sum \frac{\partial}{\partial x_i}\left(a_{ij}(x)\frac{\partial v}{\partial x_j}\right) + \sum a_i(x)\frac{\partial v}{\partial x_i} + a_o(x)v$$

where

(1.2) $$a_{ij},\ a_i,\ a_o \in L^\infty(\Omega).$$

We shall set :

$H^1(\Omega)$ = Sobolev space of order 1 = functions v such that[1] $v, \frac{\partial v}{\partial x_i} \in L^2(\Omega)$, provided with the usual hilbertian structure,

$\|v\|$ = norm in $H^1(\Omega)$:

$|v|$ = norm in $L^2(\Omega)$, (f,v) = scalar product in $L^2(\Omega)$,

and for $u, v \in H^1(\Omega)$ we define :

(1.3) $$a(u,v) = \sum\int_\Omega a_{ij}(x)\frac{\partial u}{\partial x_j}\frac{\partial v}{\partial x_i}\,dx + \sum\int_\Omega a_i(x)\frac{\partial u}{\partial x_i}v\,dx + \int_\Omega a_o u v\,dx.$$

We define

(1.4) $H_o^1(\Omega)$ = closed subspace of $H^1(\Omega)$ of functions v which are zero on Γ.

We assume that (ellipticity hypothesis)

(1.5) $$a(v,v) \geq \alpha\|v\|^2,\quad \alpha > 0,\quad \forall v \in H_o^1(\Omega).$$ ■

The problem we want to consider first consists in finding u such that

(1.6) $$\begin{cases} u \in H_o^1(\Omega), \\ Au - f \leq 0,\ u \leq 0,\ (Au-f)u = 0 \text{ in } \Omega \end{cases}$$

where f is given in $L^2(\Omega)$.

(1) All functions are supposed to be real valued.

It is a simple matter, assuming that u is smooth enough (for instance that $u \in H^2(\Omega)$[1], to verify that (1.6) is equivalent to

(1.7)
$$\left| \begin{array}{l} a(u,v-u) \geq (f,v-u) \quad \forall\, v \in H_0^1(\Omega), \quad v \leq 0 , \\ u \in H_0^1(\Omega) \;,\; u \leq 0 \text{ in } \Omega. \end{array} \right.$$

If we introduce

(1.8)
$$K = \{v \mid v \in H_0^1(\Omega) \;,\quad v \leq 0 \quad \text{a.e. in } \Omega \}$$

then (1.7) is equivalent to

(1.9)
$$\left| \begin{array}{l} a(u,v-u) \geq (f,v-u) \quad \forall\, v \in K , \\ u \in K. \end{array} \right.$$

The set of inequalities (1.7) (or (1.9)) is what is called a V.I.

It was proved in [7] that <u>under the hypothesis</u> (1.5) <u>there exists a unique solution of</u> (1.9).

In Section 2 below, we shall give a proof of this result using <u>a penalty argument</u>.

Remark 1.1.

The uniqueness of the solution (if it exists) is obvious ; indeed if u, $\hat{u}$, are two solutions, and if we choose $v = \hat{u}$ (resp. u) in the V.I. satisfied by u (resp. $\hat{u}$), and if we add up, we find that :

$$a(u - \hat{u},\, u - \hat{u}) \leq 0$$

hence $u - \hat{u} = 0$ by virtue of (1.5). ∎

Remark 1.2.

The solution of (1.9) defines <u>a weak solution of</u> (1.6).

The next problem is to study <u>the regularity</u> of the solution of (1.9). The situation here is entirely different from the case of, say, the usual Dirichlet problem : even with coefficients in $C^\infty(\bar{\Omega})$ and with $f \in C^\infty(\bar{\Omega})$, u <u>does not belong to</u> $C^\infty(\bar{\Omega})$. ∎

Remark 1.3.

A "general" V.I. associated with the operator A is (1.9) where K <u>is an arbitrary closed</u> (non empty) <u>convex set in</u> $V = H_0^1(\Omega)$.

If $K = V$, (1.9) reduces to the Dirichlet problem. ∎

Remark 1.4.

The result of existence and uniqueness of u is valid for a general V.I. Cf. [7]. We restrict ourselves here to the case (1.8). ∎

---

(1) In general $H^m(\Omega) = \{v \mid D^p v \in L^2(\Omega) \quad \forall\, p \,,\, |p| \leq m\}$.

2. PROOF OF EXISTENCE BY A PENALTY ARGUMENT.

The plan of the proof is as follows :

(i) we introduce the penalized equation

$$Au_\varepsilon + \frac{1}{\varepsilon} u_\varepsilon^+ = f \quad , \quad u_\varepsilon \in H_0^1(\Omega) \; , \tag{2.1}$$

(where $v^+ = \sup(v,0)$) ;

(ii) we prove that, as $\varepsilon \to 0$, u_ε tends to a solution of (1.9).

We shall also give

(iii) an error estimate for $u-u_\varepsilon$.

Part (i) : Existence (and uniqueness) of a solution of the penalized equation (2.1).

The uniqueness is straightforward, since the operator $v \longrightarrow v^+$ is monotone, i.e.

$$(u^+ - v^+, u - v) \geqslant 0.$$

The existence follows from the fact that the operator $v \longrightarrow Av + \frac{1}{\varepsilon} v^+$ is monotone from $H_0^1(\Omega) = V \longrightarrow V'$ = dual space of V when $L^2(\Omega)$ is identified with its dual, and coercive :

$$(Av + \frac{1}{\varepsilon} v^+, v) \geqslant \alpha \|v\|^2 + \frac{1}{\varepsilon} |v^+|^2 \; ; \tag{2.2}$$

therefore one can apply general results from Minty and Browder (cf. e.g. [6] and the bibliography therein).

A very elementary proof can be given as follows ; one introduces a family of subspaces $V_h \subset V$ (1) (where h is a parameter which tends to zero), that satisfies the following assumptions :

$$V_h \text{ is finite dimensional ,} \tag{2.3}$$

$$\forall v \in V, \text{ there exists } v_h \in V_h \text{ such that } \|v-v_h\| \longrightarrow 0 \text{ as } h \to 0. \tag{2.4}$$

We then consider the equation

$$a(u_h,v) + \frac{1}{\varepsilon}(u_h^+,v) = (f,v) \quad \forall v \in V_h, \quad u_h \in V_h \; . \tag{2.5}$$

There exists a (unique) u_h satisfying (2.5), by applying the Brower's fixed point theorem.

---

(1) The constructive proof given now is presented so as to cover the method of finite elements, the next step being then to construct spaces V_h (and this is actually the main point of interest !).

Taking $v = u_h$ in (2.5) we see that

$$\alpha\|u_h\|^2 + \frac{1}{\varepsilon}|u_h^+|^2 \leq (f,u_h) ,$$

hence in particular

(2.6) $$\|u_h\| \leq C$$

(where here and in what follows, C denotes various constants).

Therefore we can extract a subsequence, still denoted by u_h, such that

(2.7) $$u_h \longrightarrow w \quad \text{in } V \text{ weakly as } h\to 0.$$

We are going to verify that $w = u_\varepsilon$. [1]

Since (Rellich-Kondrachoff) the identity mapping $V \longrightarrow L^2(\Omega)$ is compact it follows from (2.7) that

(2.8) $$u_h \longrightarrow w \quad \text{in } L^2(\Omega) \text{ (strongly)}$$

and therefore that

(2.9) $$u_h^+ \longrightarrow w^+ \text{ in } L^2(\Omega) .$$

Let now v be given in V. We introduce v_h satisfying (2.4) ; we have :

(2.10) $$a(u_h,v_h) + \frac{1}{\varepsilon}(u_h^+,v_h) = (f, v_h)$$

and since $\|v_h - v\| \to 0$, we can pass to the limit in (2.10).

We obtain

$$a(w,v) + \frac{1}{\varepsilon}(w^+,v) = (f,v)$$

so that $w = u_\varepsilon$ is solution of (2.1) [2]. ∎

Part (ii) :

We now let $\varepsilon \to 0$. We have

(2.11) $$a(u_\varepsilon,u_\varepsilon) + \frac{1}{\varepsilon}|u_\varepsilon^+|^2 = (f,u_\varepsilon)$$

hence it follows that

(2.12) $$\|u_\varepsilon\| \leq C ,$$

(2.13) $$\frac{1}{\varepsilon}|u_\varepsilon^+|^2 \leq C .$$

---

(1) One could avoid the use of this theorem by using the monotonicity of $v \to v^+$.

(2) By virtue of the uniqueness of the limit, $u_h \to u_\varepsilon$ in V weakly (and actually strongly) without extracting a subsequence.

Therefore we can extract a subsequence, still denoted by u_ε, such that

(2.14) $$u_\varepsilon \longrightarrow u \quad \text{in } V \text{ weakly.}$$

Therefore, as in part (i), $u_\varepsilon \to u$ in $L^2(\Omega)$ and $u_\varepsilon^+ \to u^+$ in $L^2(\Omega)$. But, by virtue of (2.13), $u_\varepsilon^+ \to 0$ in $L^2(\Omega)$ so that

(2.15) $$u^+ = 0 .$$

In other words, $u \in K$.

Let now v be given in K. It follows from (2.1) that :

$$(Au_\varepsilon - f,\ v-u_\varepsilon) = -\frac{1}{\varepsilon}(u_\varepsilon^+,\ v-u_\varepsilon) = \frac{1}{\varepsilon}(v^+ - u_\varepsilon^+,\ v-u_\varepsilon)$$

(since $v^+ = 0$), hence

$$(Au_\varepsilon - f,\ v-u_\varepsilon) = a(u_\varepsilon,\ v-u_\varepsilon) - (f, v-u_\varepsilon) \geqslant 0$$

i.e.

$$a(u_\varepsilon, v) - (f,\ v-u_\varepsilon) \geqslant a(u_\varepsilon, u_\varepsilon).$$

Letting $\varepsilon \to 0$, we obtain

$$a(u,v) - (f,v-u) \geqslant \text{lim. inf. } a(u_\varepsilon,u_\varepsilon) \geqslant a(u,u)$$

i.e. u is a solution of the V.I. ∎

Part (iii) : Error estimate for $u - u_\varepsilon$.

We now take the scalar product of (2.1) with u_ε^+ ; we observe that – due to the fact that A is a second-order operator

(2.16) $$a(u,u^+) = a(u^+,u^+) ;$$

it follows that

(2.17) $$a(u_\varepsilon^+, u_\varepsilon^+) + \frac{1}{\varepsilon}|u_\varepsilon^+|^2 = (f, u_\varepsilon^+)$$

so that

(2.18) $$\frac{1}{\varepsilon}|u_\varepsilon^+| \leqslant |f|$$

(2.19) $$\| u_\varepsilon^+ \| \leqslant C\sqrt{\varepsilon} ,$$

where C depends on $|f|$, $\|a_{ij}\|_{L^\infty}$, $\|a_i\|_{L^\infty}$, $\|a_0\|_{L^\infty}$, α .

We are going to show :

(2.20) $$\left| \begin{array}{l} \|u-u_\varepsilon\| \leqslant C\sqrt{\varepsilon} \quad \text{where } C \text{ depends only on} \\ |f| ,\ \|a_{ij}\|_{L^\infty} ,\ \|a_i\|_{L^\infty} ,\ \|a_0\|_{L^\infty} ,\ \alpha . \end{array} \right.$$

Since $u-u_\varepsilon = u+u_\varepsilon^- - u_\varepsilon^+$ and since we have (2.19) it suffices to prove that

(2.21) $$\|u + u_\varepsilon^-\| \leqslant C\sqrt{\varepsilon}\ .$$

We take $v = -u_\varepsilon^-$ in (1.7) and we take the scalar product of (2.1) with $u + u_\varepsilon^-$; adding up, we obtain :

$$a(u_\varepsilon - u,\ u + u_\varepsilon^-) + \frac{1}{\varepsilon}(u_\varepsilon^+,\ u+u_\varepsilon^-) \geqslant 0$$

hence it follows (since $(u_\varepsilon^+, u_\varepsilon^-) = 0$) that

(2.22) $$a(u+u_\varepsilon^-,\ u+u_\varepsilon^-) + \frac{1}{\varepsilon}(u_\varepsilon^+\ ,\ -u) \leqslant -a(u_\varepsilon^+,\ u+u_\varepsilon^-)\quad ;$$

since $(u_\varepsilon^+\ ,\ -u) \geqslant 0$, (2.22) implies

$$a(u+u_\varepsilon^-\ ,\ u+u_\varepsilon^-) \leqslant -a(u_\varepsilon^+\ ,\ u+u_\varepsilon^-)$$

hence

$$\|u+u_\varepsilon^-\| \leqslant C\ \|u_\varepsilon^+\|\ ;$$

using (2.19), (2.21) follows. ∎

Remark 2.1.

The estimate (2.20) depends on the fact that A is a second order operator and also on the particular structure of K (one can construct penalized equations associated to general V.I. ; cf. [6]). ∎

Remark 2.2.

It follows of course from (2.20) that $u_\varepsilon \to u$ in V strongly as $\varepsilon \to 0$. ∎

Remark 2.3.

The above existence proof gives a method which could be used for numerical purposes. But a more efficient method consists in approximating (1.9) by its finite dimensional analoguous.

(2.23) $$\left|\begin{array}{ll} a(u_h,\ v-u_h) \geqslant (f,\ v-u_h) & \forall\ v \in K_h\ , \\ u_h \in K_h & \end{array}\right.$$

where K_h = subset of V_h of functions $v_h \leqslant 0$ a.e. in Ω , and where

V_h = space of finite elements piecewise linear.

The numerical solution of (2.23) uses an iterative method with projection on K_h (cf. [4]). ∎

Remark 2.4.

It follows from (2.18) that $Au_\varepsilon = f - \frac{1}{\varepsilon}u_\varepsilon^+$ remains, as $\varepsilon \to 0$, in a bounded set of $L^2(\Omega)$, so that :

(2.24) $$Au \in L^2(\Omega).$$

If the coefficients a_{ij} satisfy

(2.25) $$a_{ij} \in W^{1,\infty}(\Omega) \; ; \text{ i.e. } \frac{\partial a_{ij}}{\partial x_k} \in L^{\infty}(\Omega) \qquad \forall k$$

then (2.24) together with $u \in H_0^1(\Omega)$ implies <u>the regularity result</u>

(2.26) $$u \in H^2(\Omega).$$

One <u>cannot</u> in general obtain an H^3 regularity, no matter how much the dataes are regular.

For a complete study of the regularity of the solution of a number of V.I. we refer to [2] [3]. ■

3. SOME OTHER V.I.

3.1. $K = \{v \mid v \in H_0^1(\Omega) \; , \; \psi_1 \leqslant v \leqslant \psi_2 \text{ a.e. in } \Omega\}.$

(Of course one should assume $\psi_1 \leqslant 0 \leqslant \psi_2$ on Γ in order K not to be empty).

One can introduce a penalized equation as follows :

(3.1) $$\left| \begin{array}{l} Au_\varepsilon + \frac{1}{\varepsilon}(u_\varepsilon - \psi_2)^+ - \frac{1}{\varepsilon}(u_\varepsilon - \psi_1)^- = f, \\ u_\varepsilon \in H_0^1(\Omega). \end{array} \right.$$

If

(3.2) $$\psi_i \in H^1(\Omega) \; , \; A\psi_i \in L^2(\Omega) \; ,$$

then one proves along similar lines to what has been done in Section 2, that

(3.3) $$\|u - u_\varepsilon\| \leqslant C\sqrt{\varepsilon} .$$

Hence C depends on $|f|$, $\|a_{ij}\|_{L^\infty}$, $\|a_i\|_{L^\infty}$, $\|a_0\|_{L^\infty}$, $|A\psi_i|$, α. ■

3.2. $$\left| \begin{array}{l} K = \{v \mid v \in H_0^1(\Omega) \; , \; v \leqslant \psi \text{ a.e. on } E \subset \Omega \; , \\ E = \text{set of} \geqslant 0 \text{ measure} \} \end{array} \right.$$

A penalized equation associated to the corresponding V.I. is

(3.4) $$Au_\varepsilon + \frac{1}{\varepsilon}(u_\varepsilon - \psi)^+ \chi_E = f \; , \quad u_\varepsilon \in H_0^1(\Omega) \; ,$$

where χ_E = characteristic function of E.

One proves again that $u_\varepsilon \to u$ in $H_0^1(\Omega)$ as $\varepsilon \to 0$, u being the solution of the corresponding V.I. but the estimate $\|u - u_\varepsilon\| \leqslant \sqrt{\varepsilon}$ seems dubious.

If we assume that

(3.5) $$\left\{ \begin{array}{l} E \text{ is an open set of } \Omega \; , \; \bar{E} \subset \Omega \; , \\ \partial E \text{ smooth} \end{array} \right.$$

and that

(3.6) $$\psi \in H^1(E) \; ,$$

we introduce (following H. Brézis) φ solution of

$$(3.7)\quad \left|\begin{array}{l} A\varphi = f \quad \text{in} \quad F = \complement_\Omega \bar{E}. \\ \varphi = \psi \quad \text{on} \quad \partial E, \\ \varphi = 0 \quad \text{on} \quad \Gamma \end{array}\right.$$

and we define

$$(3.8)\qquad \Psi = \varphi \ \text{on} \ F, \ \psi \ \text{on} \ E.$$

Then, if u denotes the solution of the V.I., we have

$$Au = f \ \text{in} \ F,$$

$$u \leqslant \psi \ \text{on} \ \partial E, \quad u = 0 \ \text{on} \ \Gamma,$$

so that according to the maximum principle,

$$(3.9)\qquad u \leqslant \varphi \ \text{on} \ F.$$

Let us introduce

$$(3.10)\qquad \hat{K} = \{ v | v \in V, \quad v \leqslant \Psi \ \text{a.e. in} \ \Omega \}.$$

<u>The solution</u> u <u>of the V.I. relative to</u> K <u>is also the solution of the V.I. relative to</u> $\hat{K}$.

We can then introduce <u>another penalized equation</u> [1] associated to the V.I., namely,

$$(3.11)\qquad A\hat{u}_\varepsilon + \frac{1}{\varepsilon}(\hat{u}_\varepsilon - \Psi)^+ = f \quad , \quad \hat{u}_\varepsilon \in H_0^1(\Omega).$$

One has then

$$(3.12)\qquad \|u - \hat{u}_\varepsilon\| \leqslant C\sqrt{\varepsilon}$$

if one assumes that $A\Psi \in L^2(\Omega)$. ■

These remarks lead to the <u>following general question</u> :

(3.13) | Construction of penalized equations associated to V.I. whose solutions u_ε give "the best" approximation of the solution u of the V.I., and estimation of the error $\|u - u_\varepsilon\|$. ■

---

(1) There is of course <u>no uniqueness</u> of "the" penalized equation associated to a V.I. !

4. AN APPLICATION OF DYNAMIC PROGRAMMING.

A general problem is as follows : let $u(\Omega)$ be the solution of a V.I. in the domain Ω ; assuming that all data are given in, say, $\mathbb{R}^n$, we want to study variations of $u(\Omega)$ with respect to Ω.

This seems to be an open question.

For the case of equations $(K = V)$ it is a classical problem, the main result being due to Hadamard [5] for the variation of the Green's function (for Dirichlet's problem). Recent extensions have been given in [9] [10] (cf. also the bibliography therein) still for equations.

For V.I. there seems to be two possible approaches :

1) to use "explicit" formulaes giving the solution of some V.I. (cf. the beginning of Chapter 2 for these formulaes ; we shall not pursue here the possible use of these formulaes for the variation of $u(\Omega)$) ;

2) to use "dynamic programming" : it was proved in [1] (cf. also [8]) that one can recover the Hadamard's formula by this method, in case of equations. We show here what kind of "functional equation" one is lead to when dealing with V.I.∎

Notations :

ν = normal to Γ, $\|\nu\| = 1$, directed toward the exterior of Ω ;

Γ_λ = surface spanned by $x - \lambda\theta(x)\nu(x)$, $x \in \Gamma$,

where

- $x \to \theta(x)$ is a given continuous > 0 function on Γ , λ is given > 0 small enough ;
- Ω_λ = region included in Ω with $\partial\Omega_\lambda = \Gamma_\lambda$;
- $a(u,v) = \int_\Omega \text{grad } u.\ \text{grad } v\, dx$ (1)
- $a_\lambda(u,v) = \int_{\Omega_\lambda} \text{grad } u.\ \text{grad } v\, dx.$
- $(f,v)_\lambda = \int_{\Omega_\lambda} fv\, dx.$

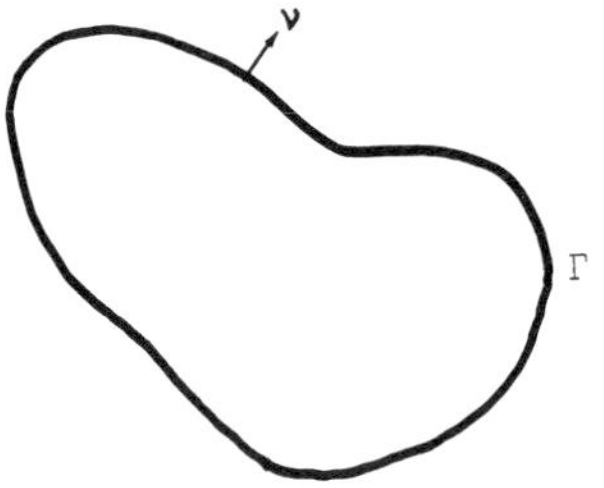

(1) All what follows could be extended to general symmetric coercive bilinear forms on $H_0^1(\Omega)$.

Let f be given in $L^2(\Omega)$ and let u be the solution of

$$(4.1) \quad \begin{cases} -\Delta u - f \leqslant 0 \ , \ u \leqslant 0 \ , \ (-\Delta u - f)u = 0 \text{ in } \Omega \ , \\ u \in H_0^1(\Omega) \ ; \end{cases}$$

u is the unique element which minimizes (1)

$$(4.2) \qquad \frac{1}{2} a(\varphi,\varphi) - (f,\varphi) \quad , \quad \varphi \in H_0^1(\Omega) \ , \ \varphi \leqslant 0.$$

We define

$$(4.3) \qquad F(\lambda,f,w) = \inf_{\substack{\varphi \leqslant 0 \text{ in } \Omega_\lambda, \\ \varphi = w \text{ on } \Gamma_\lambda \ (w \leqslant 0)}} \left[\frac{1}{2} a_\lambda(\varphi,\varphi) - (f,\varphi)_\lambda \right] .$$

We are going to verify - in a formal manner - that

$$(4.3) \qquad \frac{\partial F}{\partial \lambda}(o,f,o) = \frac{1}{2} \int_\Gamma \theta \left(\left(\frac{\partial F}{\partial w}(o,f,o) \right)^+ \right)^2 d\Gamma \ , \quad \forall f \ . \quad \blacksquare$$

Remark 4.1.

In [1][8] one deduces the Hadamard's formula from the similar equation to (4.3) which, in case $K = H_0^1(\Omega)$ (Dirichlet's problem) is :

$$(4.4) \qquad \frac{\partial F}{\partial \lambda}(o,f,o) = \frac{1}{2} \int_\Gamma \theta \left(\frac{\partial F}{\partial w}(o,f,o) \right)^2 d\Gamma \ , \quad \forall f \ . \quad \blacksquare$$

Proof of (4.3).

We write

$$(4.5) \qquad F(o,f,o) = \inf \left[\frac{1}{2} a_\lambda(\varphi,\varphi) - (f,\varphi)_\lambda + \frac{1}{2} \int_{\Omega-\Omega_\lambda} |\text{grad}\varphi|^2 dx - \int_{\Omega-\Omega_\lambda} f\varphi \, dx \right] ,$$
$$\varphi \leqslant 0 \text{ in } \Omega \ , \ \varphi = 0 \text{ on } \Gamma \ .$$

If we introduce (assuming φ to be smooth)

$$(4.6) \qquad \psi = \frac{\partial \varphi}{\partial \nu} \ ,$$

we have necessarily :

(1) This is general whenever A is symmetric, in which case the solution of the V.I. is immediate.

(4.7) $$\psi \geqslant 0 ,$$

and for λ "small" we have

$$\frac{1}{2}\int_{\Omega-\Omega_\lambda} |\mathrm{grad}\varphi|^2 dx = \frac{1}{2}\int_\Gamma \lambda\theta\, |\psi|^2\, d\Gamma + \lambda\, o(\lambda) \;, \quad \int_{\Omega-\Omega_\lambda} f,\varphi\, dx = \lambda\, o(\lambda) \;;$$

therefore (4.5) gives :

(4.8) $$F(o,f,o) = \inf_{\varphi,\psi}\left[\frac{1}{2}\, a_\lambda(\varphi,f) - (f,\varphi)_\lambda + \frac{\lambda}{2}\int_\Gamma \theta\,\psi^2\, d\Gamma + \lambda\, o(\lambda)\right] .$$

But approximately on Γ_λ we have

$$\varphi = -\lambda\theta\,\psi$$

so that ("optimality principle") (4.8) gives

$$\begin{aligned} F(o,f,o) &= \inf_{\psi\geqslant 0}\left[F(\lambda,f,-\lambda\theta\,\psi) + \frac{\lambda}{2}\int_\Gamma \theta\,\psi^2\, d\Gamma + \lambda\, o(\lambda)\right] \\ &= F(o,f,o) + \lambda\,\frac{\partial F}{\partial\lambda}(o,f,o) + \inf_{\psi\geqslant 0}\Big[-\lambda\big(\frac{\partial F}{\partial w}(o,f,o),\theta\psi\big) + \\ &\quad + \frac{\lambda}{2}\int_\Gamma \theta\,\psi^2 d\Gamma + \lambda\, o(\lambda)\Big] . \end{aligned}$$

After simplification, we divide by λ and we let $\lambda \to 0$; we obtain

$$0 = \frac{\partial F}{\partial\lambda}(o,f,o) + \inf_{\psi\geqslant 0}\left[-\big(\frac{\partial F}{\partial w}(o,f,o),\theta\psi\big) + \frac{1}{2}\int_\Gamma \theta\psi^2\, d\Gamma\right]$$

hence (4.3) follows. ∎

REFERENCES (CHAPTER I).

[1] Bellman, R. and Osborn, H. (1958). Dynamic programming and the variations of Green's functions. J. Math. and Mech. 7, Nb. 1, pp. 81-85.

[2] Brézis, H. Problèmes unilatéraux. Thèse. (1972) J. de Mathématiques Pures et Appliquées, 51, pp. 1-68.

[3] Brézis, H. and Stampacchia, G. (1968). Bulletin Société Math. de France, 96, pp. 153-180.

[4] Glowinski, R., Lions, J.L., and Trémolières, R. (1976). Book. Dunod, Ed.

[5] Hadamard, J. (1968). Mémoire sur le problème d'analyse relatif à l'équilibre des plaques élastiques encastrées. Oeuvres de J. Hadamard. C.N.R.S., Paris.

[6] Lions, J.L. (1969). Sur quelques méthodes de résolution des problèmes aux limites non linéaires. Dunod Gauthier Villars, Paris.

[7] Lions, J.L. and Stampacchia, G. (1967). Variational Inequalities. C.P.A.M. 20.

[8] Luré, K.A. (1975). Optimal control in problems of mathematical physics. Moscow. (In Russian).

[9] Murat, F., and Simon, J. To appear.

[10] Palmerio, B. and Dervieux, A. (1975). Une formule de Hadamard dans des problèmes d'identification de domaines. C.R.Ac. Sc. Paris.

II. STOPPING TIMES AND SINGULAR PERTURBATIONS

ORIENTATION

In this chapter, we first give, following [1] [3], an interpretation of the solution of some of the V.I. introduced in Chapter I in terms of the optimal cost function of a stopping time problem. This permits to give a result of singular perturbations for a family of V.I. (Section 2). These results are taken from [6]. In Section 4 we give some preliminary indications on a very interesting paper of MOSSOLOV and MIASNIKOV [9] about singular perturbations for V.I. of a different nature. Problems of singular perturbations for other (and actually simpler) classes of V.I. are considered in [6] [7].

1. OPTIMAL STOPPING TIMES AND V.I.

We state here without proofs some of the results established in [1] [2] [3].

Let $y_x(s)$ be the solution of the stochastic differential equation (Ito's differential equation) :

(1.1) $$\left| \begin{array}{l} dy = g(y)ds + \sigma(y)dw(s), \quad s > 0 , \\ y(o) = x \quad , \quad y_x(s) \in \mathbb{R}^n \end{array} \right.$$

where

(1.2) $$\left| \begin{array}{l} \sigma(x) \text{ is } C^1 \text{ in } \mathbb{R}^n , \ g(y) \text{ is also } C^1 \ (^1) \\ \sigma(x) \in \mathcal{L}(\mathbb{R}^n ; \mathbb{R}^n) , \ \sigma(x)^+ = \sigma(x) , \ \sigma(x) \geq \alpha_o I , \\ w(s) = \text{standard Wiener process in } \mathbb{R}^n. \end{array} \right.$$

If $x \in \Omega$, we denote by $\tau(x)$ the smallest $s > 0$ such that $y_x(s) \notin \Omega$.

Let θ be an arbitrary stopping time for y, $\theta \leq \tau$.

We consider the cost function

(1.3) $$J_x(\theta) = E\left[\int_0^\theta e^{-\alpha s} f(y_x(s))ds\right] , \quad \theta \leq \tau$$

where $\alpha > 0$, where f is given in Ω, and we define the optimal cost function as

(1.4) $$u(x) = \inf_{\theta \leq \tau} J_x(\theta).$$

One proves [1] [3] that u can be characterized as the solution of the V.I. (of the type introduced in Chapter I)

(1.5) $$\left\{ \begin{array}{l} Au - f \leq 0 , \ u \leq 0 , \ (Au - f)u = 0 \text{ in } \Omega , \\ u = 0 \text{ on } \Gamma \end{array} \right.$$

---

(1) These hypothesis are unnecessarily strong. Cf. for instance [1] and the bibliography therein.

where A is given by

$$(1.6) \qquad Au = -\sum_{i,j} a_{ij}(x) \frac{\partial^2 u}{\partial x_i \partial x_j} - \sum_i g_i(x) \frac{\partial u}{\partial x_i} + \alpha u$$

where

$$a_{ij}(x) = \frac{1}{2}(\sigma(x)\sigma(x))_{i,j}$$

and where $g(x) = \{g_i(x)\}$. ∎

Remark 1.1.

In (1.6) the operator A appears in a non divergence form. This leads to the natural problem of extending the theory of V.I. to operators of type (1.6) with non smooth coefficients, say :

$$a_{ij} \in C^0(\bar{\Omega}) \text{ (but not in } C^1(\bar{\Omega}))$$

and for convex sets of the type

$$K = \{v \mid v \in H_0^1(\Omega) \text{ , } v \leqslant 0 \text{ a.e. in } \Omega \text{ , or } v \leqslant \psi \text{ a.e. in } \Omega\}.$$

This is done in [1] ; it would be interesting to see if similar results can be obtained for other sets K. ∎

Remark 1.2.

One can of course - in case of hypothesis (1.2) - write A in divergence form :

$$(1.7) \qquad \left| \begin{aligned} Au &= -\sum \frac{\partial}{\partial x_i}\left(a_{ij}(x) \frac{\partial}{\partial x_j}\right) + \sum a_i(x) \frac{\partial}{\partial x_i} + \alpha u \text{ ,} \\ a_i &= -g_i + \sum_j \frac{\partial a_{ij}}{\partial x_j} \text{ .} \end{aligned} \right.$$

One does not necessarily have $a(v,v) \geqslant \alpha_1 \|v\|^2 \quad \forall\, v \in H_0^1(\Omega) \text{ , } \alpha_1 > 0$, but conditions

$$\sum a_{ij}(x)\, \xi_i\, \xi_j \geqslant \alpha_0 |\xi|^2 \text{ , } \alpha_0 \text{ , } a > 0$$

are sufficient for the results of Chapter I to hold true. ∎

Remark 1.3.

An interpretation of the solution u of the V.I. (1.5) is known, by formula (1.4), only whenever A is symmetric in its principal part. ∎

Remark 1.4.

Let us recall at this point that

$$(1.8) \qquad J_x(\tau) = w(x)$$

is the solution of the Dirichlet's problem

$$(1.9) \qquad Aw = f \text{ , } \quad w \in H_0^1(\Omega) \text{ .}$$ ∎

Remark 1.5.

We refer to [1] for other V.I. whose solution can be interpreted as the optimal cost function of a suitable stopping time problem.

It is an open question to determine the class of "all" V.I. of elliptic type whose solution can be interpreted by a stopping time problem. ∎

Remark 1.6.

We also refer to [1], Volume 2, and to [10] for the interpretation of V.I. of similar type with Neumann (or others) boundary conditions. ∎

2. SINGULAR PERTURBATIONS AND OPTIMAL STOPPING TIME.

Let us consider the situation of Section 1 with

(2.1) $$\sigma = \varepsilon I .$$

Let $y_x^\varepsilon(s)$ be the solution of the corresponding Ito's differential equation :

(2.2) $$dy^\varepsilon(s) = g(y^\varepsilon)ds + \varepsilon\, dw(s), \quad y^\varepsilon(o) = x ,$$

and let us set

(2.3) $$u_\varepsilon(x) = \inf_{\theta \leqslant \tau^\varepsilon} E \int_0^\theta e^{-\alpha s} f(y_x^\varepsilon(s))ds . \text{ (}^1\text{)}$$

According to Section 1, u_ε is characterized by the solution of the V.I.

(2.4) $$\left| \begin{array}{l} -\frac{\varepsilon^2}{2} \Delta u_\varepsilon + Bu_\varepsilon - f \leqslant 0 , \quad u_\varepsilon \leqslant 0 , \\ (-\frac{\varepsilon^2}{2} \Delta u_\varepsilon + Bu_\varepsilon - f)u_\varepsilon = 0 \text{ on } \Omega , \quad u_\varepsilon \in H_0^1(\Omega) , \end{array} \right.$$

where

(2.5) $$B u = - \sum g_i(x) \frac{\partial v}{\partial x_i} + \alpha v.$$

<u>We can now study the behaviour of u_ε as $\varepsilon \to 0$.</u> ∎

Let us first consider the <u>deterministic case</u> $\varepsilon = 0$; $y_x(s) = y_x^0(s)$ is the solution of the ordinary differential equation

(2.6) $$\frac{dy}{ds} = g(y) , \quad y(o) = x ;$$

let $\tau(x) = \inf \{s | s > 0 , \ y_x(s) \notin \Omega \}$, and let u be defined as

(2.7) $$u(x) = \inf \int_0^\theta e^{-\alpha s} f(y_x(s))ds, \qquad \theta \leqslant \tau.$$

(^{1}) $\tau^\varepsilon(x)$ = exit time of $\overline{\Omega}$ for y_x^ε .

Then u can be characterized as the solution of the V.I.

(2.8) $Bu - f \leqslant 0$, $u \leqslant 0$, $(Bu-f)u = 0$ in Ω ,

(2.9) $u = 0$ for $x \in \Gamma$ and such that $\tau(x) = 0$.

Remark 2.1.

Condition (2.9) should be treated a little more carefully ; cf. [2] for details. Roughly speaking, (2.9) means that u should be zero whenever $g.\nu \geqslant 0$.■

Remark 2.2.

One can solve directly the V.I. (2.8)(2.9) as follows :

First, by changing u into $\exp(-\rho(x))u$, with ρ suitably chosen, one can always assume that

(2.10) $$\alpha + \frac{1}{2}\Sigma \frac{\partial g_i}{\partial x_i} \geqslant C > 0.$$

We then consider - assuming (2.10) to be satisfied - the penalized equation

(2.11) $$\begin{cases} Bu_\eta + \frac{1}{\eta} u_\eta^+ = f \quad , \quad \eta > 0 , \\ u_\eta = 0 \text{ on the set } g.\nu \geqslant 0 \text{ of } \Gamma. \end{cases}$$

This equation admits a unique solution - (Observe that :

(2.12) $$(Bu_\eta , u_\eta) \geqslant C\,|u_\eta|^2).$$

We have :

(2.13) $$|u_\eta| \leqslant C \text{ as } \eta \to 0 ,$$

and taking the scalar product of (2.11) with u_η^+ , we obtain :

(2.14) $$(Bu_\eta , u_\eta^+) + \frac{1}{\eta}\,|u_\eta^+|^2 = (f, u_\eta^+) \quad ;$$

since $(Bu_\eta, u_\eta^+) = (Bu_\eta^+ , u_\eta^+) \geqslant C\,|u_\eta^+|^2$, (2.14) implies

(2.15) $$\frac{1}{\eta}\,|u_\eta^+| \leqslant C ;$$

therefore (2.11) implies

(2.16) $$|Bu_\eta| \leqslant C .$$

Moreover, since $(Bu_\eta , u_\eta) = \int_\Omega (\alpha + \frac{1}{2}\Sigma \frac{\partial g_i}{\partial x_i})\, u_\eta^2\, dx + \frac{1}{2}\int_\Gamma (-g\nu)\, u_\eta^2\, d\Gamma$,

it follows (using the fact that $u_\eta = 0$ if $g.\nu \geqslant 0$) :

(2.17) $$u_\eta \sqrt{|g\,\nu|} \text{ is bounded in } L^2(\Gamma) \text{ as } \eta \to 0.$$

Therefore we can extract a subsequence, still denoted by u_η, such that

$$(2.18)\quad \begin{cases} u_\eta \to u & \text{in } L^2(\Omega) \text{ weakly,} \\ Bu_\eta \to Bu & \text{in } L^2(\Omega) \text{ weakly,} \end{cases}$$

$$(2.19)\quad u_\eta \sqrt{|g\nu|} \to \xi \text{ in } L^2(\Gamma) \text{ weakly.}$$

It follows from (2.18) and (2.19) that

$$(2.20)\quad \begin{cases} u = 0 \text{ if } g.\nu \geq 0 , \\ \xi = u\sqrt{|g\nu|} . \end{cases}$$

If v is any function $\in L^2(\Omega)$, $v \leq 0$ a.e., we have

$$(Bu_\eta - f , v-u_\eta) \geq 0$$

i.e.

$$(2.21)\quad (Bu_\eta,v)-(f, v-u_\eta) \geq (Bu_\eta, u_\eta) = \int_\Omega (\alpha + \frac{1}{2}\Sigma \frac{\partial g_i}{\partial x_i}) u_\eta^2 \, dx + \frac{1}{2}\int_\Gamma (-g\nu)u_\eta^2 \, d\Gamma$$

and

$$\lim .\inf (Bu_\eta,u_\eta) \geq \int_\Omega (\alpha + \frac{1}{2}\Sigma \frac{\partial g_i}{\partial x_i}) u^2 dx + \frac{1}{2}\int_\Gamma (-g\nu)\, u^2 \, d\Gamma = (Bu,u)$$

so that (2.21) implies

$$(2.22)\quad (Bu - f, v-u) \geq 0 \quad \forall\, v \leq 0 , \quad v \in L^2(\Omega).$$

Therefore, it will be proven that u is a solution [1] of (2.8)(2.9) such that u, $Bu \in L^2(\Omega)$ if we check that $u \leq 0$. But for every $v \in L^2(\Omega)$,

$$(2.23)\quad (u_\eta^+ - v^+ , u_\eta - v) \geq 0 .$$

Using (2.15), (2.23) implies

$$(2.24)\quad -(v^+, u-v) \geq 0.$$

Let $\varphi \in L^2(\Omega)$ and let λ be > 0. We take $v < u+\lambda\varphi$ in (2.24), hence

$$\lambda((u+\lambda\varphi)^+ , \varphi) \geq 0$$

i.e.

$$((u+\lambda\varphi)^+, \varphi) \geq 0 .$$

We let $\lambda \to 0$, and we obtain $(u^+ ,\varphi) \geq 0 \;\; \forall\, \varphi \in L^2(\Omega)$, i.e. $u^+ = 0$, and the proof is completed. ∎

---

(1) The uniqueness is immediate.

The result of singular perturbations is now as follows :

(2.25) when $\varepsilon \to 0$, one has

$u_\varepsilon \to$ in $L^2(\Omega)$ and a.e.

The proof, given in [2], relies entirely on the interpretation of u_ε and of u by (2.3) and (2.7).

It is therefore an "indirect" proof ; a "direct" proof has been obtained by H.BREZIS [11] using rather delicate a priori estimates ; we give, in the following section, a direct proof in a very particular case ; we refer to H. BREZIS, loc. cit. and to [8], for the general case.

3. A DIRECT PROOF.

We consider, to simplify the exposition, a two-dimensional case. We assume that :

(3.1) $$\Omega =]0,1[\times]0,1[\quad,$$

(3.2) $$Bu = \frac{\partial u}{\partial x_1} + \alpha u\quad.$$

The condition (2.9) becomes :

(3.3) $$u(0,x_2) = 0.$$

We assume that

(3.4) $$f\;,\;\frac{\partial f}{\partial x_2} \in L^2(\Omega)\;.$$

Then, as $\epsilon \to 0$,

(3.5) $$u_\varepsilon \to u \quad\text{in}\quad L^2(\Omega)\quad\text{weakly}\;,$$

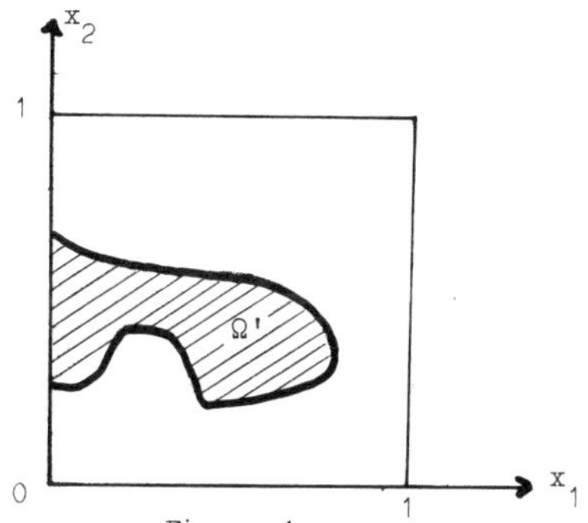

Figure 1

(3.6) $\frac{\partial u_\varepsilon}{\partial x_i} \to \frac{\partial u}{\partial x_i}$ in $L^2(\Omega')$ weakly for every set Ω' as indicated on Figure 1.

For the proof, we introduce $w_{\varepsilon_\eta} = w$, solution of the penalized equation.

(3.7) $$-\frac{\varepsilon^2}{2}\Delta w + Bw + \frac{1}{\eta} w^+ = f,$$

$$w = 0 \text{ on } \Gamma\;.$$

We know that $w \to u_\varepsilon$ in $H_0^1(\Omega)$ as $\eta \to 0$ and we immediately obtain that

(3.8) $$|w| \leqslant C\;,\quad \varepsilon|\text{grad } w| \leqslant C\;,\quad \frac{1}{\eta}|w^+| \leqslant C\;,$$

where the C's do not depend on ε and on η. Therefore :

(3.9) $$|u_\varepsilon| \leqslant C,\quad \varepsilon|\text{grad} u_\varepsilon| \leqslant C$$

and in order to prove (3.5)(3.6) it suffices to prove the following a priori estimates : let φ be $C^1(\overline{\Omega})$, with support in a set $\overline{\Omega}'$ (as indicated on Fig.1) ; then :

$$|\varphi \frac{\partial w}{\partial x_1}| \leq C , \tag{3.10}$$

$$|\varphi \frac{\partial w}{\partial x_2}| \leq C . \tag{3.11}$$

Proof of (3.10) :

We set $g = f - \frac{1}{\eta} w^+$, and we multiply (3.7) by $\varphi^2 \frac{\partial w}{\partial x_1}$.

We obtain :

$$\left| \begin{array}{l} - \frac{\varepsilon^2}{2} \int_\Gamma \frac{\partial w}{\partial x_1} \varphi^2 \frac{\partial w}{\partial x_1} d\Gamma + \frac{\varepsilon^2}{2} \int_\Omega \text{grad } w \text{ grad}(\varphi^2 \frac{\partial w}{\partial x_1})dx + |\varphi \frac{\partial w}{\partial x_1}|^2 + \alpha(\varphi w, \varphi \frac{\partial w}{\partial x_1}) = \\ = (g, \varphi^2 \frac{\partial w}{\partial x_1}) . \end{array} \right. \tag{3.12}$$

But

$\frac{\varepsilon^2}{2} \int_\Omega \text{grad } w \text{ grad}(\varphi^2 \frac{\partial w}{\partial x_1})dx = \frac{\varepsilon^2}{4} \sum_{i=1}^{2} \int_{\mathcal{O}} \varphi^2 \frac{\partial}{\partial x_1} (\frac{\partial w}{\partial x})^2 dx + O(1)$ (by virtue of (3.9)) $= - \frac{\varepsilon^2}{4} \int_0^1 \varphi^2 (\frac{\partial w}{\partial x_1}(o,x_2))^2 + O(1)$.

The first term in (3.12) equals $\frac{\varepsilon^2}{2} \int_0^1 \varphi^2 (\frac{\partial w}{\partial x_1}(o,x_2))^2 dx_2$ so that (3.12) gives

$$\frac{\varepsilon^2}{4} \int_0^1 \varphi^2 (\frac{\partial w}{\partial x_1}(o,x_2))^2 dx_2 + |\varphi \frac{\partial w}{\partial x_1}|^2 + \alpha(\varphi w, \varphi \frac{\partial w}{\partial x_1}) = (g\varphi, \varphi \frac{\partial w}{\partial x_1}) + O(1),$$

hence (3.10) follows. ■

Proof of (3.11) :

It is enough to prove the following : let ψ be a smooth function of x_2, with compact support in $]0,1[$; then

$$| \psi \frac{\partial w}{\partial x_2} | \leq C. \tag{3.13}$$

To simplify the writing, let us set $\partial = \frac{\partial}{\partial x_2}$; applying ∂ to (3.7) gives

$$- \frac{\varepsilon^2}{2} \Delta \partial w + \frac{\partial}{\partial x_1} \partial w + \alpha \partial w + \frac{1}{\eta} \partial w^+ = \partial f . \tag{3.14}$$

We take the scalar product of (3.14) with $\psi^2 \partial w$.

But $\partial w = 0$ if $x_1 = 0$ or if $x_1 = 1$ and $\psi(o) = \psi(1) = 0$; therefore we obtain :

$$\left| \begin{array}{l} \frac{\varepsilon^2}{2} \int_\Omega \text{grad } \partial w . \text{grad } (\psi^2 \partial w) \, dx + \int_\Omega \frac{\psi^2}{2} \frac{\partial}{\partial x_1} (\partial w)^2 dx + \\ + \alpha |\psi \partial w|^2 + \frac{1}{\eta} (\psi \partial w^+, \psi \partial w) = (\psi \partial f, \psi \partial w) . \end{array} \right. \tag{3.15}$$

But $\int_\Omega \frac{\psi^2}{2} \frac{\partial}{\partial x_1} (\partial w)^2 dx = 0$ so that (3.15) reduces to :

(3.16) $$\frac{\varepsilon^2}{2}\sum\int_\Omega \frac{\partial}{\partial x_i}(\partial w)\left(\psi^2 \frac{\partial}{\partial x_i}\partial w + 2\psi\frac{\partial\psi}{\partial x_i}\partial w\right)dx + \alpha|\psi\partial w|^2 + \frac{1}{\eta}|\psi\partial w^+|^2 =$$

$$= (\psi\ \partial f\ ,\ \psi\partial w).$$

The first term in (3.16) is equal to

$$\frac{\varepsilon^2}{2}\sum\int_\Omega \psi^2\left(\frac{\partial}{\partial x_i}\partial w\right)^2 dx + \varepsilon^2\int_\Omega \psi\,\psi'\,\partial w\,\frac{\partial}{\partial x_2}\partial w\ dx \geqslant$$

$$\geqslant \frac{\varepsilon^2}{2}\int_\Omega \psi^2\left(\frac{\partial}{\partial x_2}\partial w\right)^2 dx - \frac{\varepsilon^2}{2}\int_\Omega \psi^2\left(\frac{\partial}{\partial x_2}\partial w\right)^2 dx -$$

$$- \frac{\varepsilon^2}{2}\int_\Omega (\psi')^2(\partial w)^2 dx = O(1)$$ (by virtue of (3.8)), so that

(3.16) implies (3.13). ∎

4. SINGULAR PERTURBATIONS IN VISCO-PLASTICITY.

4.1. POSITION OF THE PROBLEM.

We consider in $\mathbb{R}^n$ an open set Ω,

$\Omega = \complement\,\overline{\omega}$, ω as on Fig. 2 ,

$\partial\omega = \partial\Omega = \Gamma$ being a smooth closed curve.

The formal problem [1] we consider is the following :

for v given such that

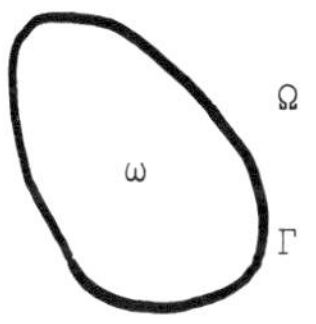

Figure 2

(4.1) v is "small" at infinity,

(4.2) grad $v \in L^1 \cap L^2(\Omega)$ [2],

and for $\varepsilon > 0$, we define

(4.3) $$J_\varepsilon(v) = \frac{\varepsilon}{2}\int_\Omega |\text{grad } v|^2 dx + \int_\Omega |\text{grad } v|\,dx$$

and we are looking for

(4.4) $\inf J_\varepsilon(v)$, v satisfying (4.1)(4.2) and $v = 1$ on Γ. ∎

Remark 4.1.

Without condition (4.1) the problem is trivial : one would have

$$\inf J_\varepsilon(v) = 0 = J_\varepsilon(1).$$ ∎

---

(1) It is formal, for the time being, since one has to make precise the statement (4.1).

(2) We denote in this way the product space $(L^1(\Omega) \cap L^2(\Omega))^2$.

Remark 4.2.

If we set

(4.5) $$a(u,v) = \int_\Omega \text{grad } u. \text{ grad } v \, dx,$$

(4.6) $$j(v) = \int_\Omega |\text{grad } v| \, dx,$$

then (4.4) is equivalent to the V.I.

(4.7) $$\left| \begin{array}{l} \varepsilon a(u_\varepsilon, v-u_\varepsilon) + j(v)-j(u_\varepsilon) \geq 0 \qquad \forall \, v \text{ satisfying } (4.1)(4.2), \\ u_\varepsilon \text{ satisfying } (4.1)(4.2). \end{array} \right.$$ ■

Remark 4.3.

Problem (4.4) is considered in [9] ; it arises in connection with the problem of longitudinal motion of a cylinder in a viscoplastic medium ; other aspects of this question are considered in [9]. ■

The problem studied in [9] consists in finding an asymptotic expansion for $\inf J_\varepsilon(v)$.

4.2. SOME FUNCTION SPACES.(1)

We introduce :

$C_c^\infty(\overline{\Omega})$ = space of C^∞ functions in $\overline{\Omega}$, with compact support in $\overline{\Omega}$.

On the space $C_c^\infty(\overline{\Omega})$ we introduce :

(4.8) $$|||\varphi||| = \int_\Omega |\text{grad } \varphi| \, dx,$$

(4.9) $$[\varphi] = \left(\int_\Omega |\text{grad } \varphi|^2 \, dx\right)^{1/2} + |||\varphi|||.$$

These quantities are obviously norms. We define :

(4.10) $$\left| \begin{array}{l} V = \text{completion of } C_c^\infty(\overline{\Omega}) \text{ for } |||\varphi|||, \\ V_0 = \text{completion of } C_c^\infty(\overline{\Omega}) \text{ for } [\varphi]. \end{array} \right.$$

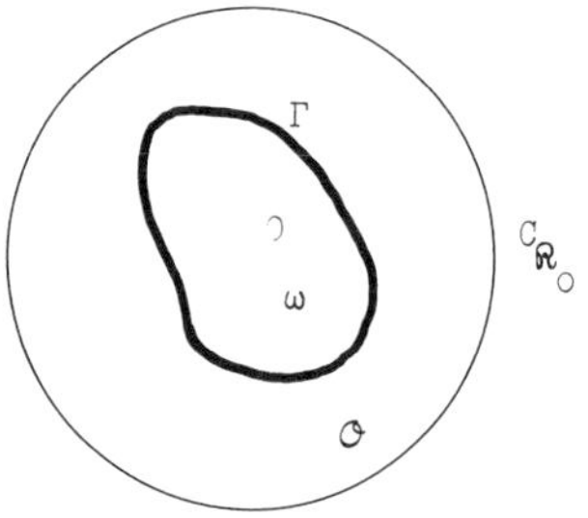

Figure 3.

We obviously have :

(4.11) $$V_0 \subset V \ .$$

We are going to show that

(4.12) $$V \subset L^1_{loc}(\Omega)$$

and that, in some sense, one has (4.1) $\forall \, v \in V$.

We do not restrict the generality by assuming that $0 \in \omega$.

Let C_R denote the circle of center 0 and radius R ; we take R large enough

(1) These considerations are not introduced in [9].

so that $C_{\mathcal{R}} \subseteq \Omega$. We have, $\forall \varphi \in C_c^\infty(\overline{\Omega})$:

(4.13) $$\int_{C_{\mathcal{R}}} |\varphi| dC_{\mathcal{R}} \leqslant \int_{\Omega_{\mathcal{R}}} |\text{grad } \varphi| \; dx,$$

where $\Omega_{\mathcal{R}} = \{x \mid |x| > \mathcal{R}\}$.

Indeed, using polar coordinates, $\varphi(x_1,x_2) = \tilde{\varphi}(r,\theta)$, one has

$$\tilde{\varphi}(\mathcal{R},\theta) = -\int_{\mathcal{R}}^\infty \frac{\partial\tilde{\varphi}}{\partial r}(r,\theta)\, dr$$

hence

$$\mathcal{R}|\tilde{\varphi}(\mathcal{R},\theta)| \leqslant \int_{\mathcal{R}}^\infty \frac{\mathcal{R}}{r}\left|\frac{\partial\tilde{\varphi}}{\partial r}(r,\theta)\right| r\, dr \leqslant \int_{\mathcal{R}}^\infty \left|\frac{\partial\tilde{\varphi}}{\partial r}(r,\theta)\right| r\, dr$$

so that :

$$\int_0^{2\pi} \mathcal{R}\left|\tilde{\varphi}(\mathcal{R},\theta)\right| d\theta \leqslant \int_0^{2\pi} d\theta \int_{\mathcal{R}}^\infty \left|\frac{\partial\tilde{\varphi}}{\partial r}(r,\theta)\right| r\, dr$$

hence (4.13) follows. ∎

Assuming Γ smooth enough, one also shows that (cf. notations on Fig. 3)

(4.14) $$\int_\Gamma |\varphi| d\Gamma \leqslant \int_{\mathcal{O}} |\text{grad } \varphi| \; dx + \int_{C_{\mathcal{R}_0}} |\varphi| \; dC_{\mathcal{R}_0}.$$

Using (4.13)(4.14), it follows that, for any open set $\mathcal{G}$ bounded, $\mathcal{G} \subset \overline{\Omega}$, $\Gamma \subset \mathcal{G}$, one has

$$\int_{\mathcal{G}} |\varphi| dx \leqslant c_{\mathcal{G}} \int_\Omega |\text{grad } \varphi| dx.$$

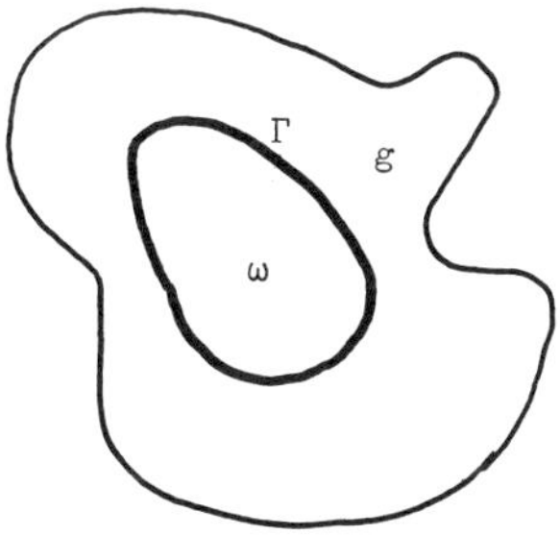

Figure 4.

Therefore (4.12) follows and inequalities (4.13)(4.14) hold true for $\varphi \in V$.
Therefore (4.13) implies that $\forall\, v \in V$, one has :

(4.16) $$\int_{C_{\mathcal{R}}} |v| dC_{\mathcal{R}} \xrightarrow{\ \ } 0 \quad \text{as} \quad \mathcal{R} \to \infty$$

which makes (4.1) precise (and shows $1 \notin V$). ∎

Remark 4.4.

The restrictions of $v \in V$ to $\mathcal{G}$ span the space $W^{1,1}(\mathcal{G})$ which consists of functions $w \in L^1(\mathcal{G})$ such that $\frac{\partial w}{\partial x_i} \in L^1(\mathcal{G})\ \forall\, i$.

One can therefore define $v|_\Gamma \in L^1(\Gamma)$ (by using (4.14) ; it follows from [5] that when v spans V, $v|_\Gamma$ spans exactly $L^1(\Gamma)$. ∎

Remark 4.5.

Since $[\varphi] \geqslant |||\varphi|||$, one has (4.11) and, in particular, one has (4.6) $\forall v \in V_0$.

The restrictions of $v \in V_0$ to $\mathcal{O}$ span the space $H^1(\mathcal{O})$ and the traces $v|_\Gamma$ span $H^{1/2}(\Gamma)$. Cf. [4] for the study of spaces of this kind. ∎

The precise statement of problem (4.4) is now to find

$$\inf J_\varepsilon(v) \ , \quad v \in V_0 \ , \ v = 1 \text{ on } \Gamma. \tag{4.17}$$

For every $\varepsilon > 0$ this problem admits a unique solution. The uniqueness follows from the strict convexity of J_ε ; let v_n be a minimizing sequence, $[v_n]$ is bounded (ε fixed > 0) and therefore we can extract a subsequence such that :

$$(4.18) \quad \left| \begin{array}{l} v_n \to u \text{ in } H^1(\mathcal{O}) \text{ weakly,} \\ \text{for every } \mathcal{O} \text{ as on Fig. 4.} \end{array} \right.$$

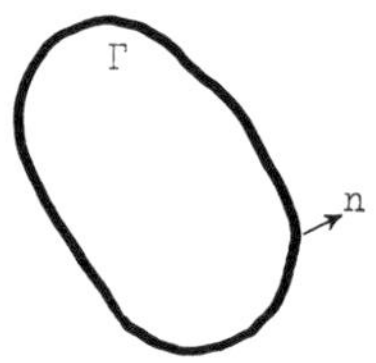

Figure 5

Let $\mathcal{M}(\Omega)$ be the space of summable measures on Ω ; one can assume that

$$\frac{\partial v_n}{\partial x_i} \longrightarrow \mu_i \text{ in } \mathcal{M}(\Omega) \text{ weak star,}$$

and by virtue of (4.18)

$$\mu_i = \frac{\partial u}{\partial x_i} \in L^1(\Omega).$$

We have also :

$$\frac{\partial u}{\partial x_i} \in L^2(\Omega) \text{ and } u=1 \text{ on } \Gamma,$$

so that u is solution of the problem.

We shall denote by u_ε the solution of (4.17). We want now to study the function :

$$\varepsilon \longrightarrow J_\varepsilon(u_\varepsilon)$$

as

$$\varepsilon \longrightarrow 0.$$

∎

4.3. ASYMPTOTIC EXPANSION.

We follow now [9]. We normalize the length ds on Γ so that $0 \leqslant s \leqslant 1$, the total length being L. We take coordinates s, n such that (we denote by x, y the coordinates x_1, x_2)

$$x = x(s) + y'(s)n \ , \quad y = y(x) - x'(s)n \ , \tag{4.19}$$

and we assume that everything is smooth so that, in particular, the curvature $k(s)$ of Γ satisfies

$$k(s) \geqslant k_0 > 0 \ . \tag{4.20}$$

In this new system of coordinates, one has :

$$(4.21) \quad J_\varepsilon(v) = L \int_0^1 \int_0^\infty \left\{ \frac{\varepsilon}{2} \frac{1}{(1+kn)^2} \left[\left(\frac{\partial v}{\partial s}\right)^2 + (1+kn)^2 \left(\frac{\partial v}{\partial n}\right)^2 \right] + \right.$$
$$\left. + \frac{1}{1+kn} \left[\left(\frac{\partial v}{\partial s}\right)^2 + (1+kn)^2 \left(\frac{\partial v}{\partial n}\right)^2 \right]^{1/2} \right\} (1+kn) \, dx \, dn.$$

By a method which is standard in the theory of singular perturbations, we now keep in (4.21) only the normal derivative to the boundary. We introduce in this way :

$$H_\varepsilon(v) = L \int_0^1 \int_0^\infty \left[\frac{\varepsilon}{2} \left(\frac{\partial v}{\partial n}\right)^2 + \left|\frac{\partial v}{\partial n}\right| \right] (1+kn)\, ds\, dn. \tag{4.22}$$

As it has been done for J_ε, one verifies the existence and uniqueness of a function u_ε^* such that :

$\frac{\partial u_\varepsilon^*}{\partial n} \sqrt{1+kn} \in L^2\,(o,1)\times(0,\infty)$

$\frac{\partial u_\varepsilon^*}{\partial n}\,(1+kn) \in L^1\,(o,1)\times(0,\infty)$

u_ε^* is small at infinity (1),

$u_\varepsilon^*\,(s,o) = 1$,

$H_\varepsilon(u_\varepsilon^*)$ is minimum, or, what amounts to the same thing, u_ε^* is solution of the V.I.

$$\int_0^1 \int_0^\infty \varepsilon \frac{\partial u_\varepsilon^*}{\partial n} \left(\frac{\partial v}{\partial n} - \frac{\partial u_\varepsilon^*}{\partial n} \right)(1+kn)\, ds\, dn + \int_0^1 \int_0^\infty \left[\left|\frac{\partial v}{\partial n}\right| (1+kn) - \left|\frac{\partial u_\varepsilon^*}{\partial n}\right| (1+kn) \right] ds\, dn \geqslant 0 \tag{4.23}$$

$\forall\, v$ such that $v(s,o) = 1$.

We are going to verify that u_ε^* is explicitely given as follows : we define $\gamma = \gamma_\varepsilon(s)$ by :

$$\varepsilon = k(s) \int_0^\gamma \left(\frac{\gamma - n}{1+kn} \right) dn \tag{4.24}$$

(one easily checks that one uniquely defines γ in this manner) ; then

$$u_\varepsilon^*(s,n) = \left| \begin{array}{l} 1 - \displaystyle\int_0^n \frac{k(\gamma-\lambda)}{\varepsilon(1+k\lambda)}\, d\lambda \quad \text{for } n \leqslant \gamma, \\ 0 \quad \text{for } n > \gamma . \end{array} \right. \tag{4.25}$$

Indeed one easily checks that u_ε satisfies the "Euler equation"

$$-\frac{\partial}{\partial n} \left[\varepsilon(1+kn) \frac{\partial u_\varepsilon^*}{\partial n} + (1+kn) \frac{\dfrac{\partial u_\varepsilon^*}{\partial n}}{\left|\dfrac{\partial u_\varepsilon^*}{\partial n}\right|} \right] = 0 . \tag{4.26}$$

---

(1) Actually u_ε^* has a compact support.

If we now multiply (4.26) by $v - u_\varepsilon^*$, we obtain :

$$\varepsilon \int_0^1 \int_0^\infty \frac{\partial u_\varepsilon^*}{\partial n} \left(\frac{\partial v}{\partial n} - \frac{\partial u_\varepsilon^*}{\partial n} \right) (1+kn)\, ds\, dn + \int_0^1 \int_0^\infty (1+kn) \frac{\frac{\partial u_\varepsilon^*}{\partial n}}{\left| \frac{\partial u_\varepsilon^*}{\partial n} \right|} \frac{\partial v}{\partial n}\, ds\, dn -$$

$$- \int_0^1 \int_0^\infty (1+kn) \left| \frac{\partial u_\varepsilon^*}{\partial n} \right| ds\, dn = 0$$

and if X denotes the left hand side of (4.23), we obtain

$$X = \int_0^1 \int_0^\infty \left(\left| \frac{\partial v}{\partial n} \right| - \frac{\frac{\partial u_\varepsilon^*}{\partial n}}{\left| \frac{\partial u_\varepsilon^*}{\partial n} \right|} \frac{\partial v}{\partial n} \right) (1+kn)\, ds\, dn$$

and $X \geq 0$ hence (4.23) follows. ∎

One introduces next w_ε obtained by a Taylor expansion in (4.25) :

(4.27)
$$w_\varepsilon = \begin{cases} 1 - \frac{kn}{\varepsilon}\left(\sqrt{\frac{2\varepsilon}{k}} - \frac{n}{2} \right), & n \leq \sqrt{\frac{2\varepsilon}{k}}, \\ 0 \text{ for } n \geq \sqrt{\frac{2\varepsilon}{k}} . \end{cases}$$

One verifies next that

$$H_\varepsilon(u_\varepsilon^*) - H_\varepsilon(w_\varepsilon) = O(\varepsilon) ,$$

$$J_\varepsilon(w_\varepsilon) - H_\varepsilon(w_\varepsilon) = O(\varepsilon) .$$

From the definitions, it follows that

$$J_\varepsilon(w_\varepsilon) \geq J_\varepsilon(u_\varepsilon) \geq H_\varepsilon(u_\varepsilon^*) \geq H_\varepsilon(w_\varepsilon) + O(\varepsilon)$$

so that

$$J_\varepsilon(u_\varepsilon) = H_\varepsilon(w_\varepsilon) + O(\varepsilon)$$

and we finally obtain (1):

(4.28)
$$J_\varepsilon(u_\varepsilon) = L\left(1 + \frac{4\sqrt{2}}{3} \varepsilon^{1/2} \int_0^1 \sqrt{k(s)}\, ds + O(\varepsilon)\right) .$$ ∎

Remark 4.6.

In [9], the A. give the second term in the expansion of $J_\varepsilon(u_\varepsilon)$. ∎

(1) In [9], the A. find $\frac{2\sqrt{2}}{3}$ instead of $\frac{4\sqrt{2}}{3}$.

REFERENCES (CHAPTER II).

[1] Bensoussan, A. and Lions, J.L. Book to appear. Hermann Ed. Vol.1 (1976), Vol.2 (1977).

[2] Bensoussan, A. and Lions, J.L. (1975). Problèmes de temps d'arrêt optimaux et de perturbations singulières dans les Inéquations Quasi Variationnelles. Lecture Notes in Economics and Mathematical Systems. Springer (107), pp. 567-584.

[3] Bensoussan, A. and Lions, J.L. (1973). Problèmes de temps d'arrêt optimal et Inéquations Variationnelles paraboliques. Applicable Analysis. Vol. 3, pp. 267-294.

[4] Deny, J. and Lions, J.L. Les espaces du type de Beppo Levi. 5 (1953-1954), pp. 305-370.

[5] Gagliardo, E. (1957) Caratterizzazione ... Rend. Sem.Mat. Padova 27,p.284-305.

[6] Huet, D. (1968) Perturbations singulières d'Inéq. Var. 267, pp.932-934.

[7] Lions, J.L.(1973). Perturbations singulières dans les problèmes aux limites et en contrôle optimal. Lecture Notes in Mathematics. Springer. Vol. 323.

[8] Mignot, F. and Puel, J.P. (1976).Archive Rat. Mech.Analysis.

[9] Mosolov, P.P. and Miasnikov, V.P. Boundary layer in the problem of longitudinal motion of a cylinder in a visco plastic medium. P.M.M. 38 (1974), pp. 682-692.

[10] Bensoussan, A. and Lions, J.L. (1975) Diffusion Processes ... in Probabilistic Methods in Differential Equations, Lecture Notes in Mathematics, Springer, 451, pp. 8-25.

[11] Brézis, H. Personal communication.

III. VARIATIONAL INEQUALITIES OF EVOLUTION

ORIENTATION

We present in Section 1, following [1], [2], the solution of a stochastic optimal stopping time problem in terms of V.I. of evolution.
Section 2 defines the strong and weak solutions of the V.I. of evolution met in Section 1, and in Section 3, we prove an important result of Mignot-Puel [6].
Other results along the lines of this (introductory) chapter are given in [1] [2].

1. OPTIMAL STOPPING TIMES.

We consider, as in Chapter II, Section 1, the stochastic Ito's differential equation :

(1.1) $$dy = g(y)ds + \sigma(y)dw(s) , \quad y(t) = x , \quad s > t$$

whose solution is denoted by $y_{xt}(s)$.

Starting from $x \in \Omega$, at time t, we denote by τ_{xt} the smallest $s > t$ such that :

$$y_{xt}(s) \notin \Omega.$$

We consider stopping times θ such that

(1.2) $$\theta \leqslant \inf(\tau_{xt}, T) = \tau_{xt} \wedge T.$$

Given θ we define the cost function

(1.3) $$J_{xt}(\theta) = E\left[\int_t^\theta f(y_{xt}(s),s)ds + \psi(y_{xt}(\theta),\theta)\, \chi_{\theta<\tau_{xt}\wedge T}\right]$$

where

- $f(x,t)$ is a given -say continuous - function in $\bar{\Omega} \times]t \leqslant T]$,
- $\psi(x,)$ is a given -say continuous - function in $\bar{\Omega} \times]t \leqslant T]$,
- $\chi_{\lambda<\mu}$ is the characteristic function of the set where $\lambda < \mu$.

We then define the optimal cost function

(1.4) $$u(x,t) = \inf J_{xt}(\theta) , \quad \theta \leqslant \tau_{xt} \wedge T.$$

One can then prove [1] [2] that $u(x,t)$ is characterized as the solution of :

(1.5) $$\left| \begin{array}{l} -\dfrac{\partial u}{\partial t} + Au - f \leqslant 0 , \quad u - \psi \leqslant 0 , \\ \\ (-\dfrac{\partial u}{\partial t} + Au - f)(u-\psi) = 0 \quad \text{in} \quad Q = \Omega \times]0,T[\end{array} \right.$$

(t=0 does not play any particular role then), where A is given by

(1.6) $$Au = -\sum a_{ij}(x)\frac{\partial^2 u}{\partial x_i \partial x_j} - \sum g_i(x)\frac{\partial u}{\partial x_i} ,$$

with the boundary conditions

(1.7) $$u(x,t) = 0 \quad \text{if} \quad x \in \Gamma = \partial\Omega ,$$

and the "initial" condition

(1.8) $$u(x,T) = 0 \quad , \quad x \in \Omega.$$ ∎

Remark 1.1.

In what follows, we shall solve directly (1.5)(1.7)(1.8) without reference to the optimal stopping time problem.

We shall change the orientation of time and we shall use A in its "divergence" form [1], i.e. we shall consider :

(1.9) $$Au = -\sum \frac{\partial}{\partial x_i}\left(a_{ij}(x)\frac{\partial u}{\partial x_j}\right) + \sum a_i(x)\frac{\partial u}{\partial x_i}$$

and we shall look for $u = u(x,t)$ solution of

(1.10) $$\left|\begin{array}{l} \frac{\partial u}{\partial t} + Au - f \leqslant 0 \quad , u - \psi \leqslant 0 \quad , \\ \\ (\frac{\partial u}{\partial t} + Au - f)(u - \psi) = 0 \quad \text{in} \quad Q \quad , \end{array}\right.$$

with (1.7) and

(1.11) $$u(x,o) = 0.$$ ∎

Remark 1.2.

What we are going to say extends to operator A whose coefficients a_{ij}, a_i depend on x and on t . ∎

Remark 1.3.

The set (1.10)(1.7)(1.11) is a so-called V.I. of evolution. We shall give precise strong and weak formulations of this problem in Section 2 below. ∎

Remark 1.4.

We recall that if we define

(1.12) $$w(x,t) = E\left(\int_t^{\tau_{xt}\wedge T} f(y_{xt}(s),s)ds\right)$$

then w is the solution of

---

[1] A transformation which is not possible (or would introduce coefficients distributions) if the a_{ij}'s are only continuous ; we refer to [1].

$$(2.8)\quad \left| \begin{array}{l} (\frac{\partial u}{\partial t}(t),\ v-u(t)) + a(u(t),\ v-u(t)) - (f(t),\ v-u(t)) \geqslant 0 \\ \forall\ v \in K(t)\ ,\quad u(t) \in K(t)\ ; \end{array} \right.$$

we have therefore a V.I. with a convex which depends on t.

Moreover – and this is the main point – in the formulation of the problem (1.3)(1.4), there is no reason to assume any smoothness on ψ except, say, ψ continuous.

Under these circumstances, the problem does not admit in general a strong solution.

We then proceed with the definition of weak solutions.

2.2. Weak formulation.

Let us define :

$$(2.9)\qquad \hat{\mathcal{K}} = \{v \mid v \in L^2(0,T\ ;\ V)\ ,\quad v \leqslant \psi \quad \text{a.e. in } Q\ \}\ ,$$

$$(2.10)\qquad \mathcal{K} = \{v \mid v \in \hat{\mathcal{K}},\ \frac{\partial v}{\partial t} \in L^2(0,T\ ;\ V')\ \}\ ;$$

and let us make the only assumption that

$$(2.11)\qquad \mathcal{K} \neq \emptyset\ .$$

Let u be a solution of (2.8) and let v be an element of $\mathcal{K}$; if we set

$$X = \left[\int_0^T (\frac{\partial v}{\partial t},\ v-u) + a(u,v-u) - (f,v-u)\right] dt$$

we have

$$X = \left[\int_0^T (\frac{\partial u}{\partial t},\ v-u) + a(u,v-u) - (f,v-u)\right] dt + \int_0^T (\frac{\partial (v-u)}{\partial t},\ v-u) dt$$

and by virtue of (2.8) we see that

$$X \geqslant \frac{1}{2}|v(T) - u(T)|^2 - \frac{1}{2}|v(0)-u(0)|^2 \geqslant -\frac{1}{2}\ |v(0)|^2\ ,$$

i.e.

$$(2.12)\quad \left| \begin{array}{l} \int_0^T \left[(\frac{\partial v}{\partial t}\ ,v-u) + a(u,v-u) - (f,v-u)\right] dt + \frac{1}{2}|v(0)|^2 \geqslant 0 \\ \forall\ v \in \mathcal{K}\ . \end{array} \right.$$

We now define a "weak solution" as a function $u \in \hat{\mathcal{K}}$ satisfying (2.12).

The main result is now given in Section 3 below.

(1.13) $$-\frac{\partial w}{\partial t} + Aw = f \quad \text{in} \quad Q ,$$

subject to

(1.14) $$\begin{cases} w(x,t) = 0 & \text{if} \quad x \in \Gamma , \\ w(x,T) = 0 & x \in \Omega ; \end{cases}$$

(1.12) is the classical Feynmann-Kac formula, so that (1.4) can be thought of as an extension of this formula. ∎

Remark 1.5.

One can use formula (1.4) to obtain informations on the projection of the solution of (1.5)(1.7)(1.8) - for instance for singular perturbations as in Chapter II. We refer to [1]. ∎

2. STRONG AND WEAK FORMULATIONS OF THE V.I.

2.1. Strong Formulation.

We define

(2.1) $$a(u,v) = \sum \int_{\Omega} a_{ij}(x) \frac{\partial u}{\partial x_j} \frac{\partial v}{\partial x_i} dx + \sum \int_{\Omega} a_i(x) \frac{\partial u}{\partial x_i} v \, dx ;$$

formally (1.10)(1.7) is equivalent to :

(2.2) $$\begin{cases} (\frac{\partial u}{\partial t}, v-u) + a(u,v-u) - (f,v-u) \geq 0 \\ \forall v \in H_0^1(\Omega) , \quad v \leq \psi(x,t) , \quad u \leq \psi , \end{cases}$$

where in (2.2) t is fixed a.e.

By a "strong solution" we mean a function $u = u(x,t)$ such that

(2.3) $$u \in L^2(0,T ;V), \quad V = H_0^1(\Omega) ,$$

(2.4) $$\frac{\partial u}{\partial t} \in L^2(0,T ;V') ,$$

(2.5) $$u \leq \psi \quad \text{a.e.}$$

(2.6) $$u(x,0) = 0$$

and u satisfies (2.2). ∎

The main problem.

Let us set :

(2.7) $$K(t) = \{v | \ v \in H_0^1(\Omega) , \ v(x) \leq \psi(x,t) \ \text{a.e. in } \Omega\} ;$$

then (2.2) is equivalent to :

3. EXISTENCE OF A WEAK MAXIMUM SOLUTION.

We shall prove now that (1) :

(3.1) There exists, in the set of weak solutions of (2.12), a maximum solution u, i.e. which is such that, $\forall\, w$ weak solution, one has $w \leqslant u$.

Preliminary reduction.

By changing u into $e^{kt}u$ we see that we do not restrict the generality by assuming that :

$$a(v,v) \geqslant \alpha \|v\|^2 \quad , \quad \alpha > 0 \quad , \quad \forall\, v \in V . \tag{3.2}$$

The plan of the proof is as follows :

Step 1 : We introduce the penalized equation

$$\begin{cases} \dfrac{\partial u_\varepsilon}{\partial t} + Au_\varepsilon + \dfrac{1}{\varepsilon}(u_\varepsilon - \psi)^+ = f \quad , \\ u_\varepsilon = 0 \quad \text{if} \quad x \in \Gamma \, , \\ u_\varepsilon(x,o) = 0 \end{cases} \tag{3.3}$$

and we check that

$$\begin{cases} u_\varepsilon \longrightarrow \tilde{u} \quad \text{in} \quad L^2(0,T\,;V) \text{ , weakly} \\ \tilde{u} \text{ being a weak solution of } (2.12). \end{cases} \tag{3.4}$$

Step 2 : We show that :

$$u_\varepsilon \downarrow \quad \text{as} \quad \varepsilon \downarrow . \tag{3.5}$$

Step 3 : We show that given w , weak solution of (2.12), one has :

$$u_\varepsilon \geqslant w \quad \forall\, \varepsilon > 0 . \tag{3.6}$$

Of course (3.1) immediately follows, with actually the supplementary information that :

(3.7) the maximum solution is the limit of the u_ε's as $\varepsilon \to 0$.

---

(1) This result is due to Mignot and Puel [6].

<u>Proof of Step 1.</u>

One first proves the existence and uniqueness of a solution of (3.3) which is such that $u_\varepsilon \in L^2(0,T;V)$, $\frac{\partial u_\varepsilon}{\partial t} \in L^2(0,T,V')$.

In order to obtain <u>a priori estimates,</u> we take the scalar product of (3.3) with $u_\varepsilon - v_0$, $v_0 \in K$; we obtain :

$$(3.8) \qquad (\frac{\partial u_\varepsilon}{\partial t}, u_\varepsilon - v_0) + a(u_\varepsilon, u_\varepsilon - v_0) + \frac{1}{\varepsilon}((u_\varepsilon - \psi)^+, u_\varepsilon - \psi + \psi - v_0) = (f, u_\varepsilon - v_0)$$

and since $\psi - v_0 \geqslant 0$, it follows that

$$(\frac{\partial(u_\varepsilon - v_0)}{\partial t}, u_\varepsilon - v_0) + a(u_\varepsilon, u_\varepsilon - v_0) + (\frac{\partial v_0}{\partial t}, u_\varepsilon - v_0) + a(v_0, u_\varepsilon - v_0) +$$
$$+ \frac{1}{\varepsilon}|(u_\varepsilon - \psi)^+|^2 \leqslant (f, u_\varepsilon - v_0)$$

hence it follows by standard arguments that

$$(3.9) \qquad \|u_\varepsilon\|_{L^2(0,T;V)\cap L^\infty(0,T;H)} \leqslant C \quad , (H = L^2(\Omega))$$

$$(3.10) \qquad \frac{1}{\sqrt{\varepsilon}} \|(u_\varepsilon - \psi)^+\|_{L^2(Q)} \leqslant C \quad ,$$

where the C's denote constants which do not depend on ε .

We can extract a subsequence, still denoted by u_ε , such that

$$(3.11) \qquad u_\varepsilon \to \tilde{u} \text{ in } L^2(0,T;V) \text{ weakly, and in } L^\infty(0,T;L^2(\Omega)) \text{ weak star ;}$$

let v be given in K ; it follows from (3.3) that

$$(\frac{\partial u_\varepsilon}{\partial t}, v - u_\varepsilon) + a(u_\varepsilon, v - u_\varepsilon) - (f, v - u_\varepsilon) = -\frac{1}{\varepsilon}((u_\varepsilon - \psi)^+, v - u_\varepsilon) =$$
$$= \text{(since } v \leqslant \psi) = \frac{1}{\varepsilon}((v-\psi)^+ - (u_\varepsilon - \psi)^+, v - \psi - (u_\varepsilon - \psi)) \geqslant 0$$

and therefore

$$(3.12) \qquad \int_0^T \left[(\frac{\partial v}{\partial t}, v - u_\varepsilon) + a(u_\varepsilon, v - u_\varepsilon) - (f, v - u_\varepsilon)\right] dt + \frac{1}{2}|v(o)|^2 \geqslant 0.$$

Since $\lim.\inf \int_0^T a(u_\varepsilon, u_\varepsilon)dt \geqslant \int_0^T a(\tilde{u}, \tilde{u})dt$, it follows that $\tilde{u}$ satisfies (2.12).

In order to prove that $\tilde{u}$ is a weak solution, it remains only to check that

$$\tilde{u} \leqslant \psi \quad \text{a.e.} \tag{3.13}$$

We argue as in (2.23)(2.24), Chapter II. We start from the inequality

$$\int_0^T ((v-\psi)^+ - (u_\varepsilon-\psi)^+, v-u_\varepsilon)dt \geqslant 0 \tag{3.14}$$

which is valid for every $v \in L^2(Q)$; by virtue of (3.10) we can pass to the limit in (3.14) ; we obtain

$$\int_0^T ((v-\psi)^+, v-\tilde{u})dt \geqslant 0 \quad \forall v. \tag{3.15}$$

We now take $v = \tilde{u} + \lambda\varphi$, $\lambda > 0$, $\varphi \in L^2(Q)$; we obtain after dividing by λ :

$$\int_0^T ((\tilde{u} - \psi + \lambda\varphi)^+, \varphi)\, dt \geqslant 0$$

and we let $\lambda \to 0$; it comes :

$$\int_0^T ((\tilde{u} - \psi)^+, \varphi)dt \geqslant 0 \quad \forall \varphi \in L^2(Q) ,$$

and therefore $(\tilde{u} - \psi)^+ = 0$, hence (3.13) follows. ∎

Proof of Step 2. Let $\varepsilon \leqslant \hat{\varepsilon}$, $u_\varepsilon = u$, $u_{\hat{\varepsilon}} = \hat{u}$. We want to show that $u \leqslant \hat{u}$. We take the scalar product of (3.3) (resp. the penalized equation relative to $\hat{u}$) with $(u - \hat{u})^+$ resp. with $-(u - \hat{u})^+$). We obtain :

$$\left(\frac{\partial}{\partial t}(u-\hat{u}), (u-\hat{u})^+\right) + a(u-\hat{u}, (u-\hat{u})^+) + X = 0 , \tag{3.16}$$

where

$$X = \frac{1}{\varepsilon}((u-\psi)^+, (u-\hat{u})^+) - \frac{1}{\varepsilon}((\hat{u}-\psi)^+, (u-\hat{u})^+) =$$

$$= \left(\frac{1}{\varepsilon} - \frac{1}{\hat{\varepsilon}}\right)((u-\psi)^+, (u-\hat{u})^+) + Y \geqslant Y ,$$

$$Y = \frac{1}{\hat{\varepsilon}}((u-\psi)^+ - (\hat{u}-\psi)^+, (u-\hat{u})^+) .$$

We have $Y \geqslant 0$; indeed we integrate over the set where $u \geqslant \hat{u}$; on this set $u-\psi \geqslant \hat{u}-\psi$ so that $(u-\psi)^+ \geqslant (\hat{u}-\psi)^+$ and $Y \geqslant 0$. Consequently (3.16) implies that

$$\left(\frac{\partial}{\partial t}(u-\hat{u}), (u-\hat{u})^+\right) + a((u-\hat{u})^+, (u-\hat{u})^+) \leqslant 0$$

i.e.

$$\frac{1}{2}\frac{d}{dt}|(u-\hat{u})^+|^2 + a((u-\hat{u})^+, (u-\hat{u})^+) \leqslant 0 \tag{3.17}$$

hence it follows that $(u-\hat{u})^+ = 0$. ∎

Proof of Step 3.

This is the most interesting part in the proof.

Let w be a weak solution and let us introduce

$$z = w - u_\varepsilon \ , \tag{3.18}$$

$$K_1 = K - u_\varepsilon . \tag{3.19}$$

There one verifies, by an elementary computation, that

$$\begin{cases} \int_0^T \left[\left(\frac{\partial v}{\partial t}, v-z\right) + a(z, v-z) - \frac{1}{\varepsilon}((u_\varepsilon-\psi)^+, v-z) \right] dt + \frac{1}{2}|v(o)|^2 \geqslant 0 \\ \forall \ v \in K_1 \ , \ z \in K_1 . \end{cases} \tag{3.20}$$

Indeed, for every $\tilde{v} \in K$ we have

$$\int_0^T \left[\left(\frac{\partial \tilde{v}}{\partial t}, \tilde{v}-w\right) + a(w, \tilde{v}-w) - (f, \tilde{v}-w) \right] dt + \frac{1}{2}|\tilde{v}(o)|^2 \geqslant 0 \ , \tag{3.21}$$

and we deduce from (3.3) that

$$\int_0^T \left[\left(\frac{\partial u_\varepsilon}{\partial t}, \tilde{v}-w\right) + a(u_\varepsilon, \tilde{v}-w) + \frac{1}{\varepsilon}((u_\varepsilon-\psi)^+, \tilde{v}-w) \right] dt = \int_0^T (f, \tilde{v}-w) dt \ ; \tag{3.22}$$

by substraction, we obtain,

$$\int_0^T \left[\left(\frac{\partial(\tilde{v}-u_\varepsilon)}{\partial t}, \tilde{v}-w\right) + a(w-u_\varepsilon, \tilde{v}-w) - \frac{1}{\varepsilon}((u_\varepsilon-\psi)^+, \tilde{v}-w) \right] dt + \frac{1}{2}|\tilde{v}(o)|^2 \geqslant 0.$$

If we set $\tilde{v} - u_\varepsilon = v$, this is (3.20). ∎

We now prove that

$$\begin{cases} \text{if } z \text{ satisfies to (3.20) then} \\ \int_0^T \left[-\left(\frac{\partial \theta}{\partial t}, z\right) + a(z, \theta) - \frac{1}{\varepsilon}((u_\varepsilon-\psi)^+, \theta) \right] dt \leqslant 0 \\ \forall \ \theta \in \Theta . \end{cases} \tag{3.23}$$

where

$$\Theta = \{\theta \mid \theta \in L^2(0,T;V), \ \frac{\partial \theta}{\partial t} \in L^2(0,T;V'), \ \theta(T) = 0 \ , \ \theta \geqslant 0 \ \} . \tag{3.24}$$

Indeed, let v_0 be a given element of K_1 and let us define

(3.25) $$v = v_o - \lambda\theta \ , \quad \lambda > 0 \ , \quad \theta \in \mathbb{H} \ .$$

We have $v \in K_1$ and we take v given by (3.25) in (3.20).

We obtain :

(3.26) $$X\lambda^2 + Y\lambda + Z \geq 0 \qquad \forall \lambda \geq 0 \ ;$$

but $X = 0$; indeed

$$X = \int_0^T \left(-\frac{\partial\theta}{\partial t}, \theta\right) dt + \frac{1}{2}|\theta(o)|^2 = 0 \quad \text{since} \quad \theta(T) = 0 \ ;$$

therefore (3.26) implies

(3.27) $$Y \geq 0 \ .$$

But

$$Y = \int_0^T \left[-\left(\frac{\partial\theta}{\partial t}, v_o - z\right) - \left(\frac{\partial v_o}{\partial t}, \theta\right) - a(z,\theta) + \frac{1}{\varepsilon}((u_\varepsilon - \psi)^+, \theta)\right] dt - (v_o(o), \theta(o))$$

and since

$$\int_0^T \left[-\left(\frac{\partial\theta}{\partial t}, v_o\right) - \left(\frac{\partial v_o}{\partial t}, \theta\right)\right] dt - (v_o(o), \theta(o)) = 0 \ ,$$

it comes

$$Y = \int_0^T \left[\left(\frac{\partial\theta}{\partial t}, z\right) - a(z,\theta) + \frac{1}{\varepsilon}((u_\varepsilon - \psi)^+, \theta)\right] dt$$

so that (3.23) follows from (3.27).

We show now that (3.23) implies :

(3.28) $$z \leq 0$$

which is (3.6).

We define θ_η as the solution of

(3.29) $$-\eta \frac{d\theta_\eta}{dt} + \theta_\eta = z^+ \ , \quad \theta_\eta(T) = 0 \ , \quad \eta > 0.$$

We have $\theta_\eta \geq 0$ so that $\theta_\eta \in \mathbb{H}$ and we can choose $\theta = \theta_\eta$ in (3.23) ; we obtain :

$$\xi_\eta + \int_0^T \left[a(z,\theta_\eta) - \frac{1}{\varepsilon}((u_\varepsilon-\psi)^+,\theta_\eta) \right] dt \leq 0 \quad ,$$

$$\xi_\eta = \int_0^T (-(\frac{\partial\theta_\eta}{\partial t},z))dt = -\int_0^T (\frac{\partial\theta_\eta}{\partial t},z^+)dt + \int_0^T (\frac{\partial\theta_\eta}{\partial t}, z^-)\, dt =$$

$$= +\eta \int_0^T \left|\frac{\partial\theta_\eta}{\partial t}\right|^2 dt - \int_0^T (\frac{\partial\theta_\eta}{\partial t}, \theta_\eta)dt + \int_0^T \frac{1}{\eta}(\theta_\eta - z^+, z^-)\, dt =$$

$$= \eta \int_0^T \left|\frac{\partial\theta_\eta}{\partial t}\right|^2 dt + \frac{1}{2}|\theta_\eta(0)|^2 + \frac{1}{\eta}\int_0^T (\theta_\eta, z^-)dt \geqslant 0$$

(since all terms are $\geqslant 0$) , so that (3.30) implies

(3.31) $$\int_0^T \left[a(z,\theta_\eta) - \frac{1}{\varepsilon}((u_\varepsilon-\psi)^+, \theta_\eta) \right] dt \leq 0.$$

We can now let $\eta \to 0$ in (3.31). We obtain

(3.32) $$\int_0^T \left[a(z,z^+) - \frac{1}{\varepsilon}((u_\varepsilon-\psi)^+, z^+) \right] dt \leq 0.$$

But $(u_\varepsilon-\psi)^+ z^+ = 0$ since if $u_\varepsilon \geqslant \psi$ and $z = w-u_\varepsilon \geqslant 0$, then $w \geqslant u_\varepsilon \geqslant \psi$ and since $w \leqslant \psi$ one has $u_\varepsilon = \psi$.

Therefore (3.32) reduces to :

$$\int_0^T a(z,z^+)dt \leq 0$$

i.e.

$$\int_0^T a(z^+,z^+)dt \leq 0 \quad \text{hence} \quad z^+ = 0 \ .$$ ∎

Remark 3.1.

Other results for V.I. of the type studied here can be found in [3] [4] [7] and the bibliography therein. ([4] uses a result of [5]). ∎

Remark 3.2.

The preceding proof uses, in an essential manner, the fact that A is a second order operator. The study of V.I. with constraints of the type $v \leqslant \psi(x,t)$, ψ "non smooth", for, say, the operator $\frac{\partial}{\partial t} + \Delta^2$, is an open question. ∎

REFERENCES (CHAPTER III).

[1] Bensoussan, A. and Lions, J.L. Book to appear. Hermann ed. Vol. 1 (1976), Vol. 2 (1977).

[2] Bensoussan, A. and Lions, J.L. (1973) Applicable Analysis. Vol. 3 pp. 267-294.

[3] Brézis, H. C.R.Ac. Sc. Paris, t. 274 (1972), pp. 310-313.

[4] Charrier, P. and Troianiello, G.M. (1975) C.R.Ac. Sc. Paris.

[5] Hanouzet, B. and Joly, J.L. (1975) C.R.Ac. Sc. Paris.

[6] Mignot, F. and Puel, J.P. (1975) C.R.Ac. Sc. Paris.

[7] Troianiello, G.M. To appear.

W. Eckhaus (ed.), New Developments in Differential Equations

FREE BOUNDARY PROBLEMS FOR POISSON'S EQUATION

Guido Stampacchia

Scuola Normale Superiore

P I S A (Italy)

§1 - In this talk we shall give an account of a joint paper with D. Kinderlehrer which describes the formulation and solution of a free boundary value problem in the framework of the theory of variational inequalities.

We confine our attention to a problem in the plane which consists in finding a domain Ω and a function u defined in Ω satisfying there a Poisson's equation together with both assigned Dirichlet and Neumann type data on the boundary Γ of Ω. Under suitable hypotheses about the given data, we prove that there is a unique solution pair Ω,u which solves this problem and that Γ is a smooth curve.

Let $z=x_1+ix_2=\rho e^{i\theta}$, $0\leq\theta<2\pi$, denote a point in the z plane. Let us suppose that F(z) is a function in $C^2(R^2)$ which satisfies the conditions

$$\begin{aligned} &\rho^{-2}F(z) \in C^2(R^2) \\ &\inf_{R^2} \rho^{-2} F(z) > 0 \\ &F_\rho(z)\geq 0 \quad \text{for} \quad z \in R^2 \\ &F(0) = F_\rho(0) = 0 . \end{aligned} \tag{1.1}$$

Our object is to solve, in some manner, the

Problem 1 - To find a bounded Ω *and a function* u *such that*

$$-\Delta u=\rho^{-1}F_\rho \quad \text{in } \Omega \tag{1.2}$$

$$\left.\begin{aligned} &u=0 \\ &\frac{\partial u}{\partial \nu} = -F\frac{d\theta}{ds} \end{aligned}\right\} \text{ on } \Gamma \tag{1.3}$$

$$u(0) = \gamma \tag{1.4}$$

where $\Gamma = \partial\Omega$, ν *is the outward directed vector and* s *is the arc length on* Γ, F *satisfies* (1.1), *and* γ *is given.*

Supposing Ω,u to be a solution to *Problem 1*, the maximum principle for superharmonics implies that $u>0$ in Ω since $-\Delta u\geq 0$ in Ω. We assume, consequently, that $\gamma>0$ and that $u \in C(R^2)$ with $\Omega = \{z:u(z)>0\}$. Further, if Ω is a domain with smooth boundary Γ and u satisfy (1.2) in Ω and (1.3) on Γ then

$$\frac{\partial u}{\partial \nu}(z) < 0 \qquad \text{for} \quad z \in \Gamma$$

or the central angle θ is a strictly increasing function of the arc length parameter on Γ. Interpreting this situation geometrically, we conclude *if* Γ *is smooth and* u *satisfies* (1.2) *in* Ω *and* (1.3) *on* Γ , *then* Ω *is starshaped with respect to* $z=0$.

We shall solve *Problem 1* by means of a variational inequality suggested by the properties of a function $g(z)$ which satisfies

$$g_\rho = -\rho^{-1}u \ . \tag{1.5}$$

A characteristic of the present work is the logarithmic nature of a function g defined by (1.5) at $z=0$. This difficulty will be overcome by considering an unbounded obstacle.

In the following section we transform our problem to one concerning a variational inequality.

§2 - In this section we introduce a variational inequality and determine its relationship to *Problem 1*. We begin with some notations. Employing usual notations for function spaces, set $B_r = \{z:|z|<r\}$, $r>0$, and

$$\mathbb{K}_r = \{v \in H^1(B_r): v \geq \log\rho \text{ in } B_r \text{ and } v=\log r \text{ on } \partial B_r\} \ .$$

Define the bilinear form

$$a(v,\zeta) = \int_{B_r} v_{x_i}\zeta_{x_i}\,dx = \int_{B_r}\{v_\rho\zeta_\rho + \frac{1}{\rho^2}v_\theta\zeta_\theta\}\rho\,d\rho\,d\theta \qquad \text{for } v,\zeta \in H^1(B_2)$$

where we have depressed the dependence of $a(v,\zeta)$ on $r>0$.
Let

$$f \in L^p_{loc}(R^2) \quad \text{for some } p > 2 \ .$$

Problem ($\star$) - *To find a pair* $r>1$ *and* $w \in \mathbb{K}_r$ *such that*

$$w \in \mathbb{K}_r : a(w,v-w) \geq \int_{B_r} f(v-w)\,dx \qquad v \in \mathbb{K}_r \tag{2.1}$$

and the function $\tilde{w}(z)$ *defined by*

$$\tilde{w}(z) = \begin{cases} w(z) & z \in B_r \\ \log|z| & z \notin B_r \end{cases} \tag{2.2}$$

is in $C^1(R^2)$.

The existence and other properties of a solution to *Problem* ($\star$) can be easily investigated. We note that the restriction of $\tilde{w}$ to B_R for $R>r$ will be a solution of (2.1) in B_R . Since this means that (2.2) will be automatically satisfied, so that $R,\tilde{w}|_{B_R} \in \mathbb{K}_r$ is also a solution to *Problem* ($\star$), we shall not distinguish between w and $\tilde{w}$ in the sequel.

We have the following

THEOREM 1 - *Let* Ω,u *be a solution of Problem 1 where* F *satisfies* (1.1) *and* $\gamma>0$. *Suppose that* Γ *is a smooth curve. Then there exists a solution* $r,w \in \mathbb{K}_r$ *of Problem* ($\star$) *for*

$$f(z) = -\frac{1}{\gamma\rho^2}F(z)$$

such that

$$\Omega = \{z:w(z)>\log\rho\} \quad \textit{and} \quad u(z) = \gamma(1-\rho w_\rho(z)) \ .$$

§3 - According to a well known theorem, there is a solution to the variational inequality (2.1) for each $r>0$. To establish its smoothness in B_r we shall prove that it is bounded. For once this is known, the obstacle $\log\rho$ may be replaced by a smooth obstacle which equals $\log\rho$ when

$$\log \rho > -\|w\|_{L^\infty(B_r)}$$

and (2.1) may be solved in the convex $\mathbb{K}_\psi$ of $H^1(B_r)$ functions which exceed ψ in B_r and satisfy the boundary condition $v(z) = \log r$, $|z|=r$. The solution to this latter problem is known to be suitably smooth and is easily shown to be the solution of (2.1).

By standard methods we have the following

LEMMA - *Let* $f \in L^p(B_r)$ *for some* $p>2$ *and satisfy*

$$f \leq 0 \quad \text{in } B_r .$$

Then the solution w *of* (2.1) *for* f *satisfies*

$$\log r - c\|f\|_{L^p(B_r)} \leq w(z) \leq \log r \quad \text{in } B_r ,$$

where $c=c(r,p)>0$.

On the other hand we have the following

THEOREM 2 - *Let* $f \in L^p_{loc}(R^2)$ *for a* $p>2$ *satisfy*

$$\sup_{R^2} f < 0 .$$

Then there exists a solution $r, w \in \mathbb{K}_r$ *to Problem* (★). *In addition,* $w \in H^{2,p}(B_r)$.

The main step in the proof is to construct a supersolution $g(z)=h(\rho)$ to the form

$$a(w,\zeta) - \int_{B_r} f\zeta\, dx$$

for some $r>1$, which satisfies

$$h \in \mathbb{K}_r \tag{3.2}$$

$$h_\rho = \frac{1}{r} \tag{3.3}$$

If (3.2) and (3.3) are fulfilled, then, from a well known property

$$w \leq h \quad \text{in } B_r .$$

Moreover, since $\log \rho \leq w \leq h$ we conclude from (3.3) that

$$w_\rho(z) = \frac{1}{r} \quad \text{for } |z| = r$$

and, since $w=\log r$ for $|z| = r$,

$$w_\theta(z) = 0 \quad \text{for } |z| = r$$

Therefore $\tilde{w}$ defined by (2.2) is in $C^1(R^2)$.

§4 - Next step is to show that the set where the solution to *Problem* (★) exceeds $\log \rho$ in starshaped under an assumption about f. Indeed we have the following:

THEOREM 3 - *Let* $f \in L^p_{loc}(R^2)$ *satisfy* $\sup_{R^2} f<0$ *and* $\rho^{-1}(\rho^2 f)_\rho \le 0$. *Let* $r,w \in \mathbb{K}_r$ *denote the solution of Problem* (★) *for* f *and set*

$$\Omega = \{z:w(z) > \log \rho\} .$$

Then Ω *is starshaped with respect to* $z = 0$.

§5 - In this section we shall report on the smoothness of the free boundary determined by a solution to *Problem* (★) assuming that $f \in C^1(R^2)$ and that

$$\sup_{R^2} f < 0 \quad \textit{and} \quad (\rho^2 f)_\rho \le 0 \quad \textit{in} \quad R^2 .$$

Then the following theorems hold

THEOREM 4 - *Let* $f \in C^1(R^2)$ *satisfy* (5.1) *and let* $r,w \in \mathbb{K}_r$ *denote the solution to Problem* (★) *for* f . *Let*

$$\Omega = \{z:w(z)\ge\log \rho\} .$$

Then the boundary Γ *of* Ω *has the representation*

$$\Gamma : \rho = \mu(\theta) \quad , \quad 0\le\theta\le 2\pi$$

where μ *is a continuous function of bounded variation.*

THEOREM 5 - *Let* $f \in C^1(R^2)$ *satisfy* (5.1). *Let* $r,w \in \mathbb{K}_r$ *denote the solution to Problem* (★) *for* f *and* Γ *the boundary of* $\Omega = \{z:w(z)>\log \rho\}$. *Then* Γ *has a* $C^{1,\tau}$ *parametrization,* $0<\tau<1$.

At this point we can give an answer in the affermative to the question of the existence of a solution to *Problem 1* with the

THEOREM 6 - *Let* $F \in C^1(R^2)$ *satisfy conditions* (1.1). *Then there exists a domain* Ω *and a function* $u \in H^{1,\infty}_{loc}(R^2)$ *such that*

$$-\Delta u = \rho^{-1} F_\rho \quad \textit{in} \quad \Omega$$

$$\begin{cases} u = 0 \\ \dfrac{\partial u}{\partial \nu} = -F \dfrac{d\theta}{ds} \end{cases} \quad \text{a.e. } \textit{on} \quad \Gamma$$

$$u(0) = \gamma$$

where ν *is the outward directed normal vector and* s *is the arc lenght of* Γ *and* $\gamma>0$ *is given.*

Given F , define $f(z) = -\frac{1}{\gamma\rho^2} F(z)$ and observe that f satisfies (5.1). Denote by $z,w \in \mathbb{K}_r$ the solution to *Problem* (★) for f and define

$$u(z) = \gamma(1-\rho w_\rho(z)) \quad z \in R^2 .$$

It can be proved that $u(z)$, so defined, satisfies the conclusions of theorem 6.

REFERENCES

For the bibliography and more details we refer to

D.Kinderlehrer and G.Stampacchia. A Free Boundary Value Problem in Potential Theory *Ann.Inst. Fourier,* 25 3/4 (1975) *to appear.*

W. Eckhaus (ed.), New Developments in Differential Equations

NONLINEAR ELLIPTIC EQUATIONS WITH NONLINEAR BOUNDARY CONDITIONS

Herbert Amann
Department of Mathematics
Ruhr-Universität
Bochum , Germany

INTRODUCTION

In this paper we study mildly nonlinear elliptic boundary value problems (BVPs) of the form

$$\begin{aligned} Au &= f(x,u) \quad \text{in} \quad \Omega \ , \\ Bu &= g(x,u) \quad \text{on} \quad \Gamma \ , \end{aligned} \tag{1.1}$$

where Ω is a bounded domain in $\mathbb{R}^N$ with sufficiently smooth boundary Γ. We suppose that A is a second order, strongly uniformly elliptic differential operator and B is a boundary operator of the form

$$Bu = \frac{\partial u}{\partial \beta} + bu \quad ,$$

where β is an outward pointing, nowhere tangent vector field on Γ.

Problems of this type arise, in particular, in the study of steady state solutions of nonlinear parabolic equations of the form

$$\begin{aligned} \frac{\partial u}{\partial t} + Au &= f(x,u) \quad \text{in} \quad \Omega \times (0,\infty) \ , \\ Bu &= g(x,u) \quad \text{on} \quad \Gamma \times (0,\infty) \ , \\ u &= u_0 \quad \text{on} \quad \overline{\Omega} \end{aligned} \tag{1.2}$$

In this connection, nonlinear boundary conditions seem to be of particular importance. For the study of the stability of the solutions of the parabolic initial-boundary value problem (1.2), one has to have a good knowledge of the steady states, that is, of the solutions of the elliptic BVP (1.1). Of course, the most interesting case occurs if the elliptic BVP has several distinct solutions. (For an interesting analysis of an initial BVP of the form (1.2) in the case of one space dimension and in the presence of several distinct steady states cf. [6]).

Unlike to the situation where g is independent of u, not much seems to be known if the boundary condition depends nonlinearly on the unknown function u. Recently, the theory of monotone operators has been applied to BVPs of the form (1.1) (cf.[7,8,10]). However, in all of these papers the boundary condition is of the special form

$$\frac{\partial u}{\partial \nu} = g(u) \quad ,$$

where g is *decreasing* and ν is the co-normal with respect to the differential operator A. Moreover, the theory of monotone operators does not seem to yield proper multiplicity results.

Besides these results, there are some scattered existence theorems for nonlinear Stecklov problems of the form

$$Au = o \quad \text{in} \quad \Omega \quad ,$$

$$\frac{\partial u}{\partial \nu} = g(x,u) \quad \text{on} \quad \Gamma \quad ,$$

where A is supposed to be formally self-adjoint such that the homogeneous linear BVP possesses a nontrivial solution (cf. [9,12]). This situation will also be covered by our general results.

So far, the only general existence theorem for the BVP (1.1) seems to be a result due to the author [2] (see also [26]), namely the result that the existence of a subsolution $\bar{v}$ and of a supersolution $\hat{v}$ for (1.1) with $\bar{v} \leq \hat{v}$ guarantees the existence of a solution. In this paper we give a new (and more elegant) proof for this theorem by transforming the BVP (1.1) into an equivalent fixed point equation in $C(\bar{\Omega})$. This transformation has the advantage that it makes the problem (1.1) accessible to the powerful tools of nonlinear functional analysis. Some of these tools, namely the theory of increasing, completely continuous maps in ordered Banach spaces (cf. [5,15]) are then used to enlarge the domain of applicability of the general existence theorem by deriving simple sufficient criteria for the existence of sub- and supersolutions. In addition, we can derive a nonexistence and a general uniqueness theorem. Moreover, in order to demonstrate the power of this abstract approach, we derive a multiplicity result, namely a criterion guaranteeing the existence of at least three distinct solutions.

In the following paragraph we state the main results for the nonlinear BVP (1) which are proved in this paper. In Paragraph 3 we establish a fundamental a priori estimate for the solutions of the linear BVP $Au = v$ in Ω , $Bu = w$ on Γ , involving only a L_p-norm of the boundary term.

In Paragraph 4 we derive the equivalent fixed point equation and we prove, besides of the fundamental existence result, the above mentioned multiplicity theorem.

The last paragraph is of more abstract nature. Namely, it contains a derivation of the basic spectral properties for positive linear operators in an ordered Banach space which map every point of the positive cone either into the interior of the cone or onto zero. These results generalize the known spectral properties for strongly positive operators (cf. [14]). Moreover they are needed for the study of the linear eigenvalue problem

$$Au = mu \quad \text{in} \quad \Omega \quad ,$$

$$Bu = ru \quad \text{on} \quad \Gamma \quad .$$

This eigenvalue problem plays a considerable rôle in the solvability theory of the nonlinear BVP (1.1).

2. STATEMENT OF THE MAIN RESULT

Throughout this paper all functions are real-valued and all vector spaces are over the reals.

In the following we specify the *hypotheses* which are used throughout the remainder of this paper.

We suppose that Ω is a nonempty bounded domain in $\mathbb{R}^N$, $N \geq 2$, of class $C^{3+\alpha}$ for some $\alpha \in (o,1)$, that is, the boundary , Γ , of Ω is an (N-1)-dimensional compact $C^{3+\alpha}$-manifold such that Ω lies locally on one side of Γ .

We denote by A a linear differential operator of the form

$$Au := -\sum_{i,k=1}^{N} a_{ik}D_iD_ku + \sum_{i=1}^{N} a_iD_iu + au$$

with symmetric coefficient matrix (a_{ik}). We suppose that $a_{ik} \in C^{2+\alpha}(\overline{\Omega})$, $a_i \in C^{1+\alpha}(\overline{\Omega})$, and $a \in C^{\alpha}(\overline{\Omega})$. Moreover, A is suppose to be strongly uniformly elliptic, that is,

$$\sum_{i,k=1}^{N} a_{ik}(x)\xi^i\xi^k \geq a_0|\xi|^2$$

for some constant $\alpha_0 > o$ and every $x \in \Omega$, and $\xi = (\xi^1,\ldots,\xi^N) \in \mathbb{R}^N$.

We denote by $\beta \in C^{2+\alpha}(\Gamma,\mathbb{R}^N)$ an outward pointing, nowhere tangent vector field on Γ, and $\frac{\partial u}{\partial \beta} := \Sigma\beta^i D_i u$ denotes the directional derivative on Γ of $u \in C^1(\overline{\Omega})$ with respect to β. It should be observed that β is not supposed to be a unit vector field. Then we define a (*regular oblique derivative*) boundary operator B by

$$Bu := \frac{\partial u}{\partial \beta} + bu \quad ,$$

where $b \in C^{1+\alpha}(\Gamma)$.

Let I be a nonempty subinterval of $\mathbb{R}$. We denote by $f : \overline{\Omega} \times I \to \mathbb{R}$ a function which is α-Hölder continuous in the first variable and locally Lipschitz in the second variable. More precisely, for every compact subinterval I' of I, there exists a constant $\gamma(I')$ such that

$$|f(x,\xi) - f(y,\eta)| \leq \gamma(I')(|x-y|^{\alpha} + |\xi-\eta|)$$

for every pair (x,ξ), $(y,\eta) \in \overline{\Omega} \times I'$. Moreover, we suppose that $g : \Gamma \times I \to \mathbb{R}$ is a locally Lipschitz continuous function.

Then we consider mildly nonlinear elliptic BVP's of the form

$$\begin{aligned} Au &= f(x,u) \quad \text{in } \Omega \quad , \\ Bu &= g(x,u) \quad \text{on } \Gamma \quad . \end{aligned} \tag{2.1}$$

By a *solution* u of (2.1) we mean a classical solution, that is, a function $u \in C^2(\Omega) \cap C^1(\overline{\Omega})$ such that $u(\overline{\Omega}) \subset I$, $Au(x) = f(x,u(x))$ for $x \in \Omega$, and $Bu(x) = g(x,u(x))$ for $x \in \Gamma$.

A function u is called a *subsolution* for the BVP (2.1) if $u \in C^2(\Omega) \cap C^1(\overline{\Omega})$, $u(\overline{\Omega}) \subset I$, and

$$\begin{aligned} Au(x) &\leq f(x,u(x)) \quad \text{for } x \in \Omega \quad , \\ Bu(x) &\leq g(x,u(x)) \quad \text{for } x \in \Gamma \quad . \end{aligned}$$

A subsolution which is not a solution is called a *strict* subsolution. *Supersolutions* and *strict* supersolutions are defined by reversing the above inequality signs.

Let X be a nonempty set. If $u,v : X \to \mathbb{R}$ are two functions such that $u(x) \leq v(x)$, for every $x \in X$, then we write $u \leq v$. Moreover, $u < v$ means that $u \leq v$ but $u \neq v$. Finally, by the *order interval* $[u,v]$, between u and v, we mean the set of all functions $w : X \to \mathbb{R}$ such that $u \leq w \leq v$.

The following theorem contains the basic existence result for the BVP (2.1).

(2.1) *Theorem: Let* $\bar{v}$ *be a subsolution and let* $\hat{v}$ *be a supersolution such that* $\bar{v} \le \hat{v}$. *Then the* BVP (2.1) *has at least one solution in the order interval* $[\bar{v},\hat{v}]$.

More precisely, there exists a minimal solution $\bar{u}$ *and a maximal solution* $\hat{u}$ *in the order interval* $[\bar{v},\hat{v}]$ *such that every solution* $u \in [\bar{v},\hat{v}]$ *satisfies* $\bar{u} \le u \le \hat{u}$.

The above existence theorem is essentially contained in an earlier paper by the author (cf. [2, Theorem 3]). In the present paper we give a simple proof by reducing the BVP (2.1) to an equivalent fixed point equation in the Banach space $C(\bar{\Omega})$.

In the following we denote by $C^{1-}(X)$ the Banach space of all Lipschitz continuous functions on the compact metric space X . If X and Y are two nonempty sets and $u : X \to \mathbb{R}$ and $v : Y \to \mathbb{R}$ are two functions, then we write $(u,v) \ge (o,o)$ if $u \ge o$ and $v \ge o$. Moreover, $(u,v) > (o,o)$ means $(u,v) \ge (o,o)$ but $(u,v) \ne (o,o)$.

The second fundamental result of this paper concerns the linear eigenvalue problem (EVP)

$$\begin{aligned} Au &= \lambda m u \quad \text{in } \Omega , \\ Bu &= \lambda r u \quad \text{on } \Gamma , \end{aligned} \tag{2.2}$$

where, as a rule, we suppose that the following *hypothesis* (H) is satisfied:

$$m \in C^{\alpha}(\bar{\Omega}) , \; r \in C^{1-}(\Gamma) , \; \text{and} \; (m,r) > (o,o) .$$

There exists a constant $\mu \ge 0$ *such that* (H)

$$(a + \mu m , b + \mu r) \ge (o,o) .$$

By means of the above mentioned reduction to a fixed point equation in $C(\bar{\Omega})$ we shall prove the following important result:

(2.2) *Theorem: Let the hypothesis* (H) *be satisfied. Then the* EVP (2.2) *possesses a smallest eigenvalue* $\lambda_0(m,r)$, *the principal eigenvalue, and* $\lambda_0(m,r)$ *is positive if* $(a,b) > (o,o)$. *The* EVP (2.2) *has exactly one linearly independent eigenfunction* $u_0 \in C^2(\Omega) \cap C^1(\bar{\Omega})$ *to the eigenvalue* $\lambda_0(m,r)$, *and* u_0 *can be chosen to be everywhere positive. Moreover,* $\lambda_0(m,r)$ *is the only eigenvalue of* (2.2) *having a nonnegative eigenfunction.*

Lastly, $\lambda_0(m,r)$ *is a strictly decreasing function of* (m,r) . *More precisely, suppose that* $m_1 \in C^{\alpha}(\bar{\Omega})$ *and* $r_1 \in C^{1-}(\Gamma)$ *satisfy* $(m_1,r_1) > (m,r)$. *Then* $\lambda_0(m,r) > \lambda_0(m_1,r_1)$.

The next theorem contains some useful results concerning the solvability of the linear BVP

$$\begin{aligned} Au - \lambda m u &= c \quad \text{in } \Omega , \\ Bu - \lambda r u &= d \quad \text{on } \Gamma , \end{aligned} \tag{2.3}$$

where $\lambda \in \mathbb{R}$.

(2.3) *Theorem: Let the hypothesis* (H) *be satisfied and suppose that* $(c,d) \in C^{\alpha}(\bar{\Omega}) \times C^{1-}(\Gamma)$. *Then the linear* BVP (2.3) *has for every* $\lambda < \lambda_0(m,r)$ *exactly one solution which is everywhere positive if* $(c,d) > (o,o)$.

The BVP (2.3) *has no positive solution if either* $\lambda > \lambda_0(m,r)$ *and* $(c,d) \ge (o,o)$ *or* $\lambda = \lambda_0(m,r)$ *and* $(c,d) > (o,o)$.

On the basis of the theorems (2.1) - (2.3), which will be proved in the following paragraphs, it is easy to derive the following general existence, nonexistence, and uniqueness results.

(2.4) *Theorem: Let the hypothesis* (H) *be satisfied. Suppose that there exist nonnegative constants* γ *and* δ *and a real number* $\lambda < \lambda_0(m,r)$ *such that*

$$f(\cdot,\xi) \leq \gamma + \lambda m \xi ,$$
$$g(\cdot,\xi) \leq \delta + \lambda r \xi$$

for every $\xi \geq 0$ *, and*

$$f(\cdot,\xi) \geq -\gamma + \lambda m \xi ,$$
$$g(\cdot,\xi) \geq -\delta + \lambda r \xi \qquad (2.4)$$

for every $\xi \leq 0$. *Then the nonlinear* BVP (2.1) *has at least one solution.*

Proof: It follows from Theorem (2.3) that the linear BVP

$$\begin{aligned} Au &= \lambda mu + \gamma \quad \text{in } \Omega , \\ Bu &= \lambda ru + \delta \quad \text{on } \Gamma \end{aligned} \qquad (2.5)$$

has exactly one solution $\hat{v} \geq o$. Hence $\overline{v} := -\hat{v}$ is the only solution of the linear BVP

$$\begin{aligned} Au &= \lambda mu - \gamma \quad \text{in } \Omega , \\ Bu &= \lambda ru - \delta \quad \text{on } \Gamma . \end{aligned}$$

It is obvious that $\overline{v}$ is a subsolution and $\hat{v}$ is a supersolution for the BVP (2.1). Hence Theorem (2.1) implies the assertion. Q.E.D.

It should be observed that the above proof shows that there exists a solution in the order interval $[-\hat{v},\hat{v}]$, where $\hat{v}$ is the solution of the BVP (2.5).

We emphasize the fact that Theorem (2.4) imposes one-sided growth conditions only. For example, Theorem (2.4) implies the existence of a solution of the BVP (2.1) provided $(a,b) > (o,o)$ and $f(x,\cdot)$, $g(y,\cdot)$ are nonincreasing for every $x \in \overline{\Omega}$ and every $y \in \Gamma$, respectively, *without any growth restriction whatsoever.*

Furthermore, it is important to notice that the functions m and r can be chosen independently of each other. For example, suppose that $a \geq o$, $b \geq o$, and $f(x,\cdot)$ is nonincreasing for every $x \in \overline{\Omega}$. Then we can take $m = o$ and Theorem (2.4) implies the existence of a solution for the BVP (2.1) for every function g satisfying one-sided estimates of the above form with $\lambda < \lambda_0(o,r)$. It follows from Theorem (2.2) that $\lambda_0(o,r) > \lambda_0(m,r)$, for every $m > o$. These considerations show, that, by using sharp estimates for f , it is possible to enlarge the class of admissible functions g , and vice versa.

Clearly, if $(a,b) \geq (o,o)$, then it is always possible to replace the functions m and r by constants $\overline{\mu}$ and $\overline{\rho}$ such that $m(x) \leq \overline{\mu}$ and $r(y) \leq \overline{\rho}$ for all $x \in \overline{\Omega}$ and $y \in \Gamma$, respectively. However, since $\lambda_0(m,r) \geq \lambda_0(\overline{\mu},\overline{\rho})$ by Theorem (2.2), it may be advantageous to use nonconstant functions m and r in a given concrete situation.

Of course, there are other geometric conditions for the functions f and g which guarantee the existence of subsolutions and supersolutions for the BVP (2.1). For example, suppose that $(a,b) \geq (o,o)$ and that there exists a positive constant ξ_1 such that

$$\begin{aligned} f(x,\xi_1) &\leq o \quad \text{for } x \in \overline{\Omega} , \\ g(y,\xi_1) &\leq o \quad \text{for } y \in \Gamma . \end{aligned} \qquad (2.6)$$

Then it is obvious that the constant function $x \to \xi_1$ is a supersolution for

(2.1). Consequently, the BVP (2.1) has a solution if the inequalities (2.6) and, for $\xi \le \xi_0$, the inequalities (2.4) are satisfied, for example. We leave it to the reader to deduce similar existence results by establishing further geometric conditions of this form. In any case, we like to point out that, up to regularity assumptions, the above existence theorems contain and considerably generalize most of the known existence results for the BVP (2.1) which have been deduced by means of the theory of monotone operators (cf. Paragraph 1).

Next we prove a simple *nonexistence theorem* which implies that, in some sense, Theorem (2.4) cannot be improved upon.

(2.5) *Theorem: Let the hypothesis* (H) *be satisfied. Suppose that* $c \in C^{\alpha}(\bar{\Omega})$ *and* $d \in C^{1-}(\Gamma)$ *satisfy* $(c,d) > (0,0)$, *and let* $\lambda \ge \lambda_0(m,r)$.

Then the BVP (2.1) *has no positive solution if*

$$\begin{aligned} f(\cdot,\xi) &\ge c + \lambda m \xi , \\ g(\cdot,\xi) &\ge d + \lambda r \xi \end{aligned} \tag{2.7}$$

for every $\xi \ge 0$, *and* (2.1) *has no negative solution if*

$$\begin{aligned} f(\cdot,\xi) &\le -c + \lambda m \xi , \\ g(\cdot,\xi) &\le -d + \lambda r \xi \end{aligned}$$

for every $\xi \le 0$.

Proof: Let the condition (2.7) be satisfied and suppose that u is a positive solution of (2.1). Then it follows from (2.7), that u is a positive supersolution for the linear BVP

$$\begin{aligned} Au - \lambda m u &= c \quad \text{in } \Omega , \\ Bu - \lambda v u &= d \quad \text{on } \Gamma . \end{aligned} \tag{2.8}$$

Since zero is a strict subsolution for this BVP, Theorem (2.1) implies the existence of a positive solution of (2.8). But this contradicts Theorem (2.3). The proof for the remaining case is similar. Q.E.D.

In the following we give an existence and *uniqueness theorem* which generalizes the main result of [4] (cf. also [11,25]). In that paper the uniqueness assertion has been proved under the assumption that the function $g(y,\cdot)$ be decreasing for every $y \in \Gamma$.

(2.6) *Theorem: Let the hypothesis* (H) *be satisfied. Suppose that* $(m,r) \in C^{\alpha}(\bar{\Omega}) \times C^{1-}(\Gamma)$ *satisfies* $(m,r) > (0,0)$. *Then the* BVP (2.1) *has exactly one solution if*

$$\begin{aligned} f(\cdot,\xi) - f(\cdot,\eta) &\le \lambda m(\xi-\eta) , \\ g(\cdot,\xi) - g(\cdot,\eta) &\le \lambda r(\xi-\eta) , \end{aligned}$$

for every $\xi,\eta \in \mathbb{R}$ *with* $\eta < \xi$ *and some* $\lambda < \lambda_0(m,r)$.

Proof: Let $\gamma := \max_{x \in \bar{\Omega}} |f(x,0)|$ and $\delta := \max_{y \in \Gamma} |g(y,0)|$. Then the hypotheses of Theorem (2.4) are satisfied. Hence the BVP (2.1) has at least one solution.

Suppose that u_1 and u_2 are two solutions of (2.1). Denote by Ω_1 an arbitrary component of $\{x \in \Omega \; u_1(x) < u_2(x)\}$. Then the hypotheses imply that the function $u := u_1 - u_2$ satisfies the inequalities

$$\begin{aligned} Au - \lambda m u &\geq o \quad \text{in} \quad \Omega_1 , \\ Bu - \lambda r u &\geq o \quad \text{on} \quad \Gamma \cap \partial\Omega_1 , \\ u &\geq o \quad \text{on} \quad \Omega \cap \partial\Omega_1 . \end{aligned} \tag{2.9}$$

It follows from Theorem (2.2) that the positive eigenfunction u_0 of the EVP (2.2) satisfies the inequality (2.8). Hence, the generalized maximum principle [22, pp. 72-76] implies that $u \geq o$ in Ω_1. Consequently, $\Omega_1 = \emptyset$ and $u_1 \geq u_2$ in Ω. A similar argument shows that $u_1 \leq u_2$, that is, $u_1 = u_2$. Q.E.D.

As already mentioned earlier, the main importance of this paper lies in the fact that the BVP (2.1) can be transformed into an equivalent fixed point equation in the Banach space $C(\overline{\Omega})$. It will be shown that this can be done in such a way that the corresponding nonlinear map is increasing with respect to the natural ordering of $C(\overline{\Omega})$. This has the consequence that (with minor modification) most of the general abstract results for increasing maps (e.g. [5,15]) can be applied to the BVP (2.1). For example, it is now possible to give a detailed study of the nonlinear eigenvalue problem

$$\begin{aligned} Au &= \lambda f(x,u) \quad \text{in} \quad \Omega , \\ Bu &= \lambda g(x,u) \quad \text{on} \quad \Gamma , \end{aligned}$$

in particular with respect to the question of the existence of multiple solutions.

We shall not go into details but we leave it to the reader to carry through this program. However, for the sake of demonstration, we shall prove the following existence theorem for multiple solutions.

(2.7) *Theorem: Suppose that* $\overline{v}_1$, $\overline{v}_2$ *are subsolutions and* $\hat{v}_1$, $\hat{v}_2$ *are supersolutions for the* BVP (2.1) *such that* $\overline{v}_1 < \hat{v}_1 < \overline{v}_2 < \hat{v}_2$, *and such that* $\hat{v}_1$ *is a strict supersolution and* $\overline{v}_2$ *is a strict subsolution. Then the* BVP (2.1) *has at least three distinct solutions* u_i, $i = 1,2,3$, *such that* $\overline{v}_1 \leq u_1 < u_2 < u_3 \leq \hat{v}_2$.

The proof of this theorem will be given in Paragraph 4.

We close this paragraph by means of a simple example illustrating the preceding theorem. Namely, we claim that, for every Ω, the BVP

$$\begin{aligned} u &= o \quad && \text{in} \quad \Omega , \\ \frac{\partial u}{\partial \beta} &= 4\pi^2 \cos u + u^2 c(x) \quad && \text{on} \quad \Gamma \end{aligned}$$

has at least three distinct solutions satisfying $-2\pi \leq u_1 < u_2 < u_3 < \pi$, provided $|c(x)| \leq 1$ for every $x \in \Gamma$. To see this, it suffices to verify that the constant functions $\overline{v}_1 := -2\pi$, $\hat{v}_1 := -\pi$, $\overline{v}_2 := o$, $\hat{v}_2 := \pi$ satisfy the hypotheses of Theorem (2.7).

3. AUXILIARY RESULTS

Let Δ be an arbitrary nonempty subdomain of $\mathbb{R}^N$, $N \geq 2$, and let k be a nonnegative integer. Suppose that $o < s < 1$, and let $p > 1$. Then the Sobolev space $W_p^{k+s}(\Delta)$ is the vector space consisting of all $u \in W_p^k(\Delta)$ such that

$$|u|^p_{s,\Delta,\kappa} := \int_\Delta \int_\Delta \frac{|D^\kappa u(x) - D^\kappa u(y)|^p}{|x - y|^{N+ps}} \, dxdy < \infty$$

for every multiindex κ with $|\kappa| = k$. It is well-known that $W_p^{k+s}(\Delta)$ is a Banach space with the norm

$$\|u\|_{W_p^{k+s}(\Delta)} := \left(\|u\|^p_{W_p^k(\Delta)} + \sum_{|\kappa|=k} |u|^p_{s,\Delta,\kappa} \right)^{1/p} .$$

Since Ω is a bounded domain of class $C^{3+\alpha}$, there exist bounded open sets U_j, $j = 1,\ldots,M$, in $\mathbb{R}^N$, covering Γ, and $C^{3+\alpha}$-diffeomorphisms ϕ_j of U_j onto the open unit ball B^N of $\mathbb{R}^N$ such that $\phi_j(U_j \cap \Omega) = B^N \cap \{x \in \mathbb{R}^N \mid x^N > 0\} =: B_+^N$ and $\phi_j(U_j \cap \Gamma) = B^N \cap (\mathbb{R}^{N-1} \times \{0\}) =: \Sigma^N$ for $j = 1,\ldots,M$. Moreover, without loss of generality, we can (and will!) assume that $\phi_j \in C^{3+\alpha}(\overline{U}_j)$ and $\phi_j^{-1} \in C^{3+\alpha}(\overline{B}^N)$.

Let π_j, $j = 1,\ldots,M$, be a $C^{3+\alpha}$-partition of unity on Γ subordinate to the covering $\{U_j \cap \Gamma \mid j = 1,\ldots,M\}$. Then every function u on Γ can be decomposed in the form $u = \sum_j(\pi_j u)$, and we define $\phi_j^*(\pi_j u)$ on B^{N-1} by

$$\phi_j^*(\pi_j u)(y) := (\pi_j u)(\phi_j^{-1}(y,0)) .$$

(Observe that $\Sigma^N = B^{N-1} \times \{0\}$.)

For every $p > 1$ and $\sigma \in [0,3]$, we define $W_p^\sigma(\Gamma)$ by $W_p^\sigma(\Gamma) := \{u \in L_p(\Gamma) \mid \phi_j^*(\pi_j u) \in W_p^\sigma(B^{N-1})$, $j = 1,\ldots,M\}$. It is well-known (cf. [17]) that $W_p^\sigma(\Gamma)$ is a Banach space with the norm

$$\|u\|_{W_p^\sigma(\Gamma)} := \left(\sum_{j=1}^M \|\phi_j^*(\pi_j u)\|^p_{W_p^\sigma(B^{N-1})} \right)^{1/p}$$

Moreover, up to equivalent norms, $W_p^\sigma(\Gamma)$ is independent of the particular choice of the local coordinates $\{(U_j,\phi_j)\}$ and of the partition of unity $\{\pi_j\}$.

Let X, Y, and Z be nonempty sets, and let $f : X \times Y \to Z$ be a map. Then we denote by $F : Y^X \to Z^X$ the *Nemytskii operator* of f, that is, F assigns to every map $u : X \to Y$ the map $F(u) : X \to Z$ defined by $F(u)(x) = f(x,u(x))$, $x \in X$.

(3.1) *__Lemma__: Let* $g : \Gamma \times \mathbb{R} \to \mathbb{R}$ *be locally Lipschitz continuous. Then, for every* $p > 1$ *and every* $\sigma \in [0,1)$, *the Nemytskii operator* G *of* g *maps* $W_p^\sigma(\Gamma) \cap L_\infty(\Gamma)$ *into* $W_p^\sigma(\Gamma)$.

Proof: Let $u \in W_p^\sigma(\Gamma) \cap L_\infty(\Gamma)$ be arbitrary and let $\delta := \|u\|_{L_\infty(\Gamma)}$. Then there exists a constant γ such that

$$|g(x,\xi) - g(y,\eta)| \le \gamma(|x-y| + |\xi-\eta|) \tag{3.1}$$

for every pair (x,ξ), $(y,\eta) \in \Gamma \times [-\delta,\delta]$.

By the definition of $W_p^\sigma(\Gamma)$, it suffices to show that

$$\|\phi_j^*(\pi_j G(u))\|_{W_p^\sigma(B^{N-1})} < \infty$$

for every $j = 1,\ldots,M$. But this is an easy consequence of the estimate (3.1) and the properties of π_j. Q.E.D.

In the following we denote by t the "trace operator" which assigns to every $u \in C(\overline{\Omega})$ its boundary value $u|\Gamma$. It is well-known (cf. [21]), that t can be extended to a continuous linear surjection (again denoted by t) of $W_p^k(\Omega)$

onto $W_p^{k-1/p}(\Gamma)$, for every $p > 1$ and $k = 1,2$. In particular, for every $p > 1$, there exists a constant γ such that

$$\|t(u)\|_{W_p^{1-1/p}(\Gamma)} \le \gamma \ \|u\|_{W_p^1(\Omega)} \tag{3.2}$$

for every $u \in W_p^1(\Omega)$.

We denote by A' the adjoint operator of A , that is

$$A'u = - \sum_{i,k=1}^{N} D_i D_k (a_{ik} u) - \sum_{i=1}^{N} D_i (a_i u) + au$$

It should be observed that the uniformly elliptic differential operator A' has α-Hölder continuous coefficients on $\overline{\Omega}$.

It can be shown (cf. [20]) that there exist a function $c_0 \in C^{2+\sigma}(\Gamma)$ with $c_0(x) > 0$ for every $x \in \Gamma$ and an outward pointing, nowhere tangent vector field $\beta' \in C^{2+\alpha}(\Gamma,\mathbb{R}^N)$ such that Green's formula takes the form

$$\int_\Omega (uAv - vA'u)dx = \int_\Gamma (v\, B'u - c_0\, u\, Bv)d\sigma \tag{3.3}$$

for every $u,v \in C^2(\overline{\Omega})$, where $B'u := \frac{\partial u}{\partial \beta'} + b'u$ and $b' \in C^{1+\alpha}(\Gamma)$.

(3.2) *Lemma: Let* $g \in C(\Gamma)$ *. Then there exists a function* $u \in C^1(\overline{\Omega})$ *, satisfying* $u|\Gamma = 0$ *and* $Bu = g$ *, such that*

$$\|u\|_{W_p^1(\Omega)} \le \gamma \|g\|_{L_p(\Gamma)} \quad , \quad 1 < p < \infty \quad ,$$

where the constant γ *is independent of* g .

Proof: (a) We first consider local coordinates. The general point $y \in \mathbb{R}^N$ will be denoted by $y = (\overline{y},t)$, with $\overline{y} \in \mathbb{R}^{N-1}$ and $t \in \mathbb{R}$. Moreover,

$$Q := \{y \in B_+^N \cup \Sigma^N \mid \overline{y} + t := (y^1 + t,\dots,y^{N-1} + t) \in \Sigma^N\}$$

Let $\psi \in C(\Sigma^N)$ be given and define $v : Q \to \mathbb{R}$ by

$$v(y) := t^{2-N} \int_{\overline{y}}^{\overline{y}+t} \psi(\overline{\eta})d\overline{\eta}$$

with an obvious abbreviation for the (N-1)-fold iterated integral. Then it is easily seen that $v \in C^1(Q)$, that $v|\Sigma^N = 0$, and that $(D_N v)|\Sigma^N = \psi$.

By means of Hölder's inequality it is easily verified that

$$\int_a^{b-t} |t^{-1} \int_x^{x+t} f(\xi)d\xi|dx \le \int_a^b |f(x)|dx$$

for every $f \in C[a,b]$, $-\infty < a < b < \infty$, and every $t \in (0,b-a)$. By applying this inequality repeatedly, one easily proves that

$$\int_Q |D_i v|^p \, dy \le \gamma \int_{\Sigma^N} |\psi|^p \, d\overline{y} \quad , \tag{3.4}$$

$i = 1,\dots,N$, where γ is independent of ψ .

(b) For each $j = 1,\dots,M$, let $V_j := \phi_j^{-1}(Q)$. Then V_j is an open subset of U_j such that $V_j \cap \Gamma = U_j \cap \Gamma$. For every $u \in C^1(\bar{\Omega})$ and every $j = 1,\dots,M$,

$$\phi_j^*\left(\frac{\partial u}{\partial \beta}\right) = \sum_{k=1}^{N} m_{k,j}\, D_k[\phi_j^*(u)]$$

and it is easy to see that

$$m_{k,j} = \phi_j^*\left(\sum_{i=1}^{N} \beta^i\, D_i\, \phi_j^k\right) .$$

Hence, $m_{k,j} \in C^{2+\alpha}(\Sigma^N)$ and, due to the fact that β is nowhere tangent to Γ, the function $m_j := m_{N,j}$ does nowhere vanish.

Let $g \in C(\Gamma)$. For every $j = 1,\dots,M$ define $\psi_j \in C(\Sigma^N)$ by $\psi_j := \frac{1}{m_j}\phi_j^*(g)$. Denote by v_j the function in $C^1(Q)$ defined in part (a) by means of $\psi_j \in C(\Sigma^N)$, and let $u_j := v_j \circ \phi_j$. Then $u_j = o$ and $Bu_j = \frac{\partial u_j}{\partial \beta} + bu_j = \frac{\partial u_j}{\partial \beta} = g$ on $U_j \cap \Gamma$.

Let $e_1,\dots,e_M$ be a smooth partition of unity with respect to the compact set Γ in $\mathbb{R}^N$, subordinate to be open covering $\{V_1,\dots,V_M\}$, and let

$$u := \begin{cases} \sum_{j=1}^{M} e_j u_j & \text{in } \bar{\Omega} \cap \bigcup_{j=1}^{m} V_j \\ o & \text{in } \bar{\Omega} \setminus \bigcup_{j=1}^{M} V_j \end{cases} .$$

Then $u \in C^1(\bar{\Omega})$, $u|\Gamma = o$, and $Bu = g$ on Γ. Lastly, by using Poincaré's inequality and the estimate (3.4), it is easily verified that u satisfies the asserted estimate. Q.E.D.

After these preparations we are ready for the proof of the following a priori estimate.

(3.3) *Proposition: Suppose that* $(a,b) > (o,o)$ *, and let* $1 < p < \infty$ *. Then there exists a constant* γ *such that*

$$\|u\|_{W_p^1(\Omega)} \le \gamma(\|Au\|_{L_p(\Omega)} + \|Bu\|_{L_p(\Gamma)})$$

for every $u \in C^2(\bar{\Omega})$.

Proof: The maximum principle together with the regularity theory for elliptic equations implies that the only function $u \in W_p^2(\Omega)$ which satisfies $Au = o$ in Ω and $Bu = o$ on Γ is the zero function. Consequently the L_p-estimates (cf. [1]) imply the existence of a constant γ such that

$$\|u\|_{W_p^2(\Omega)} \le \gamma(\|Au\|_{L_p(\Omega)} + \|Bu\|_{W_p^{1-1/p}(\Gamma)})$$

for every $u \in C^2(\bar{\Omega})$.

Suppose that $u \in W_p^2(\Omega)$ satisfies $A'u = o$ in Ω and $B'u = o$ on . Then, by using the well-known fact that $C^2(\bar{\Omega})$ is dense in $W_p^2(\Omega)$ (e.g. [21]), it follows from Green's formula (3.3) that

$$\int_\Omega uAv\,dx = -\int_\Gamma t(u)c_o Bv\,d\sigma \tag{3.5}$$

for every $v \in C^2(\bar{\Omega})$. By the classical Schauder theory (e.g. [16]), the BVP

$$Av = f \quad \text{in} \quad \Omega ,$$
$$Bv = g \quad \text{on} \quad \Gamma ,$$

has a unique solution $u \in C^{2+\alpha}(\overline{\Omega})$ for every $f \in C^{\alpha}(\overline{\Omega})$ and $g \in C^{1+\alpha}(\Gamma)$. Consequently, since $C^{\alpha}(\overline{\Omega})$ is dense in $L_{p'}(\Omega)$ and $C^{1+\alpha}(\Gamma)$ is dense in $L_{p'}(\Gamma)$, $p' := p/(p-1)$, the relation (3.14) implies that

$$\int_{\Omega} ufdx = \int_{\Gamma} t(u)g \, d\sigma$$

for every $f \in L_{p'}(\Omega)$ and $g \in L_{p'}(\Gamma)$. Consequently, $u = o$. This fact, together with the regularity properties of the coefficients of (A',B') , implies the validity of the $L_{p'}$-estimate

$$\|u\|_{W^2_{p'}(\Omega)} \le \gamma_1 \|A'u\|_{L_{p'}(\Omega)} \tag{3.6}$$

for every $u \in C^2(\overline{\Omega})$ satisfying $B'u = o$.

Let $V := \{u \in C^2(\overline{\Omega}) \mid Bu = o\}$ and $V' := \{u \in C^2(\overline{\Omega}) \mid B'u = o\}$, and denote by $(\cdot,\cdot)$ the inner product in $L_2(\Omega)$. For every $u \in C(\overline{\Omega})$ let

$$\|u\|_{-k,p} := \sup_{v \in V'} \frac{|(u,v)|}{\|u\|_{W^k_{p'}(\Omega)}}$$

where $k = 1,2$. Then the estimate (3.6) and Green's formula (3.3) imply that, for every $u \in V$,

$$\|Au\|_{-2,p} = \sup_{v \in V'} \frac{|(Au,v)|}{\|v\|_{W^2_{p'}(\Omega)}} = \sup_{v \in V'} \frac{|(u,A'v)|}{\|v\|_{W^2_{p'}(\Omega)}} \ge$$
$$\gamma_1^{-1} \sup_{v \in V'} \frac{|(u,A'v)|}{\|A'v\|_{L_{p'}(\Omega)}} .$$

Since, by the Schauder theory, the BVP

$$A'v = f \quad \text{in} \quad \Omega ,$$
$$B'v = o \quad \text{on} \quad \Gamma$$

has a unique solution $u \in C^{2+\alpha}(\overline{\Omega})$ for every $f \in C^{\alpha}(\overline{\Omega})$, it follows that $A'(V')$ is dense in $L_{p'}(\Omega)$. Consequently, the above inequality implies the estimate

$$\|u\|_{L_p(\Omega)} \le \gamma_1 \|Au\|_{-2,p} \tag{3.7}$$

for every $u \in V$. Moreover, the L_p-estimates (cf. [1]) imply the existence of a constant γ_2 such that

$$\|u\|_{W^2_p(\Omega)} \le \gamma_2 \|Au\|_{L_p(\Omega)} \tag{3.8}$$

for every $u \in V$.

Let E and F be Banach spaces. Then we write $E \hookrightarrow F$ if E is continuously imbedded in F . We denote by $L(E,F)$ the Banach space of all continuous linear operators $T : E \to F$, and by $[E,F]_{1/2}$ we denote the "interpolation space to the parameter 1/2" obtained by means of "holomorphic interpolation" (cf. [19,24]).

Let $k = 1,2$. Then V^k_p denotes the closure of V in $W^k_p(\Omega)$ and V^{-k}_p denotes the completion of $C^{\infty}(\overline{\Omega})$ in the norm $\|\cdot\|_{-k,p}$. Then the a priori estimates (3.7) and (3.8) imply that A^{-1} has a continuous extension, T, such that

$$T \in L(V_p^{-2}, L_p(\Omega)) \cap L(L_p(\Omega), V_p^2) .$$

Hence, by the theory of interpolation spaces (cf. [19, 24]),

$$T \in L([V_p^{-2}, L_p(\Omega)]_{1/2} , [L_2(\Omega), V_p^2]_{1/2}) . \tag{3.9}$$

Moreover, (cf. [24, Theorem 4.1]),

$$[L_p(\Omega), V_p^2] \hookrightarrow V_p^1 \tag{3.10}$$

with dense imbedding. Hence, by duality,

$$V_p^{-1} \hookrightarrow [V_p^{-2}, L_p(\Omega)]_{1/2} \tag{3.11}$$

The relations (3.9) - (3.11) imply

$$T \in L(V_p^{-1}, V_p^1) .$$

Consequently, there exists a constant γ_3 such that

$$\|u\|_{W_p^1(\Omega)} \le \gamma_3 \|Au\|_{-1,p} \tag{3.12}$$

for every $u \in V_p^1$. (Observe that by

$$\|Au\|_{-1,p} = \sup_{v \in V'} \frac{|(u,A'v)|}{\|v\|_{W^1_{p'}(\Omega)}}$$

the right side of (3.12) has a well-defined meaning for $u \in V_p^1$.)

Let $u \in C^2(\bar{\Omega})$ be arbitrary. By Lemma (3.2) there exists a function $u_0 \in C^1(\bar{\Omega})$ satisfying

$$\left.\begin{aligned} B(u-u_0) &= 0 \\ u_0 &= 0 \end{aligned}\right\} \text{ on } \Gamma \tag{3.13}$$

and

$$\|u_0\|_{W_p^1(\Omega)} \le \gamma_4 \|Bu\|_{L_p(\Gamma)} , \tag{3.14}$$

where the constant γ_4 is independent of u . Consequently, $u-u_0 \in V_p^1$, and (3.12) implies that

$$\begin{aligned} \|u\|_{W_p^1(\Omega)} &\le \gamma_3(\|Au\|_{-1,p} + \|Au_0\|_{-1,p}) \\ + \|u_0\|_{W_p^1(\Omega)} &\le \gamma_3(\|Au\|_{-1,p} + \|Au_0\|_{-1,p}) \\ &+ \gamma_4 \|Bu\|_{L_p(\Gamma)} . \end{aligned} \tag{3.15}$$

By partial integration one finds that $\int_\Omega uAv\,dx = \int_\Omega\{\sum_{i,k=1}^N a_{ik}D_iuD_kv + u\sum_{i=1}^N b_i' D_i v + auv\}dx - \int_\Gamma c_0 u Bv\, d\sigma$. Hence there exists a constant γ_5 such such

$$|(Aw,v)| \le \gamma_5(\|w\|_{W_p^1(\Omega)} \|v\|_{W^1_{p'}(\Omega)} + \|Bw\|_{L_p(\Gamma)} \|v\|_{L_{p'}(\Gamma)}) \tag{3.16}$$

for every $v \in C^1(\bar{\Omega})$ and $w \in C^2(\bar{\Omega})$ satisfying $w|\Gamma = o$. Since $W_{p'}^{1-1/p'}(\Gamma) \hookrightarrow L_{p'}(\Gamma)$, inequality (3.16) implies the estimate

$$|(Aw,v)| \leq \gamma_6(\|w\|_{W_p^1(\Omega)} + \|Bw\|_{L_p(\Gamma)}) \|v\|_{W_{p'}^1(\Omega)}$$

for every $w \in C^2(\bar{\Omega})$ with $w|\Gamma = o$. Consequently,

$$\|Aw\|_{-1,p} \leq \gamma_6(\|w\|_{W_p^1(\Omega)} + \|Bw\|_{L_p(\Gamma)}) \tag{3.17}$$

for every $w \in W_p^1(\Omega)$ with $t(w) = o$.

By applying (3.17) to the function u_o and using (3.13) and (3.14), we obtain the existence of a constant γ_7 such that

$$\|Au_o\|_{-1,p} \leq \gamma_7 \|Bu\|_{L_p(\Gamma)} .$$

By combining this estimate with (3.15) and using the obvious fact that $\|Lu\|_{-1,p} \leq \|Lu\|_{L_p(\Omega)}$, we obtain finally the assertion. Q.E.D.

It should be remarked that the above proof of the a priori estimate follows the ideas developed in the author's paper [2]. Since the earlier proof contains a number of inaccuracies, in particular a false statement of the a priori estimate (which, however, does not affect the validity of the main results of [2]), we have decided to include the above more detailed derivation of Proposition (3.3).

It should be observed that we have proved the much stronger a priori estimate

$$\|u\|_{W_p^1(\Omega)} \leq \gamma(\|Au\|_{-1,p} + \|Bu\|_{L_p(\Gamma)}) .$$

Moreover, it should be observed that *it suffices to suppose that* Ω *belongs to the class* $C^{2+\alpha}$ *if the pair* (L,B) *is formally self-adjoint.*

Finally, we like to point out, that the basic Proposition (3.3) is implicitly contained in [18]. However, in that paper much stronger regularity hypotheses have been presupposed. Consequently it does not seem to be much easier to deduce the needed a priori inequality from [18] than to prove it directly.

4. THE REDUCTION TO AN EQUIVALENT FIXED POINT EQUATION

Suppose that $(a,b) > (o,o)$. Then, by the maximum principle and the regularity theory for elliptic equations, the linear BVP

$$\begin{aligned} Au &= v \quad \text{in } \Omega \\ Bu &= w \quad \text{on } \Gamma \end{aligned} \tag{4.1}$$

has at most one solution in $C^{2+\alpha}(\bar{\Omega})$ as well as in $W_p^2(\Omega)$, $1 < p < \infty$. Hence the Schauder estimates and the L_p-estimates take the form

$$\|u\|_{C^{2+\alpha}(\bar{\Omega})} \leq (\|Au\|_{C^{\alpha}(\bar{\Omega})} + \|Bu\|_{C^{1+\alpha}(\Gamma)}) \tag{4.2}$$

and

$$\|u\|_{W_p^2(\Omega)} \leq \gamma(\|Au\|_{L_p(\Omega)} + \|Bu\|_{W_p^{1-1/p}(\Gamma)}) \tag{4.3}$$

respectively. Here and in the following γ denotes a positive constant (not necessarily the same in different formulas) which is independent of the functions appearing in these estimates.

It is well-known (cf. [16,20]) that the BVP (4.1) has a unique solution $u =: S(v,w) \in C^{2+\alpha}(\bar{\Omega})$ for every $(v,w) \in C^{\alpha}(\bar{\Omega}) \times C^{1+\alpha}(\Gamma)$. Hence (4.2) implies that the *solution operator* S satisfies

$$S \in L(C^{\alpha}(\bar{\Omega}) \times C^{1+\alpha}(\Gamma), C^{2+\alpha}(\bar{\Omega})) .$$

Since $C(\bar{\Omega}) \hookrightarrow L_p(\Omega)$ and $C(\Gamma) \hookrightarrow L_p(\Gamma)$ for every $p \in (1,\infty)$, Proposition (3.3) implies the existence of a constant γ such that

$$\|S(v,w)\|_{W_p^1(\Omega)} \leq \gamma(\|v\|_{C(\bar{\Omega})} + \|w\|_{C(\Gamma)}) \tag{4.4}$$

for every $(v,w) \in C^{\alpha}(\bar{\Omega}) \times C^{1+\alpha}(\Gamma)$. Since the latter space is dense in $C(\bar{\Omega}) \times C(\Gamma)$, the estimate (4.4) implies that S has for every $p \in (1,\infty)$ a unique continuous extension, denoted again by S, such that

$$S \in L(C(\bar{\Omega}) \times C(\Gamma), W_p^1(\Omega)) .$$

Let I be a nonempty subinterval of $\mathbb{R}$ and let $f : \bar{\Omega} \times I \to \mathbb{R}$ and $g : \Gamma \times I \to \mathbb{R}$ be continuous maps. Denote by F and G the corresponding Nemytskii operators, respectively. Recall that by a solution of the BVP

$$\begin{aligned} Au &= f(x,u) \quad \text{in } \Omega , \\ Bu &= g(x,u) \quad \text{in } \Omega \end{aligned} \tag{4.5}$$

we mean a function $u \in C^2(\Omega) \cap C^1(\bar{\Omega})$ such that $u(\bar{\Omega}) \subset I$ and $Au = F(u)$ in Ω, and $Bu = G(u)$ on Γ.

In the following we let $I_{C(\bar{\Omega})} := \{u \in C(\bar{\Omega}) \mid u(\bar{\Omega}) \subset I\}$, and we define

$$T : I_{C(\bar{\Omega})} \to W_p^1(\Omega)$$

by $T(u) := S(F(u), G \circ t(u))$, where t denotes the trace operator. Here and in the following we denote by p an arbitrary but fixed *real number satisfying* $p > N$. This implies in particular that W_p^1 is compactly imbedded in $C(\bar{\Omega})$. Here T can be considered as a mapping of $I_{C(\bar{\Omega})}$ into $C(\bar{\Omega})$.

The following lemma is of fundamental importance for our considerations.

(4.1) *Lemma: Let* $(a,b) > (0,0)$. *Suppose that* $f : \bar{\Omega} \times I \to \mathbb{R}$ *is locally* α*-Hölder continuous and* $g : \Gamma \times I \to \mathbb{R}$ *is locally Lipschitz continuous. Then the* BVP (4.5) *is equivalent to the fixed point equation* $u = T(u)$ *in* $C(\bar{\Omega})$. *The map* $T : I_{C(\bar{\Omega})} \to C(\bar{\Omega})$ *is completely continuous, that is,* T *is continuous and maps bounded sets into compact sets.*

Proof: It is obvious that every solution of the BVP (4.5) is a fixed point of T.

Conversely, suppose that $u \in C(\bar{\Omega})$ is a fixed point of T. Then u belongs to the range of S, hence to $W_p^1(\Omega)$. Consequently, $t(u) \in W_p^{1-1/p}(\Gamma)$, and Lemma (3.1) implies that $G \circ t(u) \in W_p^{1-1/p}(\Gamma)$. Since $F(u) \in C(\bar{\Omega}) \subset L_p(\Omega)$, and since, by (4.3), S maps $L_p(\Omega) \times W_p^{1-1/p}(\Gamma)$ into $W_p^2(\Omega)$, it follows that $u \in W_p^2(\Omega)$. It is well-known (cf. [21]) that $W_p^2(\Omega)$ is continuously imbedded in $C^{1+\sigma}(\bar{\Omega})$, where $\sigma := 1-N/p$. Hence $u \in C^1(\bar{\Omega}) \cap W_p^2(\Omega)$. This implies that $F(u) \in C^{\alpha}(\Omega)$ and consequently, by the interior regularity theory for elliptic equations, $u \in C^{2+\alpha}(\Omega')$ for every subdomain Ω' such that $\Omega' \subset \Omega$. Hence, $u \in C^2(\Omega) \cap C^1(\bar{\Omega})$, and u is a solution of the BVP (4.7).

It is easily seen that the maps $F : I_{C(\overline{\Omega})} \to C(\overline{\Omega})$ and $G : I_{C(\Gamma)} \to C(\Gamma)$ are bounded and continuous, where $I_{C(\Gamma)} := \{u \in C(\Gamma) \mid u(\Gamma) \subset I\}$. Moreover, t is a continuous linear operator from $C(\overline{\Omega})$ into $C(\Gamma)$ such that $t(I_{C(\overline{\Omega})}) \subset I_{C(\Gamma)}$. Since $S \in L(C(\overline{\Omega}) \times C(\Gamma), W_p^1(\Omega))$, and since $W_p^1(\Omega)$ is compactly imbedded in $C(\overline{\Omega})$ for $p > N$, it follows that T is completely continuous from $I_{C(\overline{\Omega})}$ into $C(\overline{\Omega})$. Q.E.D.

Let X be a nonempty compact Hausdorff space. We denote by $C_+(X)$ the set of all nonnegative continuous functions on X , that is, $C_+(X) := \{u \in C(X) \mid u \geq o\}$. Clearly, $C_+(X)$ is a closed convex proper cone in $C(X)$ with nonempty interior. In fact, $u \in \text{int}\, C_+(X)$ if and only if $u(x) > o$ for every $x \in X$. Observe that $u \leq v$ if and only if $v - u \in C_+(X)$. In the following we write $u \gg v$ if $u - v \in \text{int}\, C_+(X)$.

Let D be a nonempty subset of $C(X)$. Then a map $h : D \to C(X)$ is said to be increasing if $h(u) \leq h(v)$ for every pair $u,v \in D$ satisfying $u \leq v$.

(4.2) *Lemma: Let the hypotheses of Lemma* (4.1) *be satisfied and suppose that* $f(x,\cdot)$ *and* $g(y,\cdot)$ *are increasing for every* $x \in \overline{\Omega}$ *and* $y \in \Gamma$ *, respectively. Then the map* $T : I_{C(\overline{\Omega})} \to C(\overline{\Omega})$ *is increasing.*

Suppose in addition that

$$f(x,\xi) < f(x,\eta)$$

for some $x \in \Omega$ *and every* $\xi,\eta \in I$ *with* $\xi < \eta$ *, or*

$$g(y,\xi) < g(y,\eta)$$

for some $y \in \Gamma$ *and every* $\xi,\eta \in I$ *with* $\xi < \eta$ *. Then* $T(u) - T(v) \in \text{int}\, C_+(\overline{\Omega})$ *for every pair* $u,v \in I_{C(\overline{\Omega})}$ *such that* $v \ll u$.

Proof: The maximum principle implies that every function $u \in C^2(\Omega) \cap C^1(\overline{\Omega})$ satisfying

$$Au \geq o \quad \text{in } \Omega \quad ,$$

$$Bu \geq o \quad \text{on } \Gamma \quad ,$$

either vanishes identically or belongs to in $C_+(\overline{\Omega})$. Consequently, the solution operator S maps $C_+^{\alpha}(\overline{\Omega}) \times C_+^{1+\alpha}(\Gamma) \setminus \{(o,o)\}$ into $\text{int}\, C_+(\overline{\Omega})$, where $C_+^{\alpha}(\overline{\Omega}) := C_+(\overline{\Omega}) \cap C^{\alpha}(\overline{\Omega})$ and $C_+^{1+\alpha}(\Gamma) := C_+(\Gamma) \cap C^{1+\alpha}(\Gamma)$. Since $C_+^{\alpha}(\overline{\Omega}) \times C_+^{1+\alpha}(\Gamma)$ is dense in $C_+(\overline{\Omega}) \times C_+(\Gamma)$, it follows that S maps $C_+(\overline{\Omega}) \times C_+(\Gamma)$ into $C_+(\overline{\Omega})$, that is, $S \in L(C(\overline{\Omega}) \times C(\Gamma), C(\overline{\Omega}))$ is a positive linear operator. This fact easily implies that T is increasing.

Suppose that $(u,v) \in C_+(\overline{\Omega}) \times C_+(\Gamma)$ and $(u,v) \neq (o,o)$. Then, by using mollifiers, it is easy to see that there exists a pair $(u_0,v_0) \in C^{\alpha}(\overline{\Omega}) \times C^{1+\alpha}(\Gamma)$ such that $(u,v) \geq (u_0,v_0) > (o,o)$. Hence $S(u,v) \geq S(u_0,v_0) \gg o$ and it follows that $S(u,v) \in \text{int}\, C_+(\overline{\Omega})$. Hence $S : C(\overline{\Omega}) \times C(\Gamma) \to C(\overline{\Omega})$ is strongly positive, that is, S maps $C_+(\overline{\Omega}) \times C_+(\Gamma) \setminus \{(o,o)\}$ into $\text{int}\, C_+(\overline{\Omega})$. The second part of the assertion is now obvious. Q.E.D.

After these preparations we are ready for the proof of Theorem (2.1) and Theorem (2.7).

Proof of Theorem (2.1): Without loss of generality we can assume that I equals the compact interval $[\min \overline{v}, \max \hat{v}]$. Hence the regularity assumptions for f and g imply the existence of a positive constant ω_0 such that $f(x,\xi) - f(x,\eta) \geq -\omega_0(\xi-\eta)$ and $g(y,\xi) - g(y,\eta) \geq -\omega_0(\xi-\eta)$ for every $x \in \overline{\Omega}$, $y \in \Gamma$, and every $\xi,\eta \in I$ satisfying $\eta \leq \xi$. Let $\omega := \omega_0 + \max\{\|a\|_{C(\overline{\Omega})}$,

$\|b\|_{C(\Gamma)}\}$. Then the BVP (2.1) is obviously equivalent to the BVP

$$\begin{aligned}(A+\omega)u &= f(x,u)+\omega u \quad \text{in } \Omega ,\\ (B+\omega)u &= g(\mathbf{x},u)+\omega u \quad \text{on } \Gamma .\end{aligned} \tag{4.6}$$

Moreover $\bar{v}$ is a subsolution and $\hat{v}$ is a supersolution for the BVP (4.6). Consequently, without loss of generality we can assume that $(a,b) > (o,o)$, and that $f(x,\cdot)$ and $g(y,\cdot)$ are increasing for every $x \in \bar{\Omega}$ and $y \in \Gamma$, respectively.

Therefore, by Lemma (4.1) and Lemma (4.2), the BVP (2.1) is equivalent to the fixed point equation $u = T(u)$ in $C(\bar{\Omega})$, where $T : I_{C(\bar{\Omega})} \to C(\bar{\Omega})$ is increasing and completely continuous. Since $\bar{v} \in I_{C(\bar{\Omega})}$ and $\bar{v}$ is a subsolution, it follows that

$$T(\bar{v}) - \bar{v} = S(F(\bar{v}),G(t(\bar{v}))) - S(A\bar{v},B\bar{v}) = S(F(\bar{v}) - A\bar{v},G(t(\bar{v})) - B\bar{v}) \geq o$$

and, similarly, that $T(\hat{v}) \leq \hat{v}$. Hence T maps the nonempty order interval $[\bar{v},\hat{v}] \subset I_{C(\bar{\Omega})}$ into itself. Now the existence of a fixed point follows from Schauder's fixed point theorem. The existence of a minimal and of a maximal fixed point follows by an easy iteration argument (cf. [3, Theorem 3] or [5, Theorem (6.1)]).

Q.E.D.

Proof of Theorem (2.7): Without loss of generality we can assume that $I = [\min \bar{v}_1, \max \hat{v}_2]$. Hence, similarly as in the preceding proof, we find that the BVP (2.1) is equivalent to the fixed point equation $u = T(u)$ in $C(\bar{\Omega})$, where $T : I_{C(\bar{\Omega})} \to C(\bar{\Omega})$ is increasing and completely continuous. Furthermore,

$$\bar{v}_1 \leq T(\bar{v}_1) \ , \ T(\hat{v}_1) < \hat{v}_1 \ , \ \bar{v}_2 < T(\bar{v}_2) \ , \ T(\hat{v}_2) \leq \hat{v}_2 \ .$$

By Theorem (2.1), the BVP (2.1) has a maximal solution $\hat{u}_1$ in the order interval $[\bar{v}_1,\hat{v}_1]$ and a minimal solution $\bar{u}_2$ in the order interval $[\bar{v}_2,\hat{v}_2]$. Hence the maximum principle implies that $\hat{v}_1 - \hat{u}_1$ and $\bar{u}_2 - \bar{v}_2$ belong to the interior of $C(\bar{\Omega})$, that is, $\hat{u}_1 \ll \hat{v}_1$ and $\bar{v}_2 \ll \bar{u}_2$. These conditions suffice for the validity of the proof of the general multiplicity result [5, Theorem 14.2] (or [3]), which guarantees the existence of a third fixed point in $[\bar{v}_1,\hat{v}_2]$. The fact that there are three distinct fixed points which can be linearly ordered follows from the existence of a minimal fixed point and a maximal fixed point in $[\bar{v}_1,\hat{v}_2]$.

Q.E.D.

We now consider the linear BVP (2.3) which, as a special case, contains the linear EVP (2.2). In this case we let $\omega := 1 + \mu$, where $\mu \geq o$ is an constant such that $(a + \mu m \ , \ b + \mu r) \geq (o,o)$. Then the linear BVP (2.3) is equivalent to the BVP

$$\begin{aligned}(A+\omega m)u - (\lambda+\omega)mu &= c \quad \text{in } \Omega ,\\ (B+\omega r)u - (\lambda+\omega)ru &= d \quad \text{on } \Gamma .\end{aligned} \tag{4.7}$$

Denote by S_ω the solution operator for the pair $(A + \omega m \ , \ B + \omega r)$. Since $(a + \omega m \ , \ b + \omega r) > (o,o)$, it follows from the proof of Lemma (4.2) that S_ω is a strongly positive compact linear operator from $C(\bar{\Omega}) \times C(\Gamma)$ into $C(\bar{\Omega})$. Hence the equations (4.7) are equivalent to the equation

$$u - (\lambda + \omega)S_\omega(mu,rt(u)) = S_\omega(c,d)$$

in $C(\bar{\Omega})$. Observe that $S_\omega(c,d) \gg o$ whenever $(c,d) > (o,o)$.

We denote by T_ω the linear endomorphism of $C(\bar{\Omega})$ defined by

$$T_\omega u := S_\omega(mu,rt(u)) \ .$$

Then it follows from Lemma (4.2) that T_ω is a compact positive endomorphism of $C(\bar{\Omega})$. Moreover, T_ω is *almost strongly positive* in the sense that

$$T_\omega(C_+(\bar{\Omega})\setminus \ker(T_\omega)) \subset \operatorname{int} C_+(\bar{\Omega}) .$$

Moreover, Lemma (4.2) implies that T_ω has the additional property that

$$T_\omega(\operatorname{int} C_+(\bar{\Omega})) \subset \operatorname{int} C_+(\bar{\Omega}) ,$$

that is, $\ker(T_\omega) \cap \operatorname{int} C_+(\bar{\Omega}) = \emptyset$.

Finally, since we can obviously assume that $\lambda \neq -\omega$, it follows that the linear BVP (2.3) is equivalent to linear equation

$$\rho u - Tu = v \tag{4.8}$$

in $C(\bar{\Omega})$, where $\rho := (\lambda + \omega)^{-1}$ and $v := \rho S_\omega(c,d)$.

In the following paragraph we shall study equations of the form (4.8) in arbitrary ordered Banach spaces, where T is an almost strongly positive compact endomorphism leaving invariant the interior of the positive cone. Then, due to the above considerations, Theorem (2.2) and Theorem (2.3) will turn out to be easy consequences of the following abstract results.

5. ALMOST STRONGLY POSITIVE LINEAR OPERATORS

Let E be a real Banach space. A subset $P \subset E$ is called a cone if P is closed, $P + P \subset P$, $\mathbb{R}_+P \subset P$, and $P \cap (-P) = \{o\}$. Every cone induces an ordering in E by setting "$u \leq v$ iff $v - u \in P$" . A Banach space E together with a cone P is called an ordered Banach space (OBS) and denoted by (E,P) provided E is given the ordering induced by P . Then P is called the positive cone of E and the elements in $\dot{P} := P\setminus\{o\}$ are called positive. We write $u > o$ if $u \in \dot{P}$ and $u \gg v$ iff $u - v \in \mathring{P}$, provided, of course, that P has nonempty interior, $\mathring{P}$. In the following, the reals, $\mathbb{R}$, are always identified with the OBS $(\mathbb{R},\mathbb{R}_+)$ whose positive cone, $\mathbb{R}_+ := [o,\infty)$, induces the natural ordering.

Let (E,P) and (F,Q) be OBSs. A linear operator $T : E \to F$ is called positive if $T(P) \subset Q$, strongly positive if $T(\dot{P}) \subset \mathring{Q}$, and *almost strongly positive* if $P\setminus\ker(T) \neq \emptyset$ and $T(P\setminus\ker(T)) \subset \mathring{Q}$, provided, of course, that Q has nonempty interior.

Finally we recall that the spectral radius, $r(T)$, of a continuous endomorphism T of a Banach space is defined by $r(T) := \lim_{k\to\infty} \|T^k\|^{1/k}$.

(5.1) <u>Theorem:</u> *Let (E,P) be an OBS whose positive cone has nonempty interior. Suppose that T is an almost strongly positive compact endomorphism of E .*

Then the spectral radius of T is positive and an eigenvalue of T and of the dual operator T' . The eigenvalue $r(T)$ possesses an eigenvector $u_0 \in \mathring{P}$ of T and a positive eigenfunctional ϕ_0 of T' satisfying $\phi_0(P\setminus\ker(T)) \subset \mathring{\mathbb{R}}_+$.

Proof: By assumption, there exists a positive element v such that $Tv \in \mathring{P}$. Hence there exists a positive number α such that $Tv - \alpha v \in P$, that is, $Tv \geq \alpha v$. This implies that the spectral radius $r(T)$ is positive (cf. [2, 14,15]). It follows now from the Krein-Rutman theory [14] that $r(T)$ is an eigenvalue of T and of T' having an eigenvector $u_0 \in P$ of T and a positive eigenfunctional ϕ_0 of T'_o, respectively. Hence $u_0 \in P\setminus\ker(T)$ and, consequently, $u_0 = r(T)^{-1}Tu_0 \in \mathring{P}$.

Suppose that $u \in P \setminus \ker(T)$. Then $Tu \in \mathring{P}$ and, consequently, $\langle\phi_0, Tu\rangle > 0$ (cf. [5,14]). Hence $\langle\phi_0, u\rangle = r(T)^{-1} \langle T'\phi_0, v\rangle = r(T)^{-1} \langle\phi_0, Tu\rangle > 0$, that is, $\langle\phi_0, u\rangle > 0$ for every $u \in P \setminus \ker(T)$. Q.E.D.

(5.2) <u>*Corollary:*</u> *Let the hypotheses of Theorem* (5.1) *be satisfied. Then* $r(T)$ *is the only nonzero eigenvalue of* T *possessing a positive eigenvector.*

Proof: Suppose that there exists an element $u \in \dot{P}$ and a real number $\lambda \neq 0$ such that $Tu = \lambda u$. Then $u \in P \setminus \ker(T)$. Hence $\lambda > 0$, and $u = \lambda^{-1} Tu \in \mathring{P}$. Consequently, by applying the eigenfunctional ϕ_0 , we obtain that

$$\lambda\langle\phi_0, u\rangle = \langle\phi_0, Tu\rangle = \langle T'\phi_0, u\rangle = r(T)\langle\phi_0, u\rangle .$$

Hence $\lambda = r(T)$ since $\langle\phi_0, u\rangle > 0$. Q.E.D.

(5.3) <u>*Theorem:*</u> *Let the hypotheses of Theorem* (5.1) *be satisfied. Then the equation*

$$\lambda u - Tu = v \tag{5.1}$$

has for every $\lambda > r(T)$ *exactly one solution, which is positive if* $v \in \dot{P}$. *If* $\lambda < r(T)$ *and* $v \in P$, *then* (5.1) *has no solution in* $P \setminus \ker(T)$. *Moreover, if* $\lambda = r(T)$ *and either* $v \in P \setminus \ker(T)$ *or* $-v \in P \setminus \ker(T)$, *then* (5.1) *has no solution at all.*

Proof: If $\lambda > r(T)$, then the resolvent $(\lambda - T)^{-1}$ is a positive endomorphism of E (cf. [23]). This implies the first assertion.

Suppose that $\lambda \leq r(T)$ and $v \in P$. Then, by applying the eigenfunctional ϕ_0 to the equation (5.1), we find that

$$(\lambda - r(T))\langle\phi_0, u\rangle = \langle\phi_0, v\rangle \geq 0 .$$

Hence the remaining part of the assertion is an easy consequence of the fact that $\phi_0(P \setminus \ker T) \subset \dot{\mathbb{R}}_+$. Q.E.D.

Recall that an eigenvalue λ of an endomorphism T of E is called *simple* if its algebraic multiplicity equals 1 , that is, if the vector subspace $\bigcup_{k \geq 1} \ker(\lambda - T)^k$ of E is one-dimensional.

(5.4) <u>*Theorem:*</u> *Let the hypotheses of Theorem* (5.1) *be satisfied. Then* $r(T)$ *is a simple eigenvalue of* T *and of* T' .

Proof: By the Riesz-Schauder theory, it suffices to show that $r(T)$ is a simple eigenvalue of T .

We first show that the kernel of $r(T) - T$ is one-dimensional. Suppose that $_0v$ is an arbitrary eigenvector of T to the eigenvalue $r(T)$. Then, since $u_0 \in \mathring{P}$, there exists a nonzero real number α such that $u_0 + \alpha v \in \partial P$. Suppose that $u_0 + \alpha v \in \ker(T)$. Then $r(T)(u_0 + \alpha v) = T(u_0 + \alpha v) = 0$ implies that u_0 and v are linearly dependent. Hence, if u_0 and v were linearly independent, then $u_0 + \alpha v \in P \setminus \ker(T)$ and, consequently, $r(T)(u_0 + \alpha v) = T(u_0 + \alpha v) \in \mathring{P}$, which contradicts the choice of α . This implies that $\ker(r(T) - T) = \operatorname{span}(u_0)$.

We suppose now that there exists an element $v \in E$ such that $(r(T) - T)^2 v = 0$ and $w := (r(T) - T)v \neq 0$. Then $Tw = r(T)w$ and, by the first part of the proof, we can assume that $w = u_0$. Consequently, for every $\alpha \neq 0$,

$$r(T)(u_0 + \alpha v) - T(u_0 + \alpha v) = \alpha(r(T) - T)v = \alpha u_0 ,$$

which contradicts Theorem (5.4) since $u_0 \in P \setminus \ker(T)$. Hence $\ker(r(T) - T) = \ker(r(T) - T)^2$.

Finally, let $S := r(T) - T$. Then for every $k \geq 2$, the preceding result implies that

$$\ker S^k = S^{-k}(o) = S^{-k+2}(S^{-2}(o)) = S^{-k+2}(S^{-1}(o))$$
$$= S^{-k+1}(o) = \ker S^{k-1} .$$

Hence $\ker(r(T) - T)^k = \ker(r(T) - T)$ for every $k \geq 1$, and the assertion follows. Q.E.D.

(5.5) *Theorem: Let the hypotheses of Theorem* (5.1) *be satisfied. Suppose in addition that* $P \cap \ker(T) = \emptyset$ *and that* S *is an almost strongly positive compact endomorphism of* E *such that* $Tu - Su \in P \setminus \ker(T)$ *for every* $u \in P$. *Then* $r(S) < r(T)$.

Proof: By Theorem (5.1) there exists an eigenvector $v_0 \in \mathring{P}$ of S to the eigenvalue $r(S)$. Hence $r(S)v_0 = Sv_0 = Tv_0 - (Tv_0 - Sv_0)$. By applying the eigenfunctional ϕ_0 of the operator T to this equation, we find that

$$r(S)\langle\phi_0,v_0\rangle = r(T)\langle\phi_0,v_0\rangle - \langle\phi_0,Tv_0 - Sv_0\rangle ,$$

that is

$$r(S)\langle\phi_0,v_0\rangle < r(T)\langle\phi_0,v_0\rangle$$

since $Tv_0 - Sv_0 \in P \setminus \ker(T)$. Hence the assertion follows from the fact that $\langle\phi_0,v_0\rangle > o$. Q.E.D.

Let (E,P) be an arbitrary OBS and let $e > o$ be a positive element in E . An endomorphism T of E is said to be e-positive if, for every $u \in P$, there exist positive numbers α and β such that $\alpha e \leq Tu \leq \beta e$. This definition is a special case of the more general definition of an e-positive linear operator due to M. A. Krasnosel'skii [15] . The reason for this restricted definition of an e-positive endomorphism lies in the fact, that, besides their importance for applications, these operators turn out to be closely related to the class of strongly positive operators. In fact, by endowing the vector subspace $E_e := \bigcup_{\lambda \geq o} \lambda[-e,e]$ of E with the order unit topology (that is, with the e-norm), it can be shown (e.g. [5,15]) that E_e becomes an ordered normed vector space whose positive cone P_e has nonempty interior. Moreover, T is e-positive iff $T(\dot{P}) \subset \mathring{P}_e$ (for more details cf. [5]).

This fact suggests the definition of an almost e-positive endomorphism. Namely, a linear operator $T : E \to E$ is said to be *almost e-positive* , if $P \ \ker(T) \neq \emptyset$ and if, for every $u \in P \setminus \ker(T)$, there exist positive numbers α and β such that $\alpha e \leq Tu \leq \beta e$. Then it can be shown by a careful analysis of the proofs in [15] , that almost e-positive endomorphisms of an OBS have, roughly speaking, the same spectral properties as strongly positive endomorphisms (cf. [13]). In a much more elegant way, these results can be established by using the above indicated connection with almost strongly positive endomorphisms.

Moreover, by modifying the corresponding proof in [15] it can be shown that the spectral radius of an almost strongly positive endomorphism T is the only eigenvalue of the complexification of T lying on the spectral circle. Finally it should be observed that in the above theorems (5.1) - (5.5), the completeness of E and the compactness of T have only been used (via the Krein-Rutman theorem) in order to guarantee that $r(T)$ is an eigenvalue of T . Hence these theorems can considerably be generalized.

Proof of Theorem (2.2) and Theorem (2.3): At the end of the preceding paragraph it has already been observed that the BVP (2.3) is equivalent to the equation

$$\rho u - Tu = v$$

in $C(\bar{\Omega})$, where $\rho := (\lambda + \omega)^{-1}$, $v := \rho S_\omega(c,d)$, and T is an almost strongly positive compact endomorphism of $C(\bar{\Omega})$ such that $\text{int}\, C_+(\bar{\Omega}) \cap \ker(T) = \emptyset$. Hence the assertions follow from the above general results by observing that $S_\omega(c,d) \in \text{int}\, C_+(\bar{\Omega})$ and that, by the maximum principle, the BVP (2.3) is uniquely solvable if $\rho^{-1} = \lambda + \omega \leq 0$. Q.E.D.

BIBLIOGRAPHY

[1] S. AGMON, A. DOUGLIS, and L. NIRENBERG: Estimates near the boundary for so-solutions of elliptic partial differential equations satisfying general boundary conditions, I., Comm.Pure Appl.Math. XII (1959), 623-727.

[2] H. AMANN: On the existence of positive solutions of nonlinear elliptic boundary value problems. Indiana Univ. Math. J., 21 (1971), 125-146.

[3] ———: On the number of solutions of nonlinear equations in ordered Banach spaces. J. Functional Anal., 11 (1972), 346-384.

[4] ———: A uniqueness theorem for nonlinear elliptic boundary value problems. Arch. Rat. Mech. Anal., 44 (1972), 178-181.

[5] ———: Fixed point equations and nonlinear eigenvalue problems in ordered Banach spaces. SIAM Review, to appear.

[6] D.G. ARONSON and L.A. PELETIER: Global stability of symmetric and asymmetric concentration profiles in catalyst particles. Arch. Rat. Mech. Anal., 54 (1974), 175-2o4.

[7] H. BREZIS: Problêmes unilatêraux, J. Math. Pures Appl., 51 (1972), 1-168.

[8] H. BRILL: Eine stark nichtlinear elliptische Gleichung unter einer nichtlinearen Randbedingung, Zeitschr. Angew. Math. Mech. to appear.

[9] J.M. CUSHING: Nonlinear Steklov problems on the unit circle. J. Math. Anal. Appl., 38 (1972), 766-783.

[10] J.-P. DIAZ: Un thêorême de Sturm-Liouville pour une class d'opêrateurs non lineaires maximaux monotones. J. Math. Anal. Appl., 47 (1974), 4oo-4o5.

[11] J.P.G. EWER and L.A. PELETIER: On the asymptotic behaviour of solutions of semilinear parabolic equations. SIAM J. Appl. Math., 28 (1975), 43-53.

[12] K. KLINGELHÖFER: Nonlinear harmonic boundary value problems. I. Arch. Rat. Mech. Anal., 31 (1968), 364-371.

[13] W. KNOKE: Positive Lösungen für Zwei-Punkt-Randwertprobleme. Diplomarbeit, Ruhr-Universität Bochum, 1974.

[14] M.G. KREIN and M.A. RUTMAN: Linear operators leaving invariant a cone in a Banach space. Amer. Math. Soc. Transl., Ser. 1, 1o (1962), 199-325.

[15] M.A. KRASNOSEL'SKII: "Positive Solutions of Operator Equations". Noordhoff, Groningen, 1964.

[16] O.A. LADYZHENSKAYA and N.N. URAL'TSEVA: "Linear and Quasilinear Elliptic Equations". Academic Press, New York, 1968.

[17] J.L. LIONS and E. MAGENES: Problemi ai limiti non omogenei (III). Ann. Sc. Norm. Sup. Pisa 15 (1961), 39-1o1.

[18] ———————: Problemi ai limiti non omogenei (V). Ann. Sc. Norm. Sup. Pisa 16 (1962), 1-44.

[19] ———————: "Non-Homogeneous Boundary Value Problems and Applications I." Springer Verlag, Berlin-Heidelberg-New York, 1972.

[20] C. MIRANDA: "Partial Differential Equations of Elliptic Type", Springer Verlag, Berlin-Heidelberg-New York, 1970.

[21] J. NECAS: "Les mêthodes directes en thêorie des êquations elliptiques". Academia, Editions de l'Academie Tchêcoslovaque des Sciences, Prague, 1967.

[22] M.H. PROTTER and H.F. WEINBERGER: "Maximum Principles in Differential Equations". Prentice-Hall, Englewood Cliffs, N.Y., 1967.

[23] H.H. SCHAEFER: "Topological Vector Spaces". Springer Verlag, Berlin-Heidelberg-New York, 1971.

[24] M. SCHECHTER: On L^p estimates and regularity, I. Amer. J. Math. 85 (1963), 1-13.

[25] J. SERRIN: A remark on the preceding paper of Amann. Arch. Rat. Mech. Anal. 44 (1972), 182-186.

[26] P. HESS: On the solvability of nonlinear elliptic boundary value problems. to appear: Indiana University Math. Journal.

W. Eckhaus (ed.), New Developments in Differential Equations

ON THE RANGE OF THE SUM OF NONLINEAR OPERATORS

H. BREZIS
Dept. de Mathématiques
Université P. et M. Curie
4 place Jussieu
75230 PARIS 5°

INTRODUCTION

Let A and B be two continuous functions on $\mathbb{R}$. Clearly we have :

$$R(A+B) = \bigcup_{u\in\mathbb{R}} Au + Bu \subset R(A) + R(B) = \bigcup_{v,w\in\mathbb{R}} Av + Bw \quad ,$$

and in general $R(A+B)$ is much smaller than $R(A) + R(B)$. However equality holds in two simple cases :

CASE I : A and B are both non decreasing

CASE II : A is linear and B is (non decreasing and) bounded.

Our purpose is to extend this observation to mappings in infinite dimensional spaces and to discuss some applications.

In §1 (the extension of case I) we present some results from a joint paper with A. HARAUX (Image d'une somme d'opérateurs monotones et applications, to appear in Israel J. of Math.).

In §2 (The extension of case II) we present a preliminary version of a joint work with L. NIRENBERG.

§1. THE MONOTONE + MONOTONE CASE

Let H be a real Hilbert space and let A and B be maximal monotone operators in H. In general, $R(A+B)$ could be much smaller than $R(A) + R(B)$; consider for example in $H = \mathbb{R}^2$, A = a rotation by $+\pi/2$ and B = a rotation by $-\pi/2$. Hovewer it turns out that in "many" important cases, $R(A+B)$ and $R(A) + R(B)$ are almost equal in the following sense. We say that two sets S_1 and S_2 are almost equal ($S_1 \simeq S_2$) provided $\overline{S}_1 = \overline{S}_2$ and $\text{Int } S_1 = \text{Int } S_2$.

THEOREM 1. Assume $A = \partial\varphi$ and $B = \partial\psi$ are subdifferentials of convex l.s.c. functions such that $\partial(\varphi+\psi) = \partial\varphi + \partial\psi$.
Then $R(A+B) \simeq R(A) + R(B)$.

Proof. Let $\varphi^\star$ and $\psi^\star$ denote the conjugate functions of φ and ψ respectively. We have $D(\varphi^\star) + D(\psi^\star) \subset D((\varphi+\psi)^\star)$
Indeed

$$\forall u \in H \ , \ (f,u) - \varphi(u) \leqslant \varphi^\star(f)$$
$$\forall u \in H \ , \ (g,u) - \psi(u) \leqslant \psi^\star(g)$$

Thus

$$\forall u \in H \quad (f+g,u) - (\varphi+\psi)(u) \leqslant \varphi^\star(f) + \psi^\star(g)$$

and consequently

$$(\varphi+\psi)^\star(f+g) \leqslant \varphi^\star(f) + \psi^\star(g)$$

Next observe that

$$R(A) + R(B) = D(\partial\varphi^\star) + D(\partial\psi^\star) \subset D(\varphi^\star) + D(\psi^\star) \subset D((\varphi+\psi)^\star) \ .$$

Recall that for any convex l.s.c. function ζ

$$\overline{D(\zeta)} = \overline{D(\partial\zeta)} \quad \text{and} \quad \text{Int } D(\zeta) = \text{Int } D(\partial\zeta) \ .$$

Hence

$$\overline{R(A) + R(B)} \subset \overline{D((\varphi+\psi)^\star)} = \overline{D(\partial(\varphi+\psi)^\star)} = \overline{R(\partial(\varphi+\psi))} = \overline{R(A+B)} \ .$$

Also

$$\text{Int } [R(A)+R(B)] \subset \text{Int } D((\varphi+\psi)^\star) = \text{Int } D(\partial(\varphi+\psi)^\star) = \text{Int } R(\partial(\varphi+\psi)) = \text{Int } R(A+B) \ .$$

THEOREM 2. Assume A is any maximal monotone operator and $B = \partial\psi$ with $D(B) = H$
Then $R(A+B) \simeq R(A) + R(B)$.

Proof. By a well know result of R.T. ROCKAFELLAR , $A + B$ is maximal monotone. We prove first that $R(A) + R(B) \subset \overline{R(A+B)}$. Let $f \in R(A) + R(B)$, so that $f \in Av + Bw$ (for some v and w).

Let $u_\varepsilon \in D(A)$ be the solution of

$$\text{(1)} \qquad \varepsilon u_\varepsilon + Au_\varepsilon + Bu_\varepsilon \ni f \ , \quad \varepsilon > 0 \ .$$

We are going to prove that $\sqrt{\varepsilon}\,|u_\varepsilon|$ remains bounded as $\varepsilon \to 0$. So, we will be able to conclude that $f \in \overline{R(A+B)}$.
By the monotonicity of A we have

$$\text{(2)} \qquad (Au_\varepsilon - Av \ , \ u_\varepsilon - v) \geqslant 0 \ .$$

On the other hand we have

$$(3) \qquad (Bu_\varepsilon - Bw, u_\varepsilon - v) \geqslant (Bw,v) - \psi(v) - \psi^\star(Bw)$$

Indeed

$$\psi(v) - \psi(u_\varepsilon) \geqslant (Bu_\varepsilon, v - u_\varepsilon) = (Bu_\varepsilon - Bw, v - u_\varepsilon) + (Bw, v - u_\varepsilon)$$

So

$$(Bu_\varepsilon - Bw, u_\varepsilon - v) \geqslant \psi(u_\varepsilon) - \psi(v) + (Bw, v - u_\varepsilon)$$

$$\geqslant (Bw,v) - \psi(v) - \psi^\star(Bw).$$

Adding (2) and (3) we find

$$(f - \varepsilon u_\varepsilon - f, u_\varepsilon - v) \geqslant - C, \ C \text{ independent of } \varepsilon.$$

i.e. $\quad \varepsilon|u_\varepsilon|^2 \leqslant \varepsilon|u_\varepsilon|\,|v| + C$.

Therefore $\quad \frac{1}{2}\varepsilon|u_\varepsilon|^2 \leqslant \frac{1}{2}\varepsilon|v|^2 + C$.

Next we prove that $\operatorname{Int}[R(A) + R(B)] \subset R(A+B)$.
Let $f \in \operatorname{Int}[R(A) + R(B)]$; we are going to show that the solution u_ε of (1) remains bounded as $\varepsilon \to 0$. This will enable us to infer that $u_{\varepsilon_n} \rightharpoonup u$ (weak convergence) with $Au + Bu \ni f$.
By assumption, there is some $r > 0$ such that for all $h \in H$ with $|h| < r$, $f + h \in Av_h + Bw_h$ (for some v_h and w_h).
Using now (2) and (3) with v and w replaced by v_h and w_h we get

$$(f - \varepsilon u_\varepsilon - f - h \ , \ u_\varepsilon - v_h) \geqslant C(h)$$

where C(h) is independent of ε (but depends on h).
Thus

$$(h,u_\varepsilon) \leqslant (h,v_h) + C(h) + \frac{1}{2}\varepsilon|v_h|^2 .$$

Applying the uniform boundedness principle, it follows that u_ε remains bounded as $\varepsilon \to 0$.

<u>EXAMPLE 1</u> Let $\Omega \subset R^N$ be a bounded smooth domain.
Given $f \in L^2(\Omega)$, the equation

$$(4) \qquad \begin{cases} - \Delta u + \dfrac{u}{\sqrt{1+u^2}} = f & \text{on } \Omega \\ \dfrac{\partial u}{\partial n} = 0 & \text{on } \partial\Omega \end{cases}$$

has a (unique) solution $u \in H^2(\Omega)$ iff $\frac{1}{|\Omega|}\left|\int_\Omega f(x)\,dx\right| < 1$.

First observe that if u exists we have from (4)

$$\int_\Omega \frac{u}{\sqrt{1+u^2}}\,dx = \int_\Omega f\,dx \quad \text{and so} \quad \left|\int_\Omega f\,dx\right| < |\Omega|.$$

Next we can write (4) in $H = L^2(\Omega)$ as

$$Au + Bu = f \quad \text{where}$$

$$Au = -\Delta u \quad \text{with} \quad D(A) = \{u \in H^2(\Omega)\ ;\ \frac{\partial u}{\partial n} = 0 \text{ on } \partial\Omega\}$$

$$Bu = \frac{u}{\sqrt{1+u^2}} \quad \text{with} \quad D(B) = H \quad \text{and} \quad B = \partial\psi$$

$$(\psi(u) = \int_\Omega \sqrt{1+u^2}\,dx)$$

Applying Theorem 1 <u>or</u> 2 we see that $R(A + B) \simeq R(A) + R(B)$. But, the assumption $\left|\int_\Omega f\,dx\right| < |\Omega|$ implies $f \in \operatorname{Int}[R(A) + R(B)]$, since for every $h \in L^2(\Omega)$ with $|h|_{L^2} < r$ (r small enough) we have

$$f + h = [\,(f+h) - \frac{1}{|\Omega|}\int_\Omega (f+h)\,dx] + [\frac{1}{|\Omega|}\int_\Omega (f+h)\,dx] \in R(A) + R(B)$$

By a similar argument one can treat the following

<u>EXAMPLE 2</u> Given $f \in L^2(\Omega)$, the equation

$$\begin{cases} -\Delta u - \lambda_1 u + \dfrac{1}{\sqrt{1+u^2}} = f & \text{on } \Omega \\ u = 0 \quad \text{on } \partial\Omega \end{cases}$$

where λ_1 is the first eigenvalue of $-\Delta$ (with homogenous Dirichlet B.C.) and v_1 (> 0 on Ω) is the corresponding eigen function, has a unique solution iff

$$\left|\int_\Omega f\,v_1\,dx\right| < \left|\int_\Omega v_1\,dx\right|$$

<u>REMARK</u> This kind of equation can also be solved by the "semi-coercive" methods (see the papers of SCHATZMAN and HESS)

<u>EXAMPLE 3</u> Given $f \in L^2(0,T)$, the equation

$$\begin{cases} \dfrac{du}{dt} + \dfrac{u}{\sqrt{1+u^2}} = f & \text{on } (0,T) \\ u(0) = u(T) \end{cases}$$

has a solution iff $\frac{1}{T}\left|\int_0^T f\right| < 1$.

<u>PROOF</u> Use Theorem 2 in $H = L^2(0,T)$ with

$Au = u'$, $D(A) = \{u \in H^1(0,T)\ ;\ u(0) = u(T)\}$

$Bu = \frac{u}{\sqrt{1+u^2}}$, $D(B) = H (B = \partial\psi)$.

§ 2 THE LINEAR NON MONOTONE + MONOTONE NONLINEAR CASE.

THEOREM 3. Let A be a linear (unbounded) operator in H with $\overline{D(A)} = H$ and closed graph.

Assume

(5) $N(A) = N(A^\star)$

(6) $\{u \in D(A)\ ;\ |u| + |Au| \leqslant 1\}$ is compact in H

Let B be a monotone hemicontinuous operator from H into H with $R(B)$ bounded. Then $R(A+B) \simeq R(A) + R(B)$.

Proof. It follows from (6) that $R(A)$ is closed and $\dim N(A) < \infty$. By (5), $R(A) = N(A)^\perp$ and so we can split H into a direct sum $R(A) \oplus N(A)$.

We write $u = u_1 + u_2$ with $u_1 \in R(A)$, $u_2 \in N(A)$.

Then $\tilde{A} = A_{|R(A) \cap D(A)}$ is $1 - 1$ and onto $R(A)$. $\tilde{A}^{-1}$ is a compact operator from $R(A)$ into itself. Given $f \in H$ and $\varepsilon > 0$, there is a $u_\varepsilon \in D(A)$ satisfying

(7) $\varepsilon u_{2\varepsilon} + Au_\varepsilon + Bu_\varepsilon = f$

Indeed (7) can be written as a system

$$\begin{cases} \tilde{A}u_{1\varepsilon} + B_1(u_\varepsilon) = f_1 \\ \varepsilon u_{2\varepsilon} + B_2(u_\varepsilon) = f_2 \end{cases}$$

or

$$\begin{cases} u_{1\varepsilon} = \tilde{A}^{-1}\,[\,f_1 - B_1(u_\varepsilon)] \\ u_{2\varepsilon} = \frac{1}{\varepsilon}\,[\,f_2 - B_2(u_\varepsilon)] \end{cases}$$

which has a solution by Schauder fixed point theorem (note that B is continuous from the strong to the weak topology in H). First we prove that $R(A) + R(B) \subset \overline{R(A+B)}$. Let $f \in R(A) + R(B)$, so that $f = Av + Bw$.

Clearly we have

(8) $|u_{1\varepsilon}| \leqslant C$, $|Au_{1\varepsilon}| \leqslant C$

By the monotonicity of B we obtain

$$(Bu_\varepsilon - Bw\ ,\ u_\varepsilon - w) \geqslant 0$$

i.e.

$$(f - \varepsilon u_{2\varepsilon} - Au_\varepsilon - f + Av, u_\varepsilon - w) \geqslant 0$$

Hence

$$\varepsilon|u_{2\varepsilon}|^2 \leqslant \varepsilon|u_{2\varepsilon}|\ |w| + (C + |Av|)(C + |w|)$$

and therefore $\sqrt{\varepsilon}\ |u_{2\varepsilon}|$ remains bounded as $\varepsilon \to 0$. Consequently $f \in \overline{R(A+B)}$.

Next we show that $\text{Int}\,[R(A) + R(B)] \subset R(A+B)$.

Let $f \in \text{Int}\,[R(A) + R(B)]$, so that $f + h = Av_h + Bw_h$, for all h with $|h| < r$. We have now

$$(- \varepsilon u_{2\varepsilon} - Au_\varepsilon + Av_h - h \,,\, u_\varepsilon - w_h) \geq 0$$

which implies $(h,u_{2\varepsilon}) \leq C(h)$ and therefore $u_{2\varepsilon}$ remains bounded as $\varepsilon \to 0$.

Hence $u_{\varepsilon_n} \to u$ which is a solution of $Au + Bu = f$.

Remark By a slight modification of the proof, one can show that, under the assumptions of Theorem 3 ,

$$R(A) + \text{conv}\, R(B) \simeq R(A+B).$$

THEOREM 4. Let A and B be as in Theorem 3 .

Let $f \in H$ be such that

(9) $\lim_{t\to+\infty} (B(tv),v) > (f,v)$

for all $v \in N(A)$, $v \neq 0$.

Then there is a $u \in D(A)$ solution of

(10) $Au + Bu = f$.

Conversely if (10) has a solution, then

(11) $\lim_{t\to+\infty} (B(tv),v) \geq (f,v)$ for all $v \in N(A)$.

Sketch of the proof. Observe that since B is monotone, $t \mapsto (B(tv),v)$ is nondecreasing, and $\lim_{t\to+\infty} B(tv),v)$ exists. First assume that (10) has a solution. We then have

$$(B(tv) - Bu \,,\, tv - u) \geq 0$$

or

$$(B(tv) - f + Au \,,\, tv - u) \geq 0 \,.$$

Therefore, for $v \in N(A) = N(A^\star)$

$$(B(tv),v) \geq \frac{(B(tv),u)}{t} + (f,v) - \frac{(f,u)}{t} + \frac{(Au,u)}{t}$$

and (11) follows.

Assume now that (9) holds. We have

$$\lim_{t\to+\infty} (B(tv),v) \leq \delta(v) = \sup_{a\in \text{conv}\, R(B)} (a,v) \quad .$$

Since $\dim N(A) < \infty$, (9) implies the existence of $\varepsilon > 0$ such that

(12) $\delta(v) - (f,v) \geq \varepsilon|v|$ for all $v \in N(A)$.

From (12) we deduce by a separation argument (relying on the fact that dim N(A) < ∞) that $f \in \text{Int}\,[N(A)^{\perp} + \text{conv}\, R(B)]$. Then we apply the remark following Theorem 3 to conclude that $f \in \text{Int}\,[R(A+B)]$.

Example 4. Let λ be any eigenvalue of $-\Delta$ (with homogenous Dirichlet B.C) and let N be the corresponding nullspace.
Let $\beta : \mathbb{R} \to \mathbb{R}$ be a continuous, nondecreasing, bounded function.
The equation

$$\begin{cases} -\Delta u - \lambda u + \beta(u) = f & \text{on } \Omega \\ u = 0 & \text{on } \partial\Omega \end{cases}$$

has a solution provided

$$\int_{[v>0]} \beta_+ v + \int_{[v<0]} \beta_- v > \int_\Omega fv \quad \text{for all } v \in N,\ v \not\equiv 0,$$

where $\beta \pm = \lim_{r \to \pm\infty} \beta(r)$.

Proof. Apply Theorem 4 in $H = L^2(\Omega)$ with $Au = -\Delta u - \lambda u$, $Bu = \beta(u)$ and note that (as a consequence of Lebesgue Theorem)

$$\lim_{t \to +\infty} (B(tv), v) = \int_{[v>0]} \beta_+ v + \int_{[v<0]} \beta_- v .$$

Remark. Example 4 can also be treated (even for non monotone β's) by using a Theorem of Landesman-Lazer ; see also the subsequent works of L. NIRENBERG , P. HESS etc ...

W. Eckhaus (ed.), New Developments in Differential Equations

On the asymptotic behaviour of solutions of an equation arising in population genetics

L.A. Peletier

Technische Hogeschool Delft

1. Introduction

In this paper we present some results which were obtained jointly with P.C. Fife of the University of Arizona.

Consider a population of diploid individuals, living in a one-dimentional habitat Ω. Suppose we can distinguish in a gene in this population two different types, called alleles. We shall denote them by A and B. Then the population can be divided into three classes of individuals, called genotypes. Two of these carry alleles which are alike, they are denoted by AA and BB. The third one carries both alleles and is denoted by AB.

Let $u(x,t)$ denote the frequency of allele A at place $x \in \Omega$ and time t. Then it can be shown that in a simple model, which incorporates the effects of dispersal in the habitat and selection advantages of the genotypes, the function u satisfies an equation of the form

$$u_t = u_{xx} + f(x,u) \qquad x \in \Omega, \quad t > 0 \tag{1}$$

in which the subscripts denote partial differentiation.

When the viability of genotype AB lies between the viabilities of genotypes AA and BB, an appropriate choice for f is

$$f(x,u) = g(x)u(1-u).$$

This case has been considered by Conley [2], Fleming [3] and Hoppensteadt [4].

In the present paper we shall investigate the case when the viability of AB is less than the viabilities of AA and BB. In this case, a possible choice for f is

$$f(x,u) = u(1-u)[u-a(x)] \qquad (2)$$

where the function $a : \Omega \to (0,1)$ depends on the viabilities of the three genotypes [1]. For simplicity we shall assume throughout this paper, that f is given by (2). However, it should be emphasized that our analysis applies equally well to many other choices of f.

It is clear that the functions $u \equiv 0$ and $u \equiv 1$ are solutions of equation (1). If $a(x) \equiv$ constant, equation (1) also possesses a wave front solution, i.e. a solution of the form $u(x,t) = q(x-ct)$ where c is a constant and $q(\xi)$ a monotone function such that $q(-\infty) = 0$ and $q(\infty) = 1$ [1, 5]. The sign of the wave speed c is determined by the integral $\int_0^1 f(s)ds$. To see this we observe that q satisfies the equation

$$q_{\xi\xi} + cq_{\xi} + f(q) = 0 \qquad \xi \in \mathbb{R}.$$

If we now multiply this equation by q_{ξ} and integrate the result over $(-\infty,\infty)$, we obtain

$$c\int_{-\infty}^{\infty} q_{\xi}^{2} d\xi + \int_0^1 f(q)dq = 0$$

which implies that

$$\operatorname{sgn} c = - \operatorname{sgn} \int_0^1 f(s)ds.$$

Thus, for example, if $\int_0^1 f(s)ds < 0$, then $c > 0$ and the wave moves towards positive x, i.e. for each $x \in \mathbb{R}$, $u(x,t) = q(x-ct)$ tends to zero as $t \to \infty$. Genetically this means that genotype BB is dominant and that the genotypes AA and AB are swept away.

In this paper we shall discuss a population, living in an infinitely extended habitat, in which genotype AA is dominant for large values of x and genotype BB is dominant for large values of (-x). Thus, the function

$$J(x) = \int_0^1 f(x,s)ds \tag{3}$$

is positive for large x and negative for large (-x).

2. The transition layer

By a transition layer we shall mean a monotone equilibrium solution u(x) of equation (1) such that $u(x) \to 0$ as $x \to -\infty$ and $u(x) \to 1$ as $x \to \infty$. Thus, u is a monotone solution of the two point boundary value problem

$$\text{(I)} \begin{cases} u'' + f(x,u) = 0 & -\infty < x < \infty \\ u(-\infty) = 0 \ , \ u(+\infty) = 1. \end{cases} \tag{4}$$

We shall assume that the function a, appearing in f, has the following properties:

(i) $a \in C^1(\mathbb{R})$;

(ii) $a' < 0$ on $\mathbb{R}$;

(iii) $a(-\infty) \in (\frac{1}{2}, 1)$ and $a(\infty) \in (0, \frac{1}{2})$.

Without loss of generality we may assume that $a(0) = \frac{1}{2}$. Thus, the function J(x), defined in (3) is negative on $(-\infty, 0)$ and positive on $(0, \infty)$.

THEOREM 1. *Let* f *be given by* (2), *and let the function* a *in* (2) *have the properties* (i) - (iii). *Then problem I has one and only one solution. This solution is monotone*.

To prove this result we first consider the auxiliary problem

$$(\mathrm{II}^+) \begin{cases} u'' + f(x,u) = 0 & 0 < x < \infty \\ u(0) = \alpha, \ u(\infty) = 1, \end{cases}$$

where $\alpha \in [0, 1]$. It can be shown that this problem has a unique solution $v(x, \alpha)$. This solution is monotone, its derivative at $x = 0$, $v'(0, \alpha)$ depends continuously on α and $v'(0, 0) > 0$, $v'(0, 1) = 0$.

Next we consider in a similar vein the problem

$$(\mathrm{II}^-)\quad \begin{cases} u'' + f(x,u) = 0 & -\infty < x < 0 \\ u(-\infty) = 0 \quad , \; u(0) = \alpha , \end{cases}$$

where $\alpha\in[0, 1]$. We denote the solution of this problem by $w(x, \alpha)$. It is also monotone, its derivative at $x = 0$, $w'(0, \alpha)$, is continuous on $[0, 1]$ and $w'(0, 0) = 0$, $w'(0, 1) > 0$.

It follows that there exists an $\alpha\epsilon^*(0, 1)$ such that

$$v'(0, \alpha^*) = w'(0, \alpha^*).$$

Hence the composite function

$$u(x) = \begin{cases} v(x, \alpha^*) & 0 \leq x < \infty \\ w(x, \alpha^*) & -\infty < x < 0 \end{cases}$$

is the desired solution. Because v and w are both monotone, u is monotone as well.

To prove that this solution is unique, we assume to the contrary that there exist two solutions, u_1 and u_2. By the uniqueness of solutions of problems II^+ and II^-, $u_1(0) \neq u_2(0)$. Thus, let us assume that $u_1(0) > u_2(0)$. Then, invoking uniqueness again, it is readily shown that $u_1(x) > u_2(x)$ on $\mathbb{R}$.

If we multiply equation (4) by u' and integrate over $\mathbb{R}$, we obtain, after subtracting the result for u_2 from that for u_1:

$$\int_0^1 \{f(y_1(s), s) - f(y_2(s), s)\}ds = 0, \tag{5}$$

where y_1 and y_2 are the inverse functions of u_1 and u_2.
However, because $u_1 > u_2$ on $\mathbb{R}$, $y_1 < y_2$ on $(0, 1)$. Because $f_x(x,u) = -a'(x)u(1-u) > 0$ on $(0, 1)$ this implies that

$$\int_0^1 \{f(y_1(s), s) - f(y_2(s), s)\}ds < 0,$$

which contradicts (5).

It follows that problem I can only have one solution.

3. Stability

We now investigate the stability of the transition layer $\phi(x)$, we constructed in the previous section. Thus, we consider the Cauchy problem

$$u_t = u_{xx} + f(x,u) \qquad -\infty < x < \infty,\ t > 0 \qquad (1)$$

$$u(x,0) = \psi(x) \qquad -\infty < x < \infty \qquad (6)$$

in which $\psi \in C(\mathbb{R})$, and takes on values in the interval $[0, 1]$. The existence and uniqueness of a solution of this problem was established in [6]; we denote it by $u(x,t;\psi)$.

Consider the one parameter family of functions

$$v(x, h) = \phi(x + h) \qquad h \in \mathbb{R}.$$

Note that for each $x \in \mathbb{R}$, $\partial v(x, h)/\partial h = \phi'(x+h) > 0$. Let $h > 0$. Then, writing $x + h = y$, we have

$$\begin{aligned} v_{xx} + f(x,v) &= \phi_{yy} + f(y-h, \phi(y)) \\ &= -f(y,\phi(y)) + f(y-h, \phi(y)) \\ &< 0 \end{aligned}$$

because $f_x > 0$. Therefore, if $h > 0$, $v(x,h)$ is a supersolution of problem I. Similarly, if $h < 0$, $v(x,h)$ is a subsolution of problem I. This family of sub and supersolutions enables us to prove the following result.

THEOREM 2. *Let* $u(x,t;\psi)$ *be the solution of problem* (1), (6) *in which* f *is given by* (2), *and* a *satisfies the assumptions* (i) - (iii). *Suppose there exist numbers* $h_1, h_2 \in \mathbb{R}$ *such that* $h_1 < 0 < h_2$, *and*

$$v(x, h_1) \leq \psi(x) \leq v(x, h_2) \qquad -\infty < x < \infty.$$

Then

$$u(x,t;\psi) \to \phi(x) \quad \text{as } t \to \infty$$

uniformly on $\mathbb{R}$.

Proof. Since

$$v(x, h_1) \leq \psi(x) \leq v(x, h_2) \qquad -\infty < x < \infty,$$

it follows from the maximum principle that

$$u(x,t;v(.,h_1)) \leq u(x,t;\psi) \leq u(x,t;v(.,h_2)).$$

However, it can be shown by means of an argument, similar to one used by Aronson and Weinberger [1] that the functions $u(x,t;v(.,h_i))$ $(i = 1, 2)$ both tend to a solution of problem I. Since ϕ is the only solution of problem I the result follows.

The basic tool in the proof of Theorem 2 was the family $v(x,h)$ of sub and super solutions of problem I. By using more subtle families of sub and super solutions, we can prove the following result.

THEOREM 3. *Let* $u(x,t;\psi)$ *be the solution of problem* (1), (6) *in which* f *is given by* (2) *and* a *satisfies assumptions* (i) - (iii). *Suppose*

$$\liminf_{x\to\infty} \psi(x) > \lim_{x\to\infty} a(x),$$

$$\limsup_{x\to-\infty} \psi(x) < \lim_{x\to-\infty} a(x).$$

Then

$$u(x,t;\psi) \to \phi(x) \text{ as } t \to \infty$$

uniformly on $\mathbb{R}$.

The proof of this theorem, as well as the details of the proofs of the previous theorems will appear in a subsequent paper.

REFERENCES

[1] Aronson, D.G. and Weinberger, H.F., Nonlinear diffusion in population genetics, combustion, and nerve propogation, Proc. Tulane Progr. in Partial Differential Eqns., Springer Lecture Notes in Mathematics, (446), 1975.

[2] Conley, C., An application of Wazewski's method to a nonlinear boundary value problem which arises in population genetics, Univ. of Wisconsin Math. Research Center Tech. Summary Report No. 1444, 1975.

[3] Fleming, W.H., A selection-migration model in population genetics, Journal Math. Biology (to appear).

[4] Hoppensteadt, F.C., Analysis of a stable polymorphism arising in a selection-migration model in population genetics.

[5] Kanel', Ja.I., Stabilization of solutions of the Cauchy problem for equations encountered in combustion theory, Mat. Sbornik (N.S.) 59 (101) (1962), supplement, 245 - 288.

[6] Kolmogoroff, A., Petrovsky, I., and Piscounoff, N., Étude de l'équation de la diffusion avec croissance de la quantité de matière et son application à un problème biologique, Bull. Univ. Moskou, Ser. Internat., Sec. A, 1 (1937) #6, 1 - 25.

W. Eckhaus (ed.), New Developments in Differential Equations

OPTIMAL CONTROL OF A SYSTEM GOVERNED BY THE NAVIER-STOKES EQUATIONS COUPLED WITH THE HEAT EQUATION.

C. CUVELIER
university of Delft
Delft, The Netherlands.

Introduction *)

Let Ω be an open set in $\mathbb{R}^2$. The boundary Γ of Ω, which is assumed to be regular, is divided into two parts: Γ_1 and Γ_2.

We consider the following problem of free convection (cf. LANDAU, LIFCHITZ [1]): Find two scalar functions Te = Te(x,t) = temperature, p=p(x,t)=pressure and one vector function $u=u(x,t)=\{u_1(x,t),u_2(x,t)\}=$ = velocity, defined on $\bar{\Omega} \times [0,T[$, (x is the space variable, t is the time variable) such that

$$(1)\qquad \frac{\partial Te}{\partial t} - \varkappa \Delta Te + \sum_{i=1}^{2} u_i \frac{\partial Te}{\partial x_i} = g$$

$$(2)\qquad \frac{\partial u}{\partial t} - \nu \Delta u + \sum_{i=1}^{2} u_i \frac{\partial u}{\partial x_i} + \operatorname{grad} p - \sigma Te = f$$

$$(3)\qquad \operatorname{div} u = 0$$

in $\Omega \times]0,T[$, where $\varkappa$ = thermal diffusivity, ν = kinematic viscosity, σ = constant two-component vector, g = heat sources anf f = external force.

The functions Te, u, p should satisfy the following initial and boundary conditions:

$$(4)\qquad Te(x,o) = Te_o(x), \qquad u(x,o) = u_o(x), \qquad x \in \bar{\Omega},$$

$$(5)\qquad \frac{\partial Te}{\partial n}(x,t) = v(x,t), \qquad \{x,t\} \in \Sigma_1 = \Gamma_1 \times]o,T[,$$

$$(6)\qquad \frac{\partial Te}{\partial n}(x,t) = 0, \qquad \{x,t\} \in \Sigma_2 = \Gamma_2 \times]o,T[,$$

$$(7)\qquad u(x,t) = 0, \qquad \{x,t\} \in \Sigma = \Gamma \times]o,T[,$$

$\frac{\partial}{\partial n}$ denotes the normal derivative at Γ, directed towards the exterior of Ω.

The function v is at our disposal and is called the control of the problem. Frequently, in practice, the control is subject to some constraints and we express this fact by requiring that v belongs to a set U_{ad} of admissible controls.

---

*) For standard notation see LIONS [1], [2] .

We are particularly interested in the temperature distribution at time t=T and our objective is to determine the function v in such a way that the temperature distribution at time t=T is as close as possible (in some definite sense) to a desired distribution $z_d(x)$, $x \in \Omega$.

For v belonging to the set of admissible controls U_{ad}, the system (1),...,(7) defines Te(x,T) in a unique way: Te(x,T) = Te(x,T;v). Next, we define a functional $v \to J(v)$ by

$$(8) \quad J(v) = \tfrac{1}{2}\{ \int_\Omega | Te(x,T;v) - z_d(x) |^2 dx + \alpha \int_0^T \int_{\Gamma_1} |v(x,t)|^2 d\Gamma\, dt \}$$

where α is a positive constant.

The problem of optimal control can be stated as follows:

$$\text{Find} \inf_{v \in U_{ad}} J(v).$$

Any element $v^* \in U_{ad}$ such that $J(v^*) = \inf_{v \in U_{ad}} J(v)$ is termed an optimal control.

#

The contents of this paper are as follows:

In chapter 1 we study the well-posedness of problem (1),...,(7) for fixed v. Next we prove the existence of an optimal control and we state a necessary condition for an element $v \in U_{ad}$ to be an optimal control. In order to write this necessary condition in an simpler way we introduce an adjoint system of equations. With the aid of this system we construct an iterative method which provides us, in the limit, with an element of U_{ad} which satisfies the necessary condition for optimality.

Because system (1),...,(3) is not of Cauchy-Kowaleska type (i.e. div u=0 does not contain a term $\frac{\partial p}{\partial t}$), it is not easy to solve numerically. In order to overcome, in some sense ,this difficulty, we introduce a perturbed system of equations (depending on a parameter $\varepsilon > 0$) which is of Cauchy-Kowaleska type (cf. LIONS [2], TEMAM [1] , [2]).

In chapter 2 we study the well-posedness of this perturbed system and the existence of an optimal control. As in the unperturbed case we introduce an adjoint system and define an iterative method for the construction of an element satisfying the necessary condition for optimality.

Chapter 3 is devoted to a convergence theorem in which we prove that the solution $\{Te_\varepsilon , u_\varepsilon, p_\varepsilon\}$ of the perturbed system corresponding to an optimal control $v_{\varepsilon opt}$ converges, in some topological sense, as the parameter ε tends to zero, to the solution $\{Te,u,p\}$ of the unperturbed system corresponding to one of its optimal controls v_{opt}.

In chapter 4 we give, very briefly, some of the numerical results.

1. OPTIMAL CONTROL OF THE UNPERTURBED SYSTEM OF EQUATIONS.

1.1. FORMULATION OF THE UNPERTURBED SYSTEM. NOTATION P(Te,u,p;v).

Before we give the exact formulation of our problem we introduce some function spaces:

$$\mathcal{V} = \{u \in (D(\Omega))^2 \mid \operatorname{div} u = 0\}$$
$$H = \text{closure of } \mathcal{V} \text{ in } (L_2)^2 \qquad ^{*)}$$
$$V = \text{closure of } \mathcal{V} \text{ in } (H_0^1)^2$$

Given $Te_0 \in L_2$, $u_0 \in H$, $g \in L_2(o,T;L_2)$, $f \in L_2(o,T;(L_2)^2)$, κ, ν positive constants, $T > 0$, $\sigma = \{\sigma_1, \sigma_2\}$ constant vector, $v \in L_2(\Sigma_1) = L_2(o,T;L_2(\Gamma_1))$, we can state our problem as follows:

Find a scalar function Te and a vector function u such that

$$Te \in L_2(o,T;H^1), \qquad Te' \in L_1(o,T;(H^1)'), \qquad Te(o) = Te_0, \qquad ^{**)}$$
$$u \in L_2(o,T;V), \qquad u' \in L_1(o,T;V'), \qquad u(o) = u_0,$$

and

$$\text{(9)} \quad \begin{aligned} &Te' + \kappa C_1[Te;v] + A[u,Te] = g \quad \text{in } \mathcal{D}'(]o,T[\,;(H^1)'),\\ &u' + \nu C_2[u] + B[u,u] - \sigma Te = f \quad \text{in } \mathcal{D}'(]o,T[\,;V'),\end{aligned}$$

where C_1, C_2, A and B are defined as follows:

$$C_1[Te;v] : \Psi(\in H^1) \to (\operatorname{grad} Te, \operatorname{grad}\Psi)_{(L_2)^2} - (v, \gamma_0\Psi)_{L_2(\Gamma_1)}, \qquad ^{***)}$$

$$C_2[u] : \varphi(\in (H_0^1)^2) \to (\operatorname{grad} u, \operatorname{grad}\varphi)_{(L_2)^2},$$

$$A[u,Te] : \Psi(\in H^1) \to \tfrac{1}{2}\sum_{i=1}^{2}\int_\Omega \{u_i \frac{\partial Te}{\partial x_i}\Psi - u_i\, Te \frac{\partial \Psi}{\partial x_i}\}dx,$$

$$B[u,w] : \varphi(\in (H_0^1)^2 \to \tfrac{1}{2}\sum_{i=1}^{2}\sum_{j=1}^{2}\int_\Omega \{u_i \frac{\partial w_j}{\partial x_j}\varphi_j - u_i w_j \frac{\partial \varphi_j}{\partial x_i}\}dx.$$

We will call this problem: P(Te,u,p;v).

The following theorem deals with the well-posedness of P(Te,u,p;v).

THEOREM 1.1.

Problem P(Te,u,p;v) admits, for fixed $v \in L_2(\Sigma_1)$, a unique solution $\{Te,u\}$ which is continuous from $[o,T] \to L_2 \times H$. The pressure p is the unique element of $L_\infty(o,T;L_2)/\mathbb{R}$ satisfying:

---

*) We write L_2, H_0^1, H^1 in stead of $L_2(\Omega)$, $H_0^1(\Omega)$, $H^1(\Omega)$.

**) Notation $\varphi' = \frac{\partial\varphi}{\partial t}$; X' denotes the dualspace of X.

***) γ_0 denotes the trace operator, $\gamma_0 : H^1(\Omega) \to H^{\frac{1}{2}}(\Gamma)$.

$$\langle u'(t) + \nu C_2[u(t)] + B[u(t),u(t)] - \sigma Te(t) - D[p(t)] - f(t), \varphi\rangle = 0$$

for all $\varphi \in (H_o^1)^2$, where $D[p]: \varphi(\in (H_o^1)^2) \to \int_\Omega p \operatorname{div} \varphi \, dx$ and $\langle .,. \rangle$ denotes the duality pairing of $(H_o^1)^2$ and $((H_o^1)^2)'$.

PROOF

The proof is classical and can be found in LIONS [2], TEMAM [2] or CUVELIER [1].

#

Concerning the dependence of $\{Te,u\}$ on v we have

PROPOSITION 1.1.

Let $\{Te(v), u(v)\}$ be the solution of $P(Te,u,p;v)$. Then:

$$||Te(v)||^2_{L_\infty(o,T;L_2)} + \varkappa||Te(v)||^2_{L_2(o,T;H^1)} + ||u(v)||^2_{L_\infty(o,T;H)} + \nu||u(v)||^2_{L_2(o,T;V)} \leq c_1(1+||v||^2_{L_2(\Sigma_1)}) \quad {}^{*})$$

$$\int_{-\infty}^{\infty} |\tau|^{2\gamma} \{||\hat{T}e(\tau;v)||^2_{L_2} + ||\hat{u}(\tau;v)||^2_H\} \, d\tau \leq c_2(1+||v||^2_{L_2(\Sigma_1)})$$

with $0 < \gamma < \frac{1}{4}$. ($\hat{}$ denotes the Fourier transform with respect to time).
Let $\{Te(v^i), u(v^i)\}$, $i=1,2$, be the solution of $P(Te,u,p;v^i)$, then:

$$||Te(v^1)-Te(v^2)||^2_{L_\infty(o,T;L_2)} + \varkappa||Te(v^1)-Te(v^2)||^2_{L_2(o,T;H^1)} + ||u(v^1)-u(v^2)||^2_{L_\infty(o,T;H)} + \nu||u(v^1)-u(v^2)||^2_{L_2(o,T;V)} \leq$$

$$\leq c_3||v^1-v^2||^2_{L_2(\Sigma_1)} .$$

PROOF.

see LIONS [2], TEMAM [2], CUVELIER [2].

#

1.2. THE OPTIMAL CONTROL OF $P(Te,u,p;v)$.

The function v is called the control. Let U_{ad} be a closed convex set in $L_2(\Sigma_1)$; U_{ad} is the set of admissible controls. The observation is given by $v \to Te(T;v)$. Let $z_d \in L_2$ be given; z_d is the desired state. We define the cost-function $v \to J(v)$ by (8). The control problem then is:

Find $\inf_{v \in U_{ad}} J(v)$.

THEOREM 1.2.

There exists at least one element $v_{opt} \in U_{ad}$ such that $J(v_{opt}) \leq J(v)$, for all $v \in U_{ad}$. v_{opt} is termed an optimal control.

---

*) c_j, $j=1,2,\ldots$, are constants only depending on the data, and not on v.

PROOF

Let $\{v_n\}$ be a minimizing sequence in U_{ad}, i.e. $J(v_n) \to \inf_{v \in U_{ad}} J(v)$.

Because of the coercivity of the functional J ($J(v_n) \geq \frac{1}{2}\alpha \|v_n\|^2_{L_2(\Sigma_1)}$) the sequence $\{v_n\}$ is bounded in $L_2(\Sigma_1)$ and consequently it contains a subsequence $\{v_{n'}\}$ such that

(10) $\quad v_{n'} \to w \quad$ weakly in $L_2(\Sigma_1)$.

Since U_{ad} is a closed convex set, it is weakly closed. Hence (10) implies $w \in U_{ad}$.

Next we claim that

(11) $\quad Te(T;v_{n'}) \to Te(T;w) \quad$ weakly in L_2.

This result, which is based on compactness theorems, is proved in CUVELIER [2]. From (10) and (11) it is easily seen that the functional $v \to J(v)$ is lower semi-continuous in the weak topology of $L_2(\Sigma_1)$. Thus

$$\inf_{v \in U_{ad}} J(v) = \lim_{n' \to \infty} \inf J(v_{n'}) \geq J(w)$$

Hence we must necessarily have $J(w) = \inf_{v \in U_{ad}} J(v)$, so that w is an optimal control of the problem $P(Te,u,p;.)$: $w = v_{opt}$.

#

An optimal control can be characterized by the following lemma:

LEMMA 1.1.

A necessary condition for an element $v_{opt} \in U_{ad}$ to be an optimal control is that it satisfies the following inequality:

$$(12) \quad \begin{aligned} &(Te(T;v_{opt})-z_d,\ (Te^G(v_{opt})[w(t)-v_{opt}(t)])_{t=T})_{L_2} + \\ &\quad + \alpha \int_0^T (v_{opt}(t),w(t)-v_{opt}(t))_{L_2(\Gamma_1)} dt \geq 0 \end{aligned}$$

for all $w \in U_{ad}$, where $Te^G(v)\,[w]$ denotes the Gateaux derivative of Te at the point v in the direction w.

PROOF

This lemma is a direct application of a lemma in LIONS [1].

#

In order to write inequality (12) in a more sympathetic form we introduce an adjoint system of equations.

#

1.3. FORMULATION OF AN ADJOINT SYSTEM, NOTATION $P^*(\vartheta,q,\pi;v,Te,u,p)$.

Let $\{Te,u,p\}$ be the solution of $P(Te,u,p;v)$, then we define the adjoint problem as follows:

Find a scalar function ϑ and a vector function q such that

$$\vartheta \in L_2(o,T;H^1), \quad \vartheta' \in L_1(o,T;(H^1)'), \quad \vartheta(T)=Te(T)-z_d,$$

$$q \in L_2(o,T;V), \quad q' \in L_1(o,T;V'), \quad q(T) = 0,$$

and

$$-\vartheta' + \varkappa C_1[\vartheta;o] - A[u,\vartheta] - \sigma q = 0 \quad \text{in } \mathcal{D}'(]o,T[\,;(H^1)'),$$

$$-q' + \nu C_2[q] - B[u,q] - B^*[u,q] - A^*[Te,\vartheta] = 0 \quad \text{in } \mathcal{D}'(]o,T[;V'),$$

where A^* and B^* are defined by:

$$A^*[Te,\vartheta] \;:\; \varphi(\in (H_o^1)^2) \to \tfrac{1}{2} \sum_{i=1}^{2} \int_\Omega \{ \varphi_i \frac{\partial \vartheta}{\partial x_i} Te - \varphi_i \vartheta \frac{\partial Te}{\partial x_i}\} \, dx \;,$$

$$B^*[u,q] \;:\; \varphi(\in (H_o^1)^2) \to \tfrac{1}{2} \sum_{i=1}^{2} \sum_{j=1}^{2} \int_\Omega \{\varphi_i \frac{\partial q_j}{\partial x_i} u_j - \varphi_i q_j \frac{\partial u_j}{\partial x_i}\} \, dx.$$

We denote this problem by $P^*(\vartheta,q,\pi;v,Te,u,p)$.

#

THEOREM 1.3.

Let $\{Te,u,p\}$ be the solution of $P(Te,u,p;v)$. Problem $P^*(\vartheta,q,\pi;v,Te,u,p)$ has a unique solution $\{\vartheta,q\}$ and the pressure π is the unique element in $L_\infty(o,T;L_2)/\mathbb{R}$ defined by:

$$\langle -q'(t) + \nu C_2[q(t)] - B[u(t), q(t)] - B^*[u(t),q(t)] +$$
$$- A^*[Te(t), \vartheta(t)] + D\,[\pi(t)],\varphi \rangle = 0 \qquad \text{for all } \varphi \in (H_o^1)^2.$$

PROOF See CUVELIER [2].

#

Using the solution of the adjoint system we can prove the following

LEMMA 1.2.

A necessary condition for an element $v_{opt} \in U_{ad}$ to be an optimal control is that it satisfies the following inequality:

$$\int_0^T (\varkappa\gamma_0\vartheta(t;v_{opt}) + \alpha v_{opt}(t), w(t) - v_{opt}(t))_{L_2(\Gamma_1)} dt \geqq 0, \forall w \in U_{ad}.$$

PROOF

This lemma is a direct consequence of lemma 1.1 and the fact that $\{\vartheta(v),q(v)\}$ is the adjoint state of $\{Te^G(v)\,[w]\,,u^G(v)\,[w]\}$.

#

REMARK 1.1.

We also find that $J^G(v) = \varkappa\gamma_0\vartheta(v) + \alpha v$.

#

The results may now be summarized in the following theorem:

THEOREM 1.4.

A necessary condition for an element $v_{opt} \in U_{ad}$ to be an optimal control is that $\{Te,u,p,\vartheta,q,\pi,v_{opt}\}$ satisfies

(i) $\{Te,u,p\}$ is the solution of $P(Te,u,p;v_{opt})$

(ii) $\{\vartheta,q,\pi\}$ is the solution of $P^*(\vartheta,q,\pi;v_{opt},Te,u,p)$

(iii) $\int_0^T (\varkappa\gamma_0\vartheta(t) + \alpha v_{opt}(t), w(t) - v_{opt}(t))_{L_2(\Gamma_1)}dt \geq 0$ for all $w \in U_{ad}$

≠

1.4. ITERATIVE ALGORITHM.

For the construction of an optimal control we can use, among others, a gradient method with a projection. We choose $v^{(o)}_{opt} \in U_{ad}$ arbitrarily. Once we know $v^{(m)}_{opt}$ we calculate $\{Te(v^{(m)}_{opt}), u(v^{(m)}_{opt})\}$ by solving $P(Te,u,p;v^{(m)}_{opt})$ Next we solve $P^*(\vartheta,q,\pi;v^{(m)}_{opt},Te,u,p)$ which provides us with $\{\vartheta(v^{(m)}_{opt}), q(v^{(m)}_{opt})\}$. The (m+1)-th approximation of an optimal control is now given by

$$(13) \quad v^{(m+1)}_{opt} = P_{U_{ad}}[v^{(m)}_{opt} - \rho J^G(v^{(m)}_{opt})] \; (=P_{U_{ad}}[v^{(m)}_{opt} - \rho(\alpha v^{(m)}_{opt} + \varkappa\vartheta(v^{(m)}_{opt})|_{\Sigma_1})] \;)$$

where $P_{U_{ad}}$ denotes the projection on the set U_{ad}; ρ is a positive constant and $|_{\Sigma_1}$ means the restriction to Σ_1.

≠

THEOREM 1.5.

When ρ satisfies $0 < \rho < \frac{2}{M}$ (M will be defined in lemma 1.3), the sequence $\{v^{(m)}_{opt}\}$ contains a subsequence $\{v^{(m')}_{opt}\}$ converging to an element $v^*_{opt} \in U_{ad}$ weakly in $L_2(\Sigma_1)$. The limit v^*_{opt} satisfies the necessary condition for optimality, i.e. $J^G(v^*_{opt})\,[w-v^*_{opt}] \geq 0$, $\forall\, w \in U_{ad}$.

PROOF

In the proof of this theorem we use the following two lemmas; the proofs of these lemmas can be found in CUVELIER [2].

≠

LEMMA 1.3.

The functional $v \to J(v)$ is two times Gateaux-differentiable and for all $v \in U_{ad}$ we have

$$J^{GG}(v)\,[w_1;w_2] \leq M||w_1||_{L_2(\Sigma_1)}||w_2||_{L_2(\Sigma_1)}, \text{ for all } w_1,w_2 \in L_2(\Sigma_1)$$

where $J^{GG}(v)\,[w_1;w_2]$ denotes the second derivative of J at the point v in the directions w_1 and w_2.

≠

LEMMA 1.4.

If $\{v_n\}$ is a sequence in $L_2(\Sigma_1)$ that converges weakly in $L_2(\Sigma_1)$ to an element $v^* \in L_2(\Sigma_1)$, then $\gamma_0\vartheta(v_n)$ converges strongly to $\gamma_0\vartheta(v^*)$ in $L_2(\Sigma_1)$.

$\neq$

PROOF OF THEOREM 1.5.

We give the Taylor series of $J(v_{opt}^{(m+1)})$ about the element $v_{opt}^{(m)}$ up to and including the third order term

$$(14) \quad J(v_{opt}^{(m+1)}) = J(v_{opt}^{(m)}) + J^G(v_{opt}^{(m)})[v_{opt}^{(m+1)} - v_{opt}^{(m)}] + \\ + \tfrac{1}{2} J^{GG}(v_\mu^{(m+1)})[v_{opt}^{(m+1)} - v_{opt}^{(m)};\ v_{opt}^{(m+1)} - v_{opt}^{(m)}]$$

where $v_\mu^{(m+1)} = \mu v_{opt}^{(m)} + (1-\mu)v_{opt}^{(m+1)}$ for some $\mu \in [0,1]$.

The second term of the right hand side of (14) can be estimated by:

$$J^G(v_{opt}^{(m)})\ [v_{opt}^{(m+1)} - v_{opt}^{(m)}] \leqq -\frac{1}{\rho}||v_{opt}^{(m+1)} - v_{opt}^{(m)}||^2_{L_2(\Sigma_1)}\ .$$

This is a simple consequence of (13). For the third term we use lemma 1.3. Therefore we obtain

$$J(v_{opt}^{(m+1)}) - J(v_{opt}^{(m)}) \leqq (\frac{M}{2} - \frac{1}{\rho})\,||v_{opt}^{(m+1)} - v_{opt}^{(m)}||^2_{L_2(\Sigma_1)}\ .$$

If ρ satisfies $0 < \rho < \frac{2}{M}$, the sequence $\{J(v_{opt}^{(m)})\}$ is convergent and consequently the sequence $\{v_{opt}^{(m)}\} \subset U_{ad}$ is bounded in $L_2(\Sigma_1)$. Thus there exist $v^*_{opt} \in L_2(\Sigma_1)$ and a subsequence $\{v_{opt}^{(m')}\}$ of $\{v_{opt}^{(m)}\}$ with

$$v_{opt}^{(m')} \to v^*_{opt} \quad \text{weakly in } L_2(\Sigma_1).$$

Because U_{ad} is convex and closed, we also have $v^*_{opt} \in U_{ad}$. The remaining thing to check is that v^*_{opt} satisfies the necessary condition for optimality. We remark that for all $w \in U_{ad}$ the following inequality holds.

$$(15) \quad J^G(v_{opt}^{(m)})\ [w - v_{opt}^{(m)}] \geqq J^G(v_{opt}^{(m)})\ [v_{opt}^{(m+1)} - v_{opt}^{(m)}] + \\ + \frac{1}{\rho}\int_0^T (v_{opt}^{(m)} - v_{opt}^{(m+1)},\ w - v_{opt}^{(m+1)})_{L_2(\Gamma_1)}dt$$

We denote the right hand side of (15) by $R^{(m)}$. Using remark 1.1. we transform (15) into

$$\alpha\int_0^T (v_{opt}^{(m)}, w)_{L_2(\Gamma_1)}dt + \varkappa\int_0^T (\gamma_0\vartheta(v_{opt}^{(m)}),\ w - v_{opt}^{(m)})_{L_2(\Gamma_1)}dt \geqq \\ \geqq \alpha\int_0^T (v_{opt}^{(m)},\ v_{opt}^{(m)})_{L_2(\Gamma_1)}dt + R^{(m)}\ .$$

We pass to the limit ($m=m' \to \infty$) and due to lemma 1.4. and the fact that $R^{(m)} \to 0$ as $m \to \infty$, we obtain

$$\alpha \int_0^T (v^*_{opt}, w)_{L_2(\Gamma_1)} dt + \varkappa \int_0^T (\gamma_o \vartheta(v^*_{opt}), w - v^*_{opt})_{L_2(\Gamma_1)} dt \geqq$$

$$\geqq \alpha \liminf_{m' \to \infty} ||v^{(m')}_{opt}||^2_{L_2(\Sigma_1)} \geqq \alpha ||v^*_{opt}||^2_{L_2(\Sigma_1)}, \text{for all } w \in U_{ad},$$

which is equivalent to

$$J^G(v^*_{opt}) [w - v^*_{opt}] \geqq 0, \quad \text{for all } w \in U_{ad}.$$

This completes the proof of theorem 1.5.

#

2. OPTIMAL CONTROL OF THE PERTURBED SYSTEM OF EQUATIONS.

In this chapter we perturb the original system of equations in such a way that we obtain a system which is of Cauchy-Kowaleska type. This perturbation facilitates the numerical treatment of the problem, because the condition of incompressibility of the fluid (i.e. equation (3)) seriously complicates the numerical calculations. The method of perturbation (often called the method of artificial compressibility) was first introduced by TEMAM [1], [2] and LIONS [2].

#

2.1. FORMULATION OF THE PERTURBED SYSTEM. NOTATION $P_\varepsilon(Te_\varepsilon, u_\varepsilon, p_\varepsilon; v)$.

Apart from the data of chapter 1.1. we take $p_o \in L_2$ arbitrarily. We state the perturbed problem as follows:

Find two scalar functions $Te_\varepsilon, p_\varepsilon$ and a vector function u_ε such that

$$Te_\varepsilon \in L_2(o,T;H^1), \quad Te'_\varepsilon \in L_1(o,T;(H^1)'), \quad Te_\varepsilon(o) = Te_o,$$
$$u_\varepsilon \in L_2(o,T;(H^1_o)^2), \quad u'_\varepsilon \in L_1(o,T;((H^1_o)^2)'), \quad u_\varepsilon(o) = u_o,$$
$$p_\varepsilon \in L_2(o,T;L_2), \quad p'_\varepsilon \in L_2(o,T;(H^1_o)'), \quad p_\varepsilon(o) = p_o,$$

and

$$\begin{aligned}
(16) \quad & Te'_\varepsilon + \varkappa C_1[Te_\varepsilon; v] + A[u_\varepsilon, Te_\varepsilon] = g && \text{in } \mathcal{D}'(]o,T[\;;(H^1)'), \\
& u'_\varepsilon + \nu C_2[u_\varepsilon] + B[u_\varepsilon, u_\varepsilon] - \sigma Te_\varepsilon - D[p_\varepsilon] = f && \text{in } \mathcal{D}'(]o,T[;((H^1_o)^2)'), \\
& \varepsilon p'_\varepsilon + D^*[u_\varepsilon] = 0 && \text{in } \mathcal{D}'(]o,T[;L_2),
\end{aligned}$$

where D^* is defined by:

$$D^*[u_\varepsilon] : \varphi(\in L_2) \to \int_\Omega \varphi \, \text{div}\, u_\varepsilon dx.$$

We refer to this problem as to $P_\varepsilon(Te_\varepsilon, u_\varepsilon, p_\varepsilon; v)$

#

THEOREM 2.1.

For fixed $v \in L_2(\Sigma_1)$, problem $P_\varepsilon(Te_\varepsilon, u_\varepsilon, p_\varepsilon; v)$ has a unique solution $\{Te_\varepsilon, u_\varepsilon, p_\varepsilon\}$ which is continuous from $[o,T] \to L_2 \times (L_2)^2 \times L_2$.

PROOF

See CUVELIER [2] which uses standard techniques.

#

PROPOSITION 2.1.

Let $\{Te_\varepsilon(v), u_\varepsilon(v), p_\varepsilon(v)\}$ be the solution of $P_\varepsilon(Te_\varepsilon, u_\varepsilon, p_\varepsilon; v)$. Then:

(i) $$||Te_\varepsilon(v)||^2_{L_\infty(o,T;L_2)} + \varkappa||Te_\varepsilon(v)||^2_{L_2(o,T;H^1)} + ||u_\varepsilon(v)||^2_{L_\infty(o,T;(L_2)^2)} + \nu||u_\varepsilon(v)||^2_{L_2(o,T;(H^1_o)^2)} + \varepsilon||p_\varepsilon(v)||^2_{L_\infty(o,T;L_2)} \leq c_4(1+||v||^2_{L_2(\Sigma_1)}).$$

(ii) $$\int_{-\infty}^{\infty} |\tau|^{2\gamma}\{||\hat{T}e_\varepsilon(\tau;v)||^2_{L_2} + ||\hat{u}_\varepsilon(\tau;v)||^2_{(L_2)^2} + \varepsilon||\hat{p}_\varepsilon(\tau;v)||^2_{L_2}\}\, d\tau \leq c_5(1+||v||^2_{L_2(\Sigma_1)}) \qquad 0 < \gamma < \tfrac{1}{4}$$

Let $\{Te_\varepsilon(v^i), u_\varepsilon(v^i), p_\varepsilon(v^i)\}$, $i=1,2$, be the solution of $P_\varepsilon(Te_\varepsilon, u_\varepsilon, p_\varepsilon; v^i)$, then the following estimate holds :

$$||Te_\varepsilon(v^1)-Te_\varepsilon(v^2)||^2_{L_\infty(o,T;L_2)} + \varkappa||Te_\varepsilon(v^1)-Te_\varepsilon(v^2)||^2_{L_2(o,T;H^1)} + ||u_\varepsilon(v^1)-u_\varepsilon(v^2)||^2_{L_\infty(o,T;(L_2)^2)} + \nu||u_\varepsilon(v^1)-u_\varepsilon(v^2)||^2_{L_2(o,T;(H^1_o)^2)} + \varepsilon||p_\varepsilon(v^1)-p_\varepsilon(v^2)||^2_{L_\infty(o,T;L_2)} \leq c_6||v^1-v^2||^2_{L_2(\Sigma_1)}$$

PROOF

see LIONS [2], TEMAM [2], CUVELIER [2].

2.2. THE OPTIMAL CONTROL OF $P_\varepsilon(Te_\varepsilon, u_\varepsilon, p_\varepsilon; v)$.

The function v is the control and we impose the constraint that it belongs to the closed convex set U_{ad} of $L_2(\Sigma_1)$. The observation is given by $v \to Te_\varepsilon(T;v)$ and we define the cost-function $v \to J_\varepsilon(v)$ by

$$J_\varepsilon(v) = \tfrac{1}{2}\{\int_\Omega |\, Te_\varepsilon(x,T;v) - z_d(x)|^2 dx + \alpha \int_0^T \int_{\Gamma_1} |v(x,t)|^2 d\Gamma dt\}.$$

The control problem can be stated as: Find $\inf_{v \in U_{ad}} J_\varepsilon(v)$.

THEOREM 2.2.

There exists at least one element $v_{\varepsilon opt} \in U_{ad}$ such that $J_\varepsilon(v_{\varepsilon opt}) \leq J_\varepsilon(v)$ for all $v \in U_{ad}$. $v_{\varepsilon opt}$ is called an optimal control.

A necessary condition for an element $v_{\varepsilon opt} \in U_{ad}$ to be an optimal control is that it satisfies the inequality

$$(Te_\varepsilon(T;v_{\varepsilon opt})-z_d,\ (Te^G_\varepsilon(v_{\varepsilon opt})[w(t)-v_{\varepsilon opt}(t)])_{t=T})_{L_2} + \alpha\int_0^T (v_{\varepsilon opt}(t),\ w(t)-v_{\varepsilon opt}(t))_{L_2(\Gamma_1)} dt \geq 0, \text{ for all } w \in U_{ad}.$$

PROOF

See the proofs of theorem 1.2. and lemma 1.1.

#

For the same reason as we did in chapter 1.2. we introduce an adjoint system of equations with respect to $P_\varepsilon(Te_\varepsilon, u_\varepsilon, p_\varepsilon; v)$.

#

2.3. FORMULATION OF AN ADJOINT SYSTEM. NOTATION $P^*_\varepsilon(\vartheta_\varepsilon, q_\varepsilon, \pi_\varepsilon; v, Te_\varepsilon, u_\varepsilon, p_\varepsilon)$.

Let $\{Te_\varepsilon, u_\varepsilon, p_\varepsilon\}$ be the solution of $P_\varepsilon(Te_\varepsilon, u_\varepsilon, p_\varepsilon; v)$. We define the adjoint problem as follows:

Find two scalar functions $\vartheta_\varepsilon, \pi_\varepsilon$ and a vector function q_ε such that

$$\begin{array}{lll} \vartheta_\varepsilon \in L_2(o,T;H^1), & \vartheta'_\varepsilon \in L_1(o,T;(H^1)'), & \vartheta_\varepsilon(T) = Te_\varepsilon(T) - z_d, \\ q_\varepsilon \in L_2(o,T;(H^1_o)^2), & q'_\varepsilon \in L_1(o,T;((H^1_o)^2)'), & q_\varepsilon(T) = 0, \\ \pi_\varepsilon \in L_2(o,T;L_2), & \pi'_\varepsilon \in L_2(o,T;(H^1_o)'), & \pi_\varepsilon(T) = 0, \end{array}$$

and

$$-\vartheta'_\varepsilon + \varkappa\, C_1[\vartheta_\varepsilon; o] - A[u_\varepsilon, \vartheta_\varepsilon] - \sigma q = 0 \qquad \text{in } \mathcal{D}'(]o,T[;(H^1)'),$$

$$-q'_\varepsilon + \nu\, C_2[q_\varepsilon] - B[u_\varepsilon, q_\varepsilon] - B^*[u_\varepsilon, q_\varepsilon] - A^*(Te_\varepsilon, \vartheta_\varepsilon] + D[\pi_\varepsilon] = 0$$
$$\text{in } \mathcal{D}'(]o,T[;((H^1_o)^2)'),$$

$$-\varepsilon\pi'_\varepsilon - D^*[q_\varepsilon] = 0 \qquad \text{in } \mathcal{D}'(]o,T[;L_2).$$

This problem we call $P^*_\varepsilon(\vartheta_\varepsilon, q_\varepsilon, \pi_\varepsilon; v, Te_\varepsilon, u_\varepsilon, p_\varepsilon)$.

#

THEOREM 2.3.

Let $\{Te_\varepsilon, u_\varepsilon, p_\varepsilon\}$ be the solution of $P_\varepsilon(Te_\varepsilon, u_\varepsilon, p_\varepsilon; v)$. Problem $P^*_\varepsilon(\vartheta_\varepsilon, q_\varepsilon, \pi_\varepsilon; v, Te_\varepsilon, u_\varepsilon, p_\varepsilon)$ has a unique solution $\{\vartheta_\varepsilon, q_\varepsilon, \pi_\varepsilon\}$.

PROOF

The proof is based on standard techniques; see, for example, CUVELIER [2].

#

In the following theorem we give a characterization of an optimal control (cf. theorem 1.4.).

THEOREM 2.4.

A necessary condition for an element $v_{\varepsilon opt} \in U_{ad}$ to be an optimal control is that $\{Te_\varepsilon, u_\varepsilon, p_\varepsilon, \vartheta_\varepsilon, q_\varepsilon, \pi_\varepsilon, v_{\varepsilon opt}\}$ satisfies:

(i) $\{Te_\varepsilon, u_\varepsilon, p_\varepsilon\}$ is the solution of $P_\varepsilon(Te_\varepsilon, u_\varepsilon, p_\varepsilon; v_{\varepsilon opt})$

(ii) $\{\vartheta_\varepsilon, q_\varepsilon, \pi_\varepsilon\}$ is the solution of $P^*_\varepsilon(\vartheta_\varepsilon, q_\varepsilon, \pi_\varepsilon; v_{\varepsilon opt}, Te_\varepsilon, u_\varepsilon, p_\varepsilon)$,

(iii) $\displaystyle\int_0^T (\varkappa\gamma_o\vartheta_\varepsilon(t) + \alpha v_{\varepsilon opt}(t), w(t) - v_{\varepsilon opt}(t))_{L_2(\Gamma_1)}\,dt \geqq 0$, for all $w \in U_{ad}$.

#

2.4. ITERATIVE ALGORITHM.

To construct an optimal control, we use the same iterative method as in chapter 1.4. Thus we choose $v^{(o)}_{\varepsilon opt} \in U_{ad}$ arbitrarily. Let M_ε be a constant (cf. lemma 1.3.) which is defined by

$$J^{GG}_\varepsilon(v)[w_1; w_2] \leqq M_\varepsilon ||w_1||_{L_2(\Sigma_1)} ||w_2||_{L_1(\Sigma_1)}, \quad \forall\, v, w_1, w_2 \in L_2(\Sigma_1).$$

We have the following theorem

THEOREM 2.5.

If ρ satisfies $0 < \rho < \frac{2}{M_\varepsilon}$, then the sequence $\{v^{(m)}_{\varepsilon opt}\}$, defined by

$$v^{(m+1)}_{\varepsilon opt} = P_{U_{ad}} [v^{(m)}_{\varepsilon opt} - \rho\, J^G_\varepsilon(v^{(m)}_{\varepsilon opt})]$$

$$(= P_{U_{ad}} [v^{(m)}_{\varepsilon opt} - \rho\,(\alpha\, v^{(m)}_{\varepsilon opt} + \varkappa \vartheta_\varepsilon(v^{(m)}_{\varepsilon opt})|_{\Sigma_1})])$$

contains a subsequence $\{v^{(m')}_{\varepsilon opt}\}$ converging weakly to an element $v^*_{\varepsilon opt} \in U_{ad}$ in $L_2(\Sigma_1)$. Moreover $v^*_{\varepsilon opt}$ satisfies the necessary condition for optimality: $J^G_\varepsilon(v^*_{\varepsilon opt})\,[w - v^*_{\varepsilon opt}] \geqq 0$ for all $w \in U_{ad}$.

PROOF

The proof is exactly the same as that of theorem 1.5.

#

3. PASSAGE TO THE LIMIT. $P_\varepsilon(Te_\varepsilon, u_\varepsilon, p_\varepsilon; v_{\varepsilon opt}) \to P(Te,u,p;v_{opt})$ AS $\varepsilon \to 0$.

In this chapter we prove that the solution $\{Te_\varepsilon(v_{\varepsilon opt}), u_\varepsilon(v_{\varepsilon opt}), p_\varepsilon(v_{\varepsilon opt})\}$ of the perturbed optimal control problem converges, in some topological sense, to the solution $\{Te(v_{opt}),\ u(v_{opt}),\ p(v_{opt})\}$ of the unperturbed optimal control problem.

#

3.1. PRELIMINARY RESULTS.

We state some preliminary results for a fixed control. The proof of these results can be found in CUVELIER [1], [2] .

THEOREM 3.1.

Let $\{Te_\varepsilon, u_\varepsilon, p_\varepsilon\}$ be the solution of $P_\varepsilon(Te_\varepsilon, u_\varepsilon, p_\varepsilon; v)$. This solution converges to the solution $\{Te,u,p\}$ of $P(Te,u,p;v)$ in the following topological sense:

(17)
$$\begin{array}{lll}
Te_\varepsilon(T;v) \to Te(T;v) & \text{strongly in } L_2, \\
Te_\varepsilon(v) \to Te(v) & \text{weakly-* in } L_\infty(o,T;L_2). \\
Te_\varepsilon(v) \to Te(v) & \text{strongly in } L_2(o,T;H^1), \\
u_\varepsilon(T;v) \to u(T;v) & \text{strongly in } (L_2)^2, \\
u_\varepsilon(v) \to u(v) & \text{weakly-* in } L_\infty(o,T;(L_2)^2), \\
u_\varepsilon(v) \to u(v) & \text{strongly in } L_2(o,T;(H^1_o)^2).
\end{array}$$

#

3.2. $v_{\varepsilon opt}$ RANGES IN A BOUNDED SET OF $L_2(\Sigma_1)$

LEMMA 3.1.

Any optimal control $v_{\varepsilon opt}$ belongs to a bounded set of $L_2(\Sigma_1)$, independent of ε.

COROLLARY

There exist an element $w \in U_{ad}$ and a subsequence $\{v_{\varepsilon' opt}\}$of $\{v_{\varepsilon opt}\}$ such that $v_{\varepsilon' opt} \to w$ weakly in $L_2(\Sigma_1)$ as $\varepsilon \to 0$.

PROOF OF LEMMA 3.1.

Let $j = J(v_{opt}) = \inf_{v \in U_{ad}} J(v)$ and $j_\varepsilon = J_\varepsilon(v_{\varepsilon opt}) = \inf_{v \in U_{ad}} J_\varepsilon(v)$.

Because we have (17), the equation

$$(18) \qquad \lim_{\varepsilon \to 0} J_\varepsilon(v) = J(v)$$

holds for all $v \in L_2(\Sigma_1)$. Hence $j_\varepsilon = \inf_{v \in U_{ad}} J_\varepsilon(v) \leqq J_\varepsilon(v_{opt}) \to J(v_{opt}) = j$

as $\varepsilon \to 0$, so that

$$(19) \qquad \limsup_{\varepsilon \to 0} j_\varepsilon \leqq j$$

On the other hand we have $j_\varepsilon = J_\varepsilon(v_{\varepsilon opt}) \geqq \frac{1}{2}\alpha \, ||v_{\varepsilon opt}||^2_{L_2(\Sigma_1)}$, which combined with (19) shows that $\frac{1}{2}\alpha ||v_{\varepsilon opt}||^2_{L_2(\Sigma_1)} \leqq j$.

#

3.3. $P_\varepsilon(Te_\varepsilon, u_\varepsilon, p_\varepsilon; v_{\varepsilon opt}) \to P(Te,u,p;w)$ as $\varepsilon \to 0$.

THEOREM 3.2.

Let $\{Te_\varepsilon, u_\varepsilon, p_\varepsilon\}$ be the solution of $P_\varepsilon(Te_\varepsilon, u_\varepsilon, p_\varepsilon; v_{\varepsilon opt})$. There exists a triplet of functions $\{Te, u, z\}$ such that the subsequence $\{Te_{\varepsilon'}, u_{\varepsilon'}, p_{\varepsilon'}\}$ of $\{Te_\varepsilon, u_\varepsilon, p_\varepsilon\}$, where the sequence $\{\varepsilon'\}$ is taken from the corollary of lemma 3.1., satisfies:

$Te_{\varepsilon'} \to Te$	weakly-* in $L_\infty(o,T;L_2)$, weakly in $L_2(o,T;H^1)$ and strongly in $L_2(o,T;L_2)$,
$u_{\varepsilon'} \to u$	weakly-* in $L_\infty(o,T;(L_2)^2)$, weakly in $L_2(o,T;(H^1_o)^2)$ and strongly in $L_2(o,T;(L_2)^2)$,
$\sqrt{\varepsilon'}\, p_{\varepsilon'} \to z$	weakly-* in $L_\infty(o,T;L_2)$.

Where, $\{Te,u\}$ is the solution of $P(Te,u,p;w)$ with p determined by theorem 1.1. and w by the corollary of lemma 3.1.

PROOF

The proof makes use of proposition 2.1. and is based on compactness arguments.

#

3.4. w IS AN OPTIMAL CONTROL OF THE PROBLEM $P(Te,u,p;.)$

LEMMA 3.2.

Let $\{Te_\varepsilon, u_\varepsilon, p_\varepsilon\}$ be the solution of $P_\varepsilon(Te_\varepsilon, u_\varepsilon, p_\varepsilon; v_{\varepsilon opt})$ and $\{Te,u,p\}$ the solution of $P(Te,u,p;w)$.

The subsequence $\{Te_{\varepsilon'}(T;v_{\varepsilon' opt})\}$ of $\{Te_\varepsilon(T;v_{\varepsilon opt})\}$, where $\{\varepsilon'\}$ is taken from the corollary of lemma 3.1., satisfies:

$$(20) \qquad Te_{\varepsilon'}(T;v_{\varepsilon' opt}) \to Te(T;w) \text{ weakly in } L_2$$

PROOF

We take the scalar product of (16), where we replace v by $v_{\varepsilon opt}$, with $\chi \in D(\bar{\Omega})$ and we integrate with respect to time from 0 to T. We pass to the limit $(\varepsilon' \to 0)$, which is permitted by theorem 3.2.

On the other hand we may multiply (9), where v is replaced by w, with $\chi \in D(\bar{\Omega})$ and, again, integrate with respect to time from 0 to T.

Comparing the two results completes the proof of this lemma.

#

THEOREM 3.3.

The element $w \in U_{ad}$, determined by the corollary of lemma 3.1. is an optimal control of the problem P(Te,u,p;.).

PROOF

It follows from lemma 3.2. that $\liminf_{\varepsilon' \to 0} J_{\varepsilon'}(v_{\varepsilon' opt}) \geq J(w)$ which is equivalent to

(21) $$\liminf_{\varepsilon' \to 0} j_{\varepsilon'} \geq J(w) \geq j.$$

Combining (19) and (21) we find that

(22) $$\lim_{\varepsilon \to 0} j_{\varepsilon} = j.$$

So it is legitimate to write: $w = v_{opt}$.

#

3.5. TWO SUPPLEMENTARY RESULTS

THEOREM 3.4.

The convergence of $v_{\varepsilon' opt}$, as stated in the corollary of lemma 3.1, takes place in a finer topology, viz. $v_{\varepsilon' opt} \to v_{opt}$ strongly in $L_2(\Sigma_1)$.

PROOF

In order to proof this, we introduce the following norm on $L_2(\Sigma_1)$

$$|||v|||_{\varepsilon}^2 = \tfrac{1}{2}||Te_{\varepsilon}(T;v) - Te_{\varepsilon}(T;o)||_{L_2}^2 + \tfrac{1}{2}\alpha||v||_{L_2(\Sigma_1)}^2$$

(when $\varepsilon = 0$ we set $Te_{\varepsilon} = Te$).

Using propositions 1.1. and 2.1. we can prove that this norm is equivalent to the standard norm of $L_2(\Sigma_1)$ for $\varepsilon \geq 0$ (cf. CUVELIER [2]).

We consider the following equality:

$$|||v_{\varepsilon opt}|||_{\varepsilon}^2 - |||v_{opt}|||_{o}^2 = J_{\varepsilon}(v_{\varepsilon opt}) - J(v_{opt}) + J_{\varepsilon}(o) - J(o) +$$
$$+ (Te_{\varepsilon}(T;v_{\varepsilon opt}) - Te(T;v_{opt}), z_d)_{L_2} + (Te_{\varepsilon}(T;o) - Te(T;o), z_d)_{L_2} +$$
$$- (Te_{\varepsilon}(T;v_{\varepsilon opt}), Te_{\varepsilon}(T;o))_{L_2} + (Te(T;v_{opt}), Te(T;o))_{L_2}.$$

We pass to the limit $(\varepsilon \to 0)$ via the subsequence $\{\varepsilon'\}$ of $\{\varepsilon\}$ and due to (17),(18),(20),(22) we obtain $|||v_{\varepsilon' opt}|||_{\varepsilon'} \to |||v_{opt}|||_{o}$.

Combining this result with the weak convergence of $v_{\varepsilon' opt}$ to v_{opt} concludes the proof of the theorem. #

THEOREM 3.5.

Some of the weak convergence results of theorem 3.2. are in fact strong convergence results:

$$Te_{\varepsilon'}(v_{\varepsilon' opt}) \to Te(v_{opt}) \qquad \text{strongly in } L_2(o,T;H^1),$$

$$u_{\varepsilon'}(v_{\varepsilon' opt}) \to u(v_{opt}) \qquad \text{strongly in } L_2(o,T;(H_o^1)^2),$$

Moreover

$$Te_{\varepsilon'}(T;v_{\varepsilon' opt}) \to Te(T;v_{opt}) \qquad \text{strongly in } L_2,$$

$$u_{\varepsilon'}(T;v_{\varepsilon' opt}) \to u(T;v_{opt}) \qquad \text{strongly in } (L_2)^2.$$

PROOF

Consider the following expression

$$X_\varepsilon = ||Te_\varepsilon(T;v_{\varepsilon opt})-Te(T;v_{opt})||_{L_2}^2 + 2\kappa \int_0^T ||Te_\varepsilon(t;v_{\varepsilon opt})-Te(t;v_{opt})||_{H^1}^2 dt +$$

$$+||u_\varepsilon(T;v_{\varepsilon opt})-u(T;v_{opt})||_{(L_2)^2}^2 + 2\nu \int_0^T ||(u_\varepsilon(t;v_{\varepsilon opt})-u(t;v_{opt})||_{(H_o^1)^2}^2 dt +$$

$$+ \varepsilon||p_\varepsilon(T)||_{L_2}^2$$

We pass to the limit ($\varepsilon' \to 0$) and we obtain (cf. CUVELIER [2], TEMAM [1]):

$$\lim_{\varepsilon' \to 0} X_{\varepsilon'} = \lim_{\varepsilon' \to 0} \{\int_0^T (v_{\varepsilon' opt}(t),\gamma_o Te_{\varepsilon'}(t;v_{\varepsilon' opt}))_{L_2(\Gamma_1)}\, dt +$$

$$- \int_0^T (v_{opt}(t),\gamma_o Te(t;v_{opt}))_{L_2(\Gamma_1)}\, dt\}$$

Because the trace operator is a continuous map from $L_2(o,T;H^1)$ into $L_2(\Sigma_1)$, where both spaces are equipped with the weak topology, and because $v_{\varepsilon' opt} \to v_{opt}$ strongly in $L_2(\Sigma_1)$, it follows that $\lim_{\varepsilon' \to 0} X_{\varepsilon'}=0$. This proves the theorem. # #

4. A NUMERICAL EXAMPLE.

In this chapter we describe, very briefly, a numerical example to illustrate the procedure proposed in this paper. A complete treatment of the numerical considerations can be found in CUVELIER [2],[3].

We describe the physical situation as follows: An non-isothermal fluid is contained in a square closed cavity (length =d) placed in the field of gravitation (g=gravitational acceleration). The fluid is initially at rest and at constant temperature Te_o. The fluid properties are: ν=kinematic viscosity, κ = thermal diffusivity, ρ=density, c_p=specific heat.

Our objective is to obtain at time T a temperature distribution equal to Te_1 (constant).

We normalize the equations and the domain by the following constants: $t_c = d^2\nu^{-1}$ = unit of time, $x_c = d$ = unit of length, $u_c = \nu d^{-1}$ = unit of velocity, $p_c = \rho\nu^2 d^{-1}$ = unit of pressure, $\Delta Te = Te_1 - Te_0$ = unit of temperature. The situation is as follows:

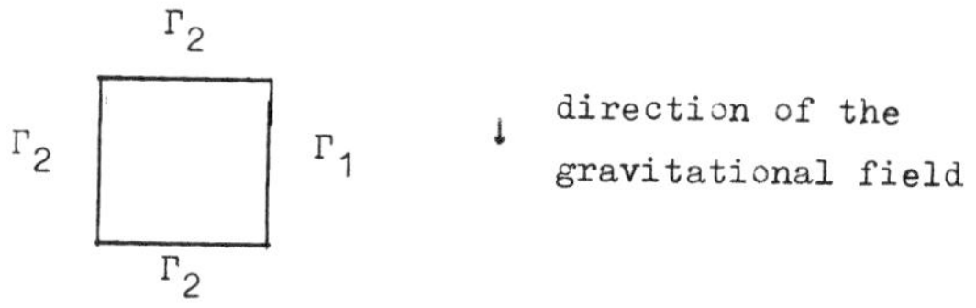

Γ_2 is that part of the boundary which is thermically insulated, while on Γ_1 heat input or output is possible without constraints (i.e. $U_{ad} = L_2(\Sigma_1)$).

The normalized problem reads as follows:

$$\Omega = \{ \{x_1,x_2\} \mid 0 < x_1 < 1,\ 0 < x_2 < 1\}, \qquad \tilde{T} = T\, t_c^{-1}$$

$$\left.\begin{aligned} &\frac{\partial Te}{\partial t} + \sum_{i=1}^{2} u_i \frac{\partial Te}{\partial x_i} = \frac{1}{Pr}\,\Delta Te \\ &\frac{\partial u}{\partial t} + \sum_{i=1}^{2} u_i \frac{\partial u}{\partial x_i} = -\operatorname{grad} p + \binom{0}{Gr} Te + \Delta u \\ &\operatorname{div}\ u = 0 \end{aligned}\right\} \quad \text{in } \Omega \times]0,\tilde{T}[$$

where $Pr = \dfrac{\rho\, c_p\, \nu}{\varkappa}$ = Prandtl number, $Gr = \dfrac{g\,\beta\,\Delta Te\, d^3}{\nu^2}$ = Grashof number

(β = coefficient of thermal expansion)

The initial conditions are $Te(x,o) = 0$, $u(x,o) = 0$, $x \in \bar{\Omega}$,

and the boundary conditions are

$$\frac{\partial Te}{\partial n} = 0 \qquad \text{on } \Gamma_2 \times]o,\tilde{T}[,$$

$$\frac{\partial Te}{\partial n} = v\,\Delta Te \qquad \text{on } \Gamma_1 \times]o,\tilde{T}[,$$

$$u = 0 \qquad \text{on } \Gamma \times]o,\tilde{T}[.$$

We fix the instant $\tilde{T}$: $\tilde{T} = 0.075$ and the desired state $Te_1 = 1$ $(= z_d)$.

The normalized cost-function becomes:

$$J(v) = \tfrac{1}{2}\{ \int_\Omega |Te(x,T;v)-1|^2 dx + \bar{\alpha} \int_0^{\tilde{T}} \int_{\Gamma_1} |v(x,t)|^2 d\Gamma\, dt\}$$

with $\bar{\alpha} = \alpha\nu^{-1} d$

For the numerical treatment we discretize, by means of a finite difference

method, the perturbed problem $P_\varepsilon(Te_\varepsilon, u_\varepsilon, p_\varepsilon; v)$ corresponding to the problem just described; furthermore we discretize the adjoint problem $P^*_\varepsilon(\vartheta_\varepsilon, q_\varepsilon, \pi_\varepsilon; v, Te_\varepsilon, u_\varepsilon, p_\varepsilon)$, the iterative algortihm and the cost-function. Considerations about the discretized problems, such as stability and convergence are treated in CUVELIER [2].
For some numerical results we choose $Pr = 0.733$ (for air), $Gr = 6850$, $\tilde{T} = 0.075$, $\varepsilon = 25 * 10^{-6}$, Δt = time-step in finite difference scheme = 0.0005, Δx = space-step = 0.1. In the iterative algortihm we fixed $\rho = 7.0$. The values of the discretized cost-function are listed below for two values of $\bar{\alpha}$.

number of iterations	$\bar{\alpha} = 10^{-4}$	$\bar{\alpha} = 10^{-5}$
1	0.50000	0.50000
2	.22660	.22629
3	.17274	.17242
4	.14916	.14872
5	.12339	.12277
6	.11562	.11491
7	.10891	.10886
8	.10487	.10402
9	.10096	.10006
10	.09770	.09673
15	.08702	.08580
20	.08105	.07971

For a discussion of these results and others see CUVELIER [2], [3].

# #

REFERENCES

C. CUVELIER [1] : Approximation numérique de la solution des équations de Navier-Stokes coupléesà celle de la chaleur (Application au problème de convection libre).Publication at the University of Delft (1973).
[2] : Doctoral thesis, Delft (to appear)
[3] : to appear.

L. LANDAU - E. LIFCHITZ [1] : Mécanique des fluides (1971)

J.L. LIONS [1] : Contrôle optimal de systèmes gouvernés par des équations aux dérivées partielles (1968).
[2] : Quelques méthodes de résolution des problèmes aux limites non linéaires (1969).

R. TEMAM [1] : Sur l'approximation de la solution des équations de Navier-Stokes par la méthode des pas fractionnaires I.

Arch.Rat.Mech.Anal. 32 (1969) p. 135-153.

R. TEMAM [2] : Cours Navier-Stokes. Cours à l'université de Maryland (1973).

# #

W. Eckhaus (ed.), New Developments in Differential Equations

SECONDARY OR DIRECT BIFURCATION OF A STEADY SOLUTION OF THE NAVIER-STOKES EQUATIONS INTO AN INVARIANT TORUS.

Gérard IOOSS
Institut de Mathématiques et Sciences Physiques
Parc Valrose 06034 - NICE (FRANCE)

I - Statement of the problem.

1) A classical experimental example.

Let us take as an illustration of our abstract results, the classical experiments on the Taylor problem. Between two concentric cylinders, rotating around their axis, there is an incompressible viscous fluid satisfying the Navier-Stokes equations. It is known that whatever are the rotation rates ω_1 and ω_2 of the inner and outer cylinder, we have a steady solution called the Couette flow, whose streamlines are concentric circles.

In fact, this solution is only observed if $|\omega_1|$ is small enough. Now, let us take $\omega_2 < 0$ and let us increase $\omega_1 > 0$. When it crosses a first critical value, in suitable conditions, it can be observed an other flow which is of cellular type and periodic in time.

If we increase again ω_1, then after the occurence of some more and more complicated flows, a true turbulent flow occurs.

This is just an example to justify the formal theories developped by E. HOPF [2] in 1942 and L. LANDAU [12] in 1944. To explain turbulence in certain situations, they had formulated an explanation based on successive bifurcations of solutions, of the Navier-Stokes equations, becoming unstable when a parameter (as ω_1) increases.

2) Navier-Stokes equations.

We have in general a system of the form

$$(1)\quad \begin{cases} \dfrac{\partial V}{\partial t} + (V.\nabla)V + \nabla p = \nu\Delta V + f , \\ \nabla.V = 0 , \end{cases} \text{ in } \Omega \qquad V\big|_{\partial\Omega} = a , \quad \text{where } \int_{\partial\Omega} a.n\, ds = 0 ,$$

where V is the velocity of the fluid at the point $(x,t)\in\Omega \times \mathbb{R}_+$, p is the presure, f is a given external force, and a is given on the boundary $\partial\Omega$ of a bounded regular domain Ω of $\mathbb{R}^2$ or $\mathbb{R}^3$.

In fact, other phenomenon such as the Bénard convection, or flows with the occurence of electromagnetic field, obeys systems of equations which have a similar structure as (1), (see [1]). In the example cited above, we have to take an Ω which is a bounded domain of periodicity of the flow in z (period ; $2\pi/\alpha$), as the experiments suggest us. The theory, that we shall developp, runs well with this sort of domain slso. Moreover, other boundary conditions are possible [3] .

3) Basic flow, perturbed equations.

Let us assume that we know a steady solution (V_0, p_0) of (1). We shall call this solution "the basic flow". In fact, following the problem, we have a characteristic parameter such as ν^{-1} or ω_1 in the example of 1). Let us denote it by λ and assume that V_0 is analytic in λ. Now we pose

$$V = V_0(\lambda) + u.$$

Hence, the perturbation u satisfies a system of the form

$$(2)\qquad \frac{du}{dt} + \mathbf{L}_\lambda u - \mathbf{M}(u) = 0,$$

where we look for $t \longmapsto u(t)$ as a continuous function taking values in the domain $\mathcal{D}$ of the linear operator $\mathbf{L}_\lambda$, with a continuous derivative in an Hilbert space H.

In the case of the system (1), we take

$$H = \{u \in [L^2(\Omega)]^3 \ ; \ \nabla.u = 0, \ u.n|_{\partial\Omega} = 0\},$$

$$\mathcal{D} = \{u \in [H^2(\Omega)]^3 \ ; \ \nabla.u = 0, \ u|_{\partial\Omega} = 0\},$$

and if we denote by Π the orthogonal projection* in $[L^2(\Omega)]^3$ onto H, we have $\forall\ u \in \mathcal{D}$

$$\mathbf{L}_\lambda u = \mathbf{\Pi}[-\nu\Delta u + (u.\nabla)V_0(\lambda) + (V_0(\lambda).\nabla)u] ,$$

$$\mathbf{M}(u) = -\mathbf{\Pi}[(u.\nabla)u].$$

---

* It is known, following [11] , [15] , that $H^\perp = \{u = \nabla\varphi, \varphi\in H_1(\Omega)\}$.

These definitions ensure us the local existence and uniqueness of the Cauchy problem (2), with $u(0) = u_0 \in \mathscr{D}$ (see [4]).
The proof is based on the facts that

i) $\{\mathbf{L}_\lambda\}_{\lambda \in D_0}$ is an holomorphic family of closed operators in H, of domain $\mathscr{D}$, where $D_0 \subset \mathbb{C}$.

ii) $\forall \lambda \in D_0$, it can be defined an holomorphic semi-group of operators in H : $\{e^{-\mathbf{L}_\lambda t}\}_{t \geq 0}$.

iii) There exists $C \geq 0$ such that $\forall \lambda \in D_0$ and $\forall\, u \in \mathscr{D}$

$$||e^{-\mathbf{L}_\lambda t}\, \mathbf{M}(u)||_{\mathscr{D}} \leq C\, t^{-\alpha} ||u||^2_{\mathscr{D}}, \text{ with } \alpha < 1,\ t \in]0,T],\ T < \infty$$

(in the case of (1) we have $\alpha = 3/4$).

4) Known results about the first bifurcation into a periodic solution.

If all points of the spectrum of $\mathbf{L}_\lambda$ have a positive real part, it is known that the basic flow V_0, i.e. the solution $u = 0$ of (2), is asymptotically stable. Let us assume that for $\lambda < \lambda_0$, the spectrum of $\mathbf{L}_\lambda$ has the preceding property, and that for $\lambda = \lambda_0$, there are two simple eigenvalues of $\mathbf{L}_{\lambda_0}$ with a null real part, and that these eigenvalues cross the imaginary axis when λ crosses λ_0. In these conditions, it is now a classical result that in general, it occurs at the neighbourhood of λ_0 (only on one side) a bifurcated periodic solution $t \mapsto u_1(\lambda,t)$ of order $|\lambda - \lambda_0|^{1/2}$, the period being analytic in λ, near λ_0.
Let us denote by $t \mapsto V_1(\lambda,t) = V_0(\lambda) + u_1(\lambda,t)$ this selfexcited periodic flow, solution of (1), and let us assume that this flow exists for $\lambda > \lambda_0$.
It is known that the basic flow V_0 losses its stability when λ crosses λ_0 and that the new periodic solution V_1 is then stable ([4], [8], [9]).

Now, our problem is

i) to look for what happens when the solution V_1 becomes unstable, λ crossing a critical value λ_1. This is the aim of § II ;

ii) to look for what happens when, instead of the preceding assumptions on the spectrum of $\mathbf{L}_{\lambda_0}$, we have four simple (conjugated) eigenvalues crossing the imaginary axis when λ crosses λ_0. This case is treated in § III.

In fact, this case is rare in nature, but it seems a good model, in the case of preceding assumptions on $\mathbf{L}_\lambda$, when very near of the imaginary axis, we have two other eigenvalues of $\mathbf{L}_{\lambda_0}$ which cross this axis for λ near λ_0.

II - Secondary bifurcation into an invariant torus.

1) The Poincaré map.

We are in the case when a self-excited periodic solution $t \mapsto V_1(\lambda,t)$ of (1) becomes unstable when λ crosses λ_1. Let us pose $V(t) = V_1(\lambda,t)+u(t)$; then the perturbation u satisfies a system of the form

$$\frac{du}{dt} = \mathscr{A}_\lambda(t)u + \mathbf{M}(u), \tag{3}$$

where $t \mapsto \mathscr{A}_\lambda(t)$ is T-periodic (we can suppose T independant of λ).

Before studying precisely the system (3), let us give one explicit example in two dimensions. Here we pose $u = (u_1,u_2)$,

$$\mathscr{A}_\lambda(t)u = ((\lambda + \cos t)u_1 - \alpha u_2,\ \alpha u_1 + (\lambda+\cos t)u_2),$$

$$\mathbf{M}(u) = (-u_1(u_1^2 + u_2^2),\ - u_2(u_1^2 + u_2^2)).$$

It can be shown on this special system, that for $\lambda \leq 0$, the 0-solution is asymptotically stable, whereas for $\lambda > 0$ the 0-solution is unstable.

Moreover, for $\lambda > 0$ we have two different situations following the fact that the coefficient α is or not a rational number. If $\alpha \in \mathbb{Q}$, then there exists a non trivial stable periodic solution (if $\alpha = P/q$, the period is $2\pi q$), whereas $\alpha \notin \mathbb{Q}$ gives a non trivial stable quasi-periodic solution ($u_1^2 + u_2^2$ is 2π-periodic, whereas the argument of (u_1,u_2) in $\mathbb{R}^2$ is $\alpha t + \theta_0$).

In the following, we study the abstract system (3) in the aim to show the existence for λ near λ_1 of an invariant two-dimensional torus in a good functional space, instead of the existence of a quasi-periodic solution which is an open problem.

For the study of the stability of the 0-solution of (3), we can consider the linearized problem :

$$\left\{\begin{aligned} &\frac{dv}{dt} = \mathscr{A}_\lambda(t)v \\ &v(\tau) = v_0 \in \mathscr{D} \end{aligned}\right. \tag{4}$$

whose solution, continuous in $\mathscr{D}$, is noted $t \mapsto v(t) = \mathbf{S}_\lambda(t,\tau)\, v_0$.

The family of operators $\{\mathbf{S}_\lambda(t,\tau)\}_{\substack{\lambda \in D_0 \\ t \geq \tau}}$ has the same regularity properties as $\{e^{-\mathbf{L}_\lambda(t-\tau)}\}_{\substack{\lambda\in D_0 \\ t\geq\tau}}$; except the semi-group property replaced by $\mathbf{S}_\lambda(t,\tau) = \mathbf{S}_\lambda(t,\eta).\ \mathbf{S}_\lambda(\eta,\tau)$ for $t \geq \eta \geq \tau$ (see [3]).

Moreover, the periodicity of $\mathcal{A}_\lambda$, gives the fundamental property

$$\text{(5)} \qquad \mathbf{S}_\lambda(t+T,0) = \mathbf{S}_\lambda(t,0).\ \mathbf{S}_\lambda(T,0).$$

Let us denote now $t \mapsto \mathcal{U}(t,\lambda,u_0)$ the solution of (3) satisfying $u(0) = u_0 \in \mathcal{D}$
The map $u_0 \mapsto \mathcal{U}(T,\lambda,u_0)$ is well defined in a neighbourhood of 0 in $\mathcal{D}$.
The derivative at the origin is the linear compact operator $\mathbf{S}_\lambda(T,0)$.
Moreover, if $||\mathcal{U}(T,\lambda,u_0)||_{\mathcal{D}}$ is small enough, we have

$$\text{(6)} \qquad \mathcal{U}\left[T,\lambda,\mathcal{U}(T,\lambda,u_0)\right] = \mathcal{U}(2T,\lambda,u_0),$$

and so on, because of the T-periodicity of the equation (3). These properties show that the knowledge of the position of the eigenvalues of $S_\lambda(T,0)$ with respect to the unit circle is essential for the study of the stability of the 0-solution of (3).
In fact we have already 1 as an eigenvalue of $\mathbf{S}_\lambda(T,0)$, an eigenvector being $\frac{\partial}{\partial t} V_1(\lambda,0)$. This is due to the fact that $\forall\ \delta \in \mathbb{R}, t \mapsto V_1(\lambda,t+\delta)$ is also solution of (1). The justification of a formal verification is given by the fact that $(\lambda,t) \mapsto \frac{\partial}{\partial t} V_1(\lambda,t)$ is shown to be analytic from $D_0 \times \mathbb{R}_+^*$ into $\mathcal{D}$([6]), hence the function $\frac{\partial}{\partial t} V_1(\lambda,.)$ can be considered as a v in (4), with $\tau = 0$.

In the aim to eliminate the eigenvalue 1, we substitute the so-called <u>Poincaré map,</u> to the previous map $u_0 \mapsto \mathcal{U}(T,\lambda,u_0)$.

<u>Assumption H.1.</u>

1 is a simple eigenvalue of $\mathbf{S}_\lambda(T,0)$ for λ near λ_1.

Hence, the projection operator $\mathbf{P}_\lambda$, which commutes with $\mathbf{S}_\lambda(T,0)$, and corresponds to the eigenvalue 1, depends analytically on λ . Let us consider $u_0 \in \mathcal{D}$ such that $\mathbf{P}_\lambda\ u_0 = 0$; if u_0 is in a good neighbourhood of 0, we can define

$$\text{(7)} \qquad u_1 = \mathcal{U}(\tau,\lambda,u_0) + V_1(\lambda,\tau) - V_1(\lambda,0),$$

where τ is near T and $\mathbf{P}_\lambda u_1 = 0$. This is our Poincaré map. The geometric meaning is indicated on the figure 1.

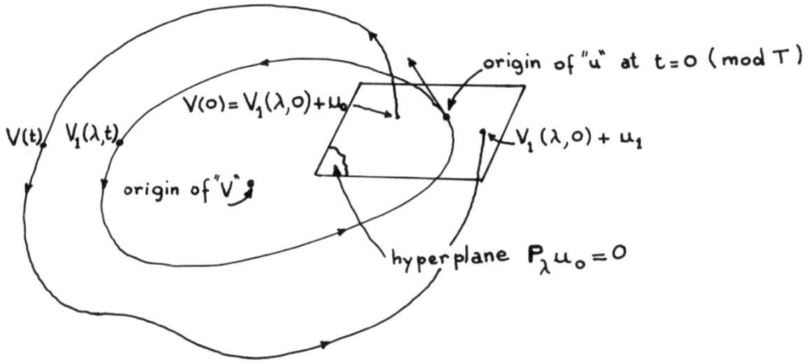

fig. 1.

2) Properties of the Poincaré map.

The determination of τ , near T, is given by the implicit function theorem applied to the equation

$$(8) \qquad \mathbf{P}_\lambda\left[\mathscr{U}(\tau,\lambda,u_0) + V_1(\lambda,\tau) - V_1(\lambda,0)\right] = 0,$$

which can be written $f(\tau,\lambda,u_0) = 0$, with $f(T,\lambda,0) = 0$,

$\frac{\partial f}{\partial \tau}(T,\lambda,0) = \frac{\partial V_1}{\partial \tau}(\lambda,0) \neq 0$. By the analyticity of f ([6]), we can find an analytic function $(\lambda,u_0) \mapsto \tau(\lambda,u_0)$ defined in $D_0 \times$ neigh. of 0 in $\mathscr{D}$, such that $\tau(\lambda,0) = T$.

The Poincaré map is then

$$(9) \qquad u_0 \longrightarrow \Phi_\lambda(u_0) = \mathscr{U}\left[\tau(\lambda,u_0),\lambda,u_0\right] + V_1\left[\lambda,\tau(\lambda,u_0)\right] - V_1(\lambda,0).$$

Lemma 1.

The map $(\lambda,u_0) \mapsto \Phi_\lambda(u_0)$ is analytic : $D_0 \times$ neigh. of 0 in $\mathscr{D} \to \mathscr{D}$ and the derivative of Φ_λ at the origin is the restriction of $\mathbf{S}_\lambda(T,0)$ in $(\mathbf{1}-\mathbf{P}_\lambda)\mathscr{D}$.

Lemma 2.

If $\Phi_\lambda^p(u_0)$ belongs to a good neighbourhood of 0 for $p \leq n$, then

$$\Phi_\lambda^p(u_0) = \mathscr{U}\left(\sum_{k=0}^{p-1} \tau_k,\lambda,u_0\right) + V_1\left(\lambda, \sum_{k=0}^{p-1} \tau_k\right) - V_1(\lambda,0),$$

where $\tau_k = \tau\left[\lambda,\Phi_\lambda^k(u_0)\right]$, $k \geq 0$.

For the proofs, see [6] . Moreover the asymptotic behaviour, of a solution $t \mapsto u(t)$ of (3), can be studied using $\Phi_\lambda^n(u_0)$, $n \to \infty$, since $\sum_{k \in \mathbb{N}} \tau_k = +\infty$.

Lemma 3.

Let the spectral radius spr $[D\, \Phi_\lambda(0)] < 1$, then the cycle V_1 is asymptotically stable.

In fact we have a more precise result in [4] :

$\exists \delta_0 > 0$, $\forall\, V(0)$ such that $||V(0) - V_1(\lambda,\alpha_0)||_{\mathscr{D}} \leq \delta_0$, then $\exists \alpha_1$ such that $||V(t) - V_1(\lambda,t+\alpha_1)||_{\mathscr{D}} \xrightarrow[t \to \infty]{} 0$ exponentially.

3) Bifurcation into a torus.

By assumption, when λ crosses λ_1, the cycle V_1 becomes unstable ; this leads to the following assumption :

H.2.

There exists two and only two conjugated simple eigenvalues of $D\, \Phi_{\lambda_1}(0)$, of moduli 1, noted ζ_0, $\bar{\zeta}_0$ such that $\zeta_0^n \neq 1$ for $n = 1,2,3,4,5$. When λ crosses λ_1, these two engenvalues cross the unit circle.

Remark.

In the case when there exists only one eigenvalue (1 or -1) on the unit circle, or two conjugated eigenvalues $(\zeta_0,\bar{\zeta}_0)$ such that $\zeta_0^n = 1$, for the operator $D\Phi_{\lambda_1}(0)$, it can then be shown that, in general (see [5]) there exists near λ_1 a new bifurcated periodic solution, of period near nT.
In this case we have to look for a non trivial fixed point of $u_0 \longmapsto \Phi_\lambda^n(u_0)$ where $\mathbf{P}_\lambda u_0 = 0$ (the period of the corresponding solution of (1) is then $\sum_{k=0}^{n-1} \tau[\lambda,\Phi_\lambda^k(u_0)] = nT + O(1)$ for λ near λ_1). This last problem is in fact put in the classical frame of stationary bifurcation problems.
Now, by the assumption H.2, and the Lemma 1 we are in the frame of a theorem of RUELLE - TAKENS [14] which says : in general there eixsts a neighbourhood of λ_1, such that, following the sign of a certain coefficient, either there exists, in a neighbourhood of 0 in $(\mathbf{1} - \mathbf{P}_\lambda)\mathscr{D}$, an invariant attracting "circle" Γ_λ for the map $\Phi_\lambda (\lambda > \lambda_1)$, or there exists an invariant repelling "circle" Γ_λ for $\Phi_\lambda(\lambda < \lambda_1)$; the "radius" of Γ_λ is of order $|\lambda - \lambda_1|^{1/2}$.

This gives us the

Theorem.

Let us assume realised the assumptions H.1 and H.2, then in general there exists a neighbourhood of λ_1 such that there is (only on one side of λ_1) a "circle" Γ_λ in $[\mathbf{1} - \mathbf{P}_\lambda]\mathscr{D}$ such that the set $\mathscr{T} = \{V(t) = V_1(\lambda,t) + u(t)$; $t \in [0, \tau(\lambda,u_0)]$, $u_0 \in \Gamma_\lambda$, $t \mapsto u(t)$ is the solution of (3) continuous in $\mathscr{D}$, with $u(0) = u_0\}$ is invariant by the dynamical system (1).
Following the sign of a certain coefficient this torus $\mathscr{T}$ occurs for $\lambda > \lambda_1$ and is attractive or it occurs for $\lambda < \lambda_1$ and is repelling.

A detailed proof can be found in [13] for the RUELLE - TAKENS theorem.
For the explicit calculation of coefficients see [6] .

III. Direct bifurcation into an invariant torus.

Let us consider the case when we have

H.1.

There are only 4 simple eigenvalues ($\pm i\omega_0$ and $\pm i\omega_1$) of $\mathbf{L}_{\lambda_0}$ on the imaginary axis, and these eigenvalues cross the imaginary axis while λ crosses λ_0.
The remaining of the spectrum stay on the right side of the complex plane for λ near λ_0.

1) Bifurcation into periodic solutions.

Let us note T the unknown period, and rescale t : $\tau = 2\pi T^{-1}t$, then $u(t) = \tilde{u}(\tau)$. We have now the system

$$
(10) \quad \begin{cases} \dfrac{d\tilde{u}}{d\tau} + \eta \mathbf{L}_\lambda \tilde{u} - \eta \mathbf{M}(\tilde{u}) = 0, \text{ with } \eta = T/2\pi, \\ \tilde{u} \in H^1(T;\mathscr{D}) \cap H^2(T;H), \end{cases}
$$

where $H^m(T,E)$ denotes the Sobolev space of nearly everywhere 2π-periodic functions such that

$$
\int_0^{2\pi} (||\tilde{u}(\tau)||^2_E + \ldots + ||\frac{d^m\tilde{u}}{d\tau^m}||^2_E)\, d\tau < +\infty.
$$

Thanks to the properties of $\mathbf{L}_\lambda$ it is shown that the linear operator $\tilde{u} \longmapsto \frac{d\tilde{u}}{d\tau} + \eta \mathbf{L}_\lambda \tilde{u}$ admits a bounded inverse in $H^1(T,H)$, if and only if ni/η is not an eigenvalue of $\mathbf{L}_\lambda$, $\forall\, n \in \mathbb{Z}$. Then for λ near λ_o, we shall obtain a bifurcation point for η near p/ω_1 or q/ω_o for a certain p or $q \in \mathbb{N}$. In the following we assume $\omega_1 > \omega_o > 0$ (no loss of generality).

If $\omega_1 \neq p\, \omega_o$, $\forall\, p \in \mathbb{N}$.

We can do exactly the same calculations as in the classical case, when there are only $\pm i\omega_o$ (or $\pm i\omega_1$) on the imaginary axis. This leads to the existence of two periodic solutions :

i) $\tilde{\mathcal{U}}_o(\tau,\varepsilon) = \sum_{n \geq 1} \varepsilon^n \mathcal{U}_o^{(n)}(\tau)$, $\mathcal{U}_o^{(1)}(\tau) = a_o\, e^{-i\tau} u^{(o)} + \bar{a}_o\, e^{i\tau} u^{(o)}$

where $\mathbf{L}_{\lambda_o} u^{(o)} = i\omega_o u^{(o)}$, $\frac{T_o}{2\pi} = \eta_o = \omega_o^{-1} + O(\lambda - \lambda_o)$, $\varepsilon = |\lambda - \lambda_o|^{1/2}$;

ii) $\tilde{\mathcal{U}}_1(\tau,\varepsilon) = \sum_{n \geq 1} \varepsilon^n \mathcal{U}_1^{(n)}(\tau)$, $\mathcal{U}_1^{(1)}(\tau) = a_1\, e^{-i\tau} u^{(1)} + \bar{a}_1\, e^{i\tau} u^{(1)}$

where $\mathbf{L}_{\lambda_o} u^{(1)} = i\omega_1 u^{(1)}$, $\frac{T_1}{2\pi} = \eta_1 = \omega_1^{-1} + O(\lambda - \lambda_o)$, $\varepsilon = |\lambda - \lambda_o|^{1/2}$;

each bifurcation is only on one side of λ_o, following the sign of a certain coefficient.

Moreover, it can be shown that these two periodic solutions* are the only ones bifurcating from λ_o (see [7]).

In the other cases we have the following results (see [7]) :

If $\omega_1 = 2\omega_o$

Either there exists only the solution $\mathcal{U}_1$ (of order $|\lambda - \lambda_o|^{1/2}$), or there exists 3 solutions : $\mathcal{U}_1$ and 2 solutions $\mathcal{U}_2$ and $\mathcal{U}_2'$ of order $(\lambda - \lambda_o)$, bifurcating from λ_o.

If $\omega_1 = 3\omega_o$

Either there exists 2 or 4 or 6 or 8 solutions of order $|\lambda - \lambda_o|^{1/2}$, bifurcating from λ_o (one of these solutions is $\mathcal{U}_1$).

If $\omega_1 = p\,\omega_o$, $p \geq 4$

There exists two solutions of order $|\lambda - \lambda_o|^{1/2}$, bifurcating from λ_o.

---

* We mean periodic solution of (2).

One of these solutions is $\mathcal{U}_1$, the other has the principal part as the previous $\mathcal{U}_0$.

2) Bifurcation into a torus.

Let us assume

H.2.

$\omega_1/\omega_0 \notin \mathbb{Q}$ or, if $\omega_1/\omega_0 \in \mathbb{Q}$ then $\frac{\omega_1}{\omega_0} = \frac{p}{q} > 1$ with $p + q \geqslant 5$.

Remark.

This contains the case when $\omega_1 = p\omega_0$, $p \geqslant 4$, and all cases when $\omega_1 \neq p\omega_0$, $\forall p \in \mathbb{N}$ (see the results of 1).

Let us now consider $u_0 \in \mathcal{D}$, in a neighbourhood of 0, as an initial condition for the system (2). We can then define the solution $t \mapsto \mathcal{U}(t,\lambda,u_0)$ which is continuous in $\mathcal{D}$ for $t \in [0,T]$, where T is chosen arbitrarily for the moment, but finite positive. The map

$$(11) \qquad u_0 \longmapsto \psi_\lambda(u_0) = \mathcal{U}(T,\lambda,u_0)$$

can then be defined in a neighbourhood of 0 in $\mathcal{D}$ and $(\lambda,u_0) \longmapsto \psi_\lambda(u_0)$ is analytic. Moreover we have the derivative at 0 :

$$D\,\psi_{\lambda_0}(0) = e^{-\mathbf{L}_{\lambda_0} T}$$

and this compact operator in $\mathcal{D}$ has 4 simple eigenvalues of moduli 1 : $e^{\pm i\omega_0 T}$, $e^{\pm i\omega_1 T}$. The other eigenvalues are of moduli less than 1.

Hence, we can use the "center-manifold theorem" (see [13]) to reduce the problem to a 4-dimensional one.

Then, in the aim to use the work of R. JOST and E. ZEHNDER [10] for the new map in a 4-dimensional space, we have to choose T such that

$$(12) \qquad \begin{cases} (S_1\omega_0 + S_2\,\omega_1)T = 2\pi\, m, \; S_i \in \mathbb{Z},\; m \in \mathbb{Z}, \\ |S_1| + |S_2| \leq 4 \end{cases}$$

leads to $S_1 = S_2 = m = 0$.

Thanks to (H.2), this is realised if we choose T such that $2\pi/T$ is not a rational combination of ω_0 and ω_1.

Now, adapting the paper [10], we obtain after a regular changing of variables, the normal form of the map (11) in $\mathbb{C}^2$: noting $z_k = r_k e^{i\varphi_k}$, $Z_k = R_k$ $^{i\psi_k}$, $k = 1,2$, the map becomes :

$$(13) \quad \begin{cases} R_k = (1 + \mu - r_k^2 + \alpha_k(\mu)\, r_{k'}^2)r_k + \Theta_4^{(k)}, & (k' \neq k) \\ \psi_k = \varphi_k + \tilde{\alpha}_k(\mu) + \sum_{j=1}^{2} d_{kj}(\mu)\, r_j^2 + \Theta_3^{(k)}, & \end{cases}$$

where $\mu = \lambda - \lambda_0$, $\alpha_k, \tilde{\alpha}_k$, d_{kj} are continuous functions in a neighbourhood of 0, and $\Theta_1^{(k)}$ are of order $(|r_1| + |r_2|)^1$, and $\tilde{\alpha}_1(0) = \omega_0 T$, $\tilde{\alpha}_2(0) = \omega_1 T$.

To write (13) we have assumed that the two periodic bifurcated solutions obtained in 1), appear for $\lambda > \lambda_0$ (this give the sign (-) in the second member in (13)). These known cycles are obtained by considering invariant "circles" of (13). Truncating (13) by suppressing the $\Theta_4^{(k)}$, we obtain the principal part of the two cycles :

$\mathcal{U}_0$: $r_1 = \mu^{1/2}$, $r_2 = 0$,

$\mathcal{U}_1$: $r_1 = 0$, $r_2 = \mu^{1/2}$.

We obtain also the principal part of an invariant torus :

i) if $\alpha_1(0)$ and $\alpha_2(0) < -1$ or if $\alpha_1(0).\ \alpha_2(0) < 1$ and $\alpha_1(0)$ and $\alpha_2(0) > -1$ we have $r_1 = \beta_1\mu^{1/2}$, $r_2 = \beta_2\ \mu^{1/2}$ with

$$\begin{cases} \beta_1^2 - \alpha_1(0)\ \beta_2^2 = 1 \\ \beta_2^2 - \alpha_2(0)\ \beta_1^2 = 1. \end{cases}$$

ii) if $\alpha_1(0).\ \alpha_2(0) > 1$, $\alpha_1(0)$ and $\alpha_2(0) > 0$ we have for $\mu < 0$ $(\lambda < \lambda_0)$

$r_1 = \beta_1(-\mu)^{1/2}$, $r_2 = \beta_2(-\mu)^{1/2}$ with

$$\begin{cases} \beta_1^2 - \alpha_1(0)\ \beta_2^2 = -1 \\ \beta_2^2 - \alpha_2(0)\ \beta_1^2 = -1. \end{cases}$$

It can be shown (using |11|) that the cycle $\mathcal{U}_0$ is stable (resp. unstable) if $\alpha_2(0) < -1$ (resp. $\alpha_2(0) > -1$), the cycle $\mathcal{U}_1$ is stable (resp. unstable) if $\alpha_1(0) < -1$ (resp. $\alpha_1(0) > -1$). The torus, when it exists, is stable (resp. unstable) if $\alpha_1(0).\ \alpha_2(0) < 1$ (resp. $\alpha_1(0).\ \alpha_2(0) > 1$).

Now, it can be shown that the invariant two-dimensional torus for the map (11) is also invariant by the dynamical system (2) (see [7]). Hence we have obtained for suitable coefficients a bifurcated invariant torus. For instance we can have the two periodic bifurcated solutions unstable, and a stable torus in $\mathcal{D}$ for $\lambda > \lambda_0$ •

BIBLIOGRAPHY

[1] G. DURAND, Thèse de 3ème cycle, Pub. Math. Orsay n°128 (1975).

[2] E. HOPF, Berichten der Math.-Phys. Kl. Sächs. Akad. Wiss. Leipzig 94, 1-22 (1942).

[3] G. IOOSS, Arch. Rat. Mech. Anal. 58-1, 35-56 (1975).

[4] G. IOOSS, Arch. Rat. Mech. Anal. 47, 301-329 (1972).

[5] G. IOOSS, Arch. Mech. Stosowanej, 26, 795-804 (1974).

[6] G. IOOSS, On the secondary bifurcation of a steady solution of systems of Navier-Stokes type (to appear).

[7] G. IOOSS, Direct bifurcation of a steady solution into an invariant torus (to appear).

[8] V.I. IUDOVICH, Prikl. Mat. Mek. 35, 638-655 (1971), and Prikl.Mat. Mek. 36, 450-459 (1972).

[9] D.D. JOSEPH and D.H. SATTINGER, Arch. Rat. Mech. Anal. 45, 79-109 (1972).

[10] R. JOST and E. ZEHNDER, Helvetica Physica Acta, 45, 258-276 (1972).

[11] O.A. LADYZHENSKAYA, The Mathematical Theory of Viscous Incompressible Flow. New-York, Gordon and Breach, 1963.

[12] L. LANDAU, C.R. Acad. Sci. U.S.S.R. 44, 311-314 (1944).

[13] O.E. LANFORD III, Lecture Notes in Math., N° 322, 159-192. Berlin Heidelberg - New-York. Springer 1973.

[14] D. RUELLE and F. TAKENS, Comm. Math. Phys. 20, 167-192 (1971).

[15] R.TEMAM., On the theory and numerical analysis of Navier-Stokes equations. Lecture Notes n°9, University of Maryland, Dept of Math. (1973).

W. Eckhaus (ed.), New Developments in Differential Equations

APPLICATIONS OF THE METHOD OF DIFFERENTIAL INEQUALITIES IN SINGULAR PERTURBATION PROBLEMS

W.A.Harris, Jr.
University of Southern California

1. Introduction

The maximum principle, a very important special case of the method of differential inequalities, is well known for its applicability in singular perturbation problems as exemplified by Eckhaus and De Jager [4] and Dorr, Parter and Shampine [3]. There is an extensive literature, see e.g. Jackson [7], on the use of differential inequalities for estimating solutions for ordinary differential equations; however, in general, not of singular perturbation type; an exception being the work of Briš [1]. Recently, the technique of differential inequalities has been effectively utilized by Howes [6] to give an elegant, unified and comprehensive treatment of some classical singular perturbed nonlinear second order boundary value problems.

In this note, based on results of Harris and Howes [5], we illustrate further the versitility and applicability of differential inequalities in singular perturbation problems by studying the model nonlinear singular perturbation problem

$$\varepsilon y'' + yy' - y = 0$$
$$y(0) = A \qquad y(1) = B$$

whose solutions exhibit a wide variety of interesting behavior; indeed, one or more solutions of the reduced problem

$$uu' - u = 0$$

may be required to approximate the solution y, and the transition may occur in an interior boundary (shock) layer.

2. Basic existence and comparison results.

We begin with the definitions of the concepts that are fundamental to our approach.

Supported in part by the United States Army under contract DAHCO4-74-6-0013. This paper was completed while visiting Rijksuniversiteit Groningen, the Netherlands.

Definition. A function $\alpha = \alpha(t)$ is called a lower solution of the differential equation

$x'' = F(t,x,x')$, $\quad F \in C([a,b] \times \mathbb{R}^2)$, $\quad$ on $[a,b]$

if $\alpha \in C^2[a,b]$ and $\alpha''(t) \geqslant F(t,\alpha(t),\alpha'(t))$ on (a,b).

Definition. A function $\beta = \beta(t)$ is called an upper solution of the differential equation

$x'' = F(t,x,x')$, $\quad F \in C([a,b] \times \mathbb{R}^2)$, $\quad$ on $[a,b]$

if $\beta \in C^2[a,b]$ and $\beta''(t) \leqslant F(t,\beta(t),\beta'(t))$ on (a,b).

Definition. The function F is said to satisfy a Nagumo condition on $[a,b]$ with respect to a pair $\alpha,\beta \in C[a,b]$ in case $\alpha(t) \leqslant \beta(t)$ on $[a,b]$ and there exists a positive continuous function Φ on $[0,\infty)$ such that $|F(t,x,x')| \leqslant \Phi(|x'|)$ for all $a \leqslant t \leqslant b$, $\alpha(t) \leqslant x \leqslant \beta(t)$, $|x'| < \infty$ and $\int^{\infty} \Phi^{-1}(s)\,s\,ds = \infty$.

With these definitions, the principle existence and comparison theorem is

Theorem (Nagumo). Let $F = F(t,x,x')$ satisfy a Nagumo condition with respect to the pair α,β which are lower and upper solutions of $x'' = F(t,x,x')$ on $[a,b]$ respectively, and let $\alpha(a) \leqslant A \leqslant \beta(a)$, $\alpha(b) \leqslant B \leqslant \beta(b)$. Then the boundary value problem, $x'' = F(t,x,x')$, $x(a) = A$, $x(b) = B$ has a solution $x \in C^2[a,b]$ with $\alpha(t) \leqslant x(t) \leqslant \beta(t)$ on $[a,b]$.

Remark. If $\alpha_1,\alpha_2/\beta_1,\beta_2$ are two lower/upper solutions, then the theorem remains valid for the pair α^+,β^- where $\alpha^+(t) = \max(\alpha_1(t),\alpha_2(t))$, $\beta^-(t) = \min(\beta_1(t),\beta_2(t))$, i.e. appropriate corners are also allowable.

Since the function $\varepsilon^{-1}(-yy' + y)$ clearly satisfies a Nagumo condition with respect to the pair α,β, $\alpha(t) = t + \min(A,B-1)$, $\beta(t) = t + \max(A,B-1)$, which are lower and upper solutions respectively for the differential equation $\varepsilon y'' + yy' - y = 0$, we have as an immediate consequence:

Theorem. For each $\varepsilon > 0$ there exists a solution $y = y(t,\varepsilon)$ of the model problem

$$\varepsilon y'' + yy' - y = 0, \qquad y(0) = A, \quad y(1) = B$$

which satisfies for $0 < t < 1$,

$$t + \min(A,B-1) \leqslant y(t,\varepsilon) \leqslant t + \max(A,B-1).$$

Having established the existence of a solution, we turn now to the construction of other lower and upper solutions of the model problem which yield not only the existence of solutions but their asymptotic behavior as $\varepsilon \to 0+$ and explicit boundary/transition layer estimates as well.

3. The model problem.

We confine our attention to four representative cases.

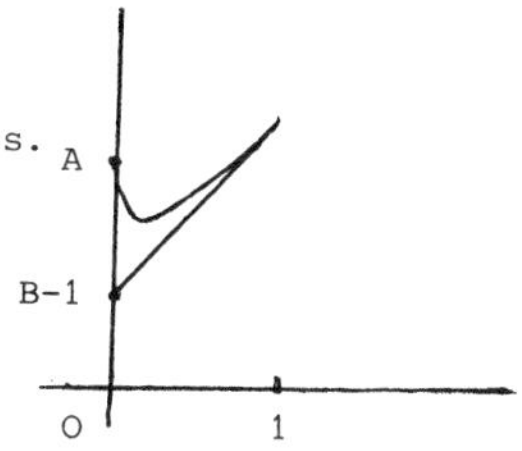

Case 1.

$A \geqslant B-1 > 0$

$\alpha(t) = t + B-1$

$\beta(t) = t + B-1 + (A-B+1) \exp\{(1-B)t/\varepsilon\}$

This is the classic case with a boundary layer of width $O(\varepsilon)$ at $t = 0$. Note the improvement of this upper solution over the upper solution $t + \max(A,B-1) = t + A$, so that we have $\beta(t) - \alpha(t) = O(\exp\{-kt/\varepsilon\})$, $k > 0$, and clearly, $\lim_{\varepsilon\to 0+} y(t,\varepsilon) = u(t) = t + B-1$, $t > 0$, where $u = u(t)$ is the solution of the reduced problem $uu' - u = 0$, $u(1) = B$.

Case 2.

$A > 0$, $0 < B < 1$.

$$\alpha(t) = \begin{cases} 0 & 0 \leqslant t \leqslant \lambda \\ t-\lambda & \lambda \leqslant t \leqslant 1 \end{cases}$$

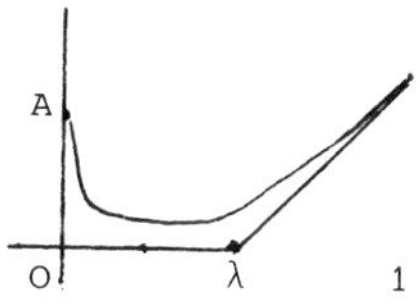

$$\beta(t) = \begin{cases} Ae^{-t/\sqrt{\varepsilon}} + \sqrt{\varepsilon} & 0 \leqslant t \leqslant \lambda \\ (t-\lambda) + Ke^{-(t-\lambda)/\sqrt{\varepsilon}} & \lambda \leqslant t \leqslant 1 \end{cases}$$

$\lambda = 1-B$, $K = \sqrt{\varepsilon} + Ae^{-\lambda/\sqrt{\varepsilon}}$

$$\lim_{\varepsilon\to 0+} y(t,\varepsilon) = u(t) = \begin{cases} 0 & 0 \leqslant t \leqslant \lambda \\ t-\lambda & \lambda \leqslant t \leqslant 1 \end{cases}, \quad t > 0$$

and we have a boundary layer of width $O(\sqrt{\varepsilon})$ at $t = 0$ and a boundary/transition layer (in $y'(t,\varepsilon)$ of width $O(\sqrt{\varepsilon})$ at $t = \lambda = 1-B$.
Note that two solutions of the reduced problem, $uu' - u = 0$, namely, $u \equiv 0$ and $u = t + B-1$ are needed to approximate the solution $y(t,\varepsilon)$ in this case.

In a similar manner, the case $0 < B < A+1 < 1$ has a solution $y(t,\varepsilon)$ for which three solutions of the reduced equation are needed for approximation, i.e.

$$\lim_{\varepsilon\to 0+} y(t,\varepsilon) = u(t) = \left\{\begin{array}{ll} t+A & 0 \leqslant t \leqslant -A \\ 0 & -A \leqslant t \leqslant 1-B \\ t+B-1 & 1-B \leqslant t \leqslant 1 \end{array}\right\} .$$

Case 3.

$0 < |A| < B-1$.

If $0 < A < B-1$, this case may be handled in a similar manner as case 1. However, if $1-B < A \leqslant 0$, the solution of the model problem must vanish on the interval $(0,1)$, we have therefore an implicit turning point, the coefficient of y' no longer has one sign, and we need a new type of lower solution. The function utilized here is natural in the sense that it is suggested by the method of matched asymptotic expansions for an inner correction to the outer solution $u(t) = t+B-1$, see e.g. Cole [2]. This illustrates how the method of differential inequalities my be utilized to establish the validity of formal approximate solutions.

An appropriate upper solution is as before

$$\beta(t) = t + B-1,$$

and an appropriate lower solution in this case is

$$\alpha(t) = t + (B-1)\tanh\left(\frac{B-1}{2\varepsilon}t - \gamma\right)$$

where γ is determined by the equation

$$\alpha(0) = (B-1)\tanh(-\gamma) = A.$$

If $A < 0$, then $\gamma > 0$, and it follows that the zero of $y(t,\varepsilon)$, $\eta(\varepsilon)$, satisfies $0 < \eta(\varepsilon) < 2\varepsilon\gamma/(B-1)$, and hence $\eta(\varepsilon) = O(\varepsilon)$.

Hence, once again, we have a boundary layer at $t = 0$ of width $O(\varepsilon)$ and

$$\lim_{\varepsilon\to 0+} y(t,\varepsilon) = u(t) = t+B-1, \quad t > 0.$$

Case 4.

$A < 0,\ B > 0,\ B > A+1,\ -(B+1) < A < 1-B$.

This is perhaps the most interesting case and indeed our results are incomplete, leading to a conjecture which we state after our discussion.

One can show that the solution $y(t,\varepsilon)$ has a unique zero, $t = \eta(\varepsilon)$, for $0 < t < 1$, and that

$$\eta(\varepsilon) \to \lambda = \tfrac{1}{2}(1-B-A) > 0 \qquad \text{as } \varepsilon \to 0+.$$

Consider, as in Case 3, the function

$$w(t) = t-\eta + K \tanh[\frac{K}{2\varepsilon}(t-\eta)]$$

for which we have

$$\varepsilon w'' + ww' - w = (t-\eta) \frac{K^2}{2\varepsilon} \operatorname{sech}^2 [\frac{K}{2\varepsilon}(t-\eta)] .$$

Clearly, on the interval $0 \leqslant t \leqslant \eta(\varepsilon)$

$$\alpha(t) = t + A$$

$$\beta(t) = (t-\eta) + K \tanh[\frac{K}{2\varepsilon}(t-\eta)]$$

are suitable lower and upper solutions, and on the interval $\eta(\varepsilon) \leqslant t \leqslant 1$

$$\alpha(t) = (t-\eta) + K \tanh[\frac{K}{2\varepsilon}(t-\eta)]$$

$$\beta(t) = t + B-1$$

are suitable lower and upper solutions, provided that $\beta(0) = -\eta + K \tanh[-\frac{K\eta}{2\varepsilon}] \geqslant A$ and $\alpha(1) = 1-\eta + K \tanh[\frac{K}{2\varepsilon}(1-\eta)] \leqslant B$, or $K < \frac{1}{2}(B-1-A)$.

It is clear that

$$\lim_{\varepsilon \to 0+} y(t,\varepsilon) = \begin{Bmatrix} t+A & 0 \leqslant t < \lambda \\ t+B-1 & \lambda < t \leqslant 1 \end{Bmatrix},$$

but the boundary layer behavior of the transition (shock) layer at $t = \lambda$ depends on how $\eta(\varepsilon)$ approaches λ as $\varepsilon \to 0+$. Our results are incomplete in this respect, but we

<u>Conjecture</u>: the unique zero, $\eta(\varepsilon)$, of the solution, $y(t,\varepsilon)$, of the model problem $\varepsilon y'' + yy' - y = 0$, $y(0) + A$, $y(1) = B$ in the case $A < 0$, $B > 0$, $B > A+1$, $-(B+1) < A < 1-B$, satisfies

$$\eta(\varepsilon) = \tfrac{1}{2}(1-B-A) + O\left(e^{-k/\varepsilon^p}\right) \quad \varepsilon \to 0+$$

for suitable positive constants k,p.

References

1. N.I.Briš, *On Boundary Value Problems for the equation* $\varepsilon y''=f(x,y,y')$ *for small* ε. Dokl.Akad.Nauk,SSSR 95 (1954), 429-432.

2. J.D.Cole, *Perturbation Methods in Applied Mathematics*. Blaisdell, Waltham, Mass. 1968.

3. F.W.Dorr, S.V.Parter and L.F.Shampine, *Applications of the Maximum Principle to Singular Perturbation Problems*. SIAM Rev. 15 (1973), 43-88.

4. W.Eckhaus and E.M.De Jager, *Asymptotic Solutions of Singular Perturbation Problems for Linear Differential Equations of Elliptic Type*. Arch.Rational Mech.Anal., 23 (1966), 26-86.

5. W.A.Harris,Jr. and F.A.Howes, *A Model Singular Perturbation Problem*. To appear.

6. F.A.Howes, *Singular Perturbations and Differential Inequalities*. Memoirs, American Math.Soc. To appear.

7. L.K.Jackson, *Subfunctions and Second-Order Differential Inequalities*. Advances in Math. 2 (1968), 307-363.

W. Eckhaus (ed.), New Developments in Differential Equations

A SINGULAR PERTURBATION PROBLEM OF TURNING POINT TYPE

P.P.N. de GROEN

1. INTRODUCTION

We study the two-point boundary value problems

$$(1^{\pm}) \qquad \varepsilon\tau u \pm xu' - \lambda u = f, \qquad u(1) = A \ \text{ and } \ u(-1) = B$$

on the real interval $[-1,+1]$; τ is the second order formal differential operator,

$$\tau u := (au')' + bu' + cu,$$

whose coefficients are C^{∞}-functions and whose principal coefficient a is strictly positive and satisfies $a(0) = 1$; ε is a small positive parameter, λ a complex "spectral" parameter and f a sufficiently smooth complex valued function.

We shall give a survey of the results on convergence for $\varepsilon \to +0$ of the eigenvalues and of the solution of problem $(1^{\pm})$, which we obtained in [3] and [4], so we will omit detailed proofs.

We begin with a study of the spectrum; we find that the eigenvalues of (1^{+}) converge to the negative integers and those of (1^{-}) to the nonpositive integers for $\varepsilon \to +0$. The behaviour of the spectrum for $\varepsilon \to +0$ depends heavily on the power of x in the coefficient of u' in $(1^{\pm})$: in the problem

$$(2^{\pm}) \qquad \varepsilon\tau u \pm x|x|^{\nu-1}u - \lambda u = f, \qquad u(\pm 1) = 0 \qquad \nu \in \mathbb{R}^{+}$$

the spectrum is real and it disappears at $-\infty$ for $\varepsilon \to +0$ if $0 < \nu < 1$, and it fills the negative part of the real axis densily for $\varepsilon \to +0$ if $\nu > 1$. Thereafter we consider convergence of the solutions of (1^{+}) and (1^{-}) for $\varepsilon \to +0$. We will mainly use Hilbert space methods, cf. [5] and [7]; we denote the inner product and norm in $L^2(-1,1)$ by $(\cdot,\cdot)$ and $||\cdot||$ and the supremum norm on $(-1,1)$ by $[\cdot]$.

In the past ten years a large number of papers has been published on the subject. Interest in the subject was raised by Ackerberg & O'Malley [1], who made formal asymptotic expansions and discovered that these expansions showed "resonance" at negative integral values of λ. By refined matching techniques Cook & Eckhaus [2] arrived at better estimates of the criterion for resonance.

From a "spectral" point of view the phenomenon of resonance is caused merely by the neighbouring eigenvalue: a solution of a boundary value problem grows beyond bound when λ tends to an eigenvalue. Independently and with totally different methods Rubenfeld & Willner [6] obtained a proof of convergence of the eigenvalues too; their proof is based on Langer's approximation method for turning point problems and it requires an enormous amount of explicit computations.

2. THE SPECTRUM

In connection with (2^+) we define the differential operator

(3) $$ {}_\nu T_\varepsilon u := \varepsilon \tau u + x|x|^{\nu-1}u', \qquad \text{for all } u \in \mathcal{D}({}_\nu T_\varepsilon), $$

$$ \mathcal{D}({}_\nu T_\varepsilon) := \{v \in L^2(-1,1) \mid v'' \in L^2(-1,1) \;\&\; v(\pm 1) = 0\}, \text{ if } \varepsilon > 0, $$

$$ \mathcal{D}({}_\nu T_0) := \{v \in L^2(-1,1) \mid x|x|^{\nu-1}u' \in L^2(-1,1) \;\&\; v(\pm 1) = 0\}. $$

The ν-dependence of $\sigma({}_\nu T_\varepsilon)$ is as follows:

THEOREM 1: *If* $0 \le \nu < 1$, *then the largest eigenvalue of* ${}_\nu T_\varepsilon$ *tends to* $-\infty$ *for* $\varepsilon \to +0$; *If* $\nu > 1$ *then* $\sigma({}_\nu T_\varepsilon)$ *becomes dense in* $(-\infty,0)$ *for* $\varepsilon \to +0$.

PROOF: Define $\omega(x) := \frac{1}{2}\int_0^x (t|t|^{\nu-1}/a(t))dt$. A function u is an eigenfunction of ${}_\nu T_\varepsilon$ if and only if $v := u \exp(\omega(\cdot)/\varepsilon)$ is an eigenfunction of the operator ${}_\nu \tilde{T}_\varepsilon$,

(4) $$ {}_\nu \tilde{T}_\varepsilon v := \exp(\omega/\varepsilon)\, {}_\nu T_\varepsilon\{v \exp(-\omega/\varepsilon)\} = $$

$$ = \varepsilon \tau v - (\tfrac{1}{2}|x|^{\nu-1} + \tfrac{1}{2}x|x|^{\nu-1}\, b/a + \tfrac{1}{4}|x|^{2\nu}/\varepsilon a)v. $$

It is easily seen that we can find a constant C such that

$$ ({}_\nu \tilde{T}_\varepsilon v,v) \le -((\tfrac{1}{2}x^{\nu-1} + \tfrac{1}{4}x^{2\nu}/\varepsilon a)\, v,v) + C(v,v). $$

If $0 < \nu < 1$ we have $\frac{1}{2}|x|^{\nu-1} + \frac{1}{4}|x|^{2\nu}/\varepsilon a > \frac{1}{4}\varepsilon^{(\nu-1)/(\nu+1)}$ and hence we find

$$ ||{}_\nu \tilde{T}_\varepsilon v - \lambda v|| \ge |\mathrm{Re}({}_\nu \tilde{T}_\varepsilon v - \lambda v,v)| \;/\; ||v|| $$

$$ \ge (\mathrm{Re}\lambda - C + \varepsilon^{(\nu-1)/(\nu+1)})||v||, $$

provided $\mathrm{Re}\lambda > -\varepsilon^{(\nu-1)/(\nu+1)} + C$. This proves that the spectrum disappears at $-\infty$ if $\varepsilon \to +0$.

If $\nu > 1$ we define for each $v \in \mathcal{D}({}_\nu\tilde{T}_\varepsilon)$ the function w,

$$w(\xi) := v(\varepsilon^{\frac{1}{2}}\xi)a^{-1}(\varepsilon^{\frac{1}{2}}\xi)\exp\{\tfrac{1}{2}\int_0^{\varepsilon^{\frac{1}{2}}\xi}(a'(s)/a(s) - b(s))ds$$

Analogous to (4) this induces the transformation of ${}_\nu\tilde{T}_\varepsilon$ into ${}_\nu\hat{T}_\varepsilon$,

$$(5)\qquad {}_\nu\hat{T}_\varepsilon w = \frac{d^2w}{d\xi^2} - P(\xi,\varepsilon)w, \qquad w(\pm\varepsilon^{-\frac{1}{2}}) = 0, \text{ for all } w \in \mathcal{D}({}_\nu\hat{T}_\varepsilon),$$

which is selfadjoint. The "potential" function P satisfies

$$P(\xi,\varepsilon) = \mathcal{O}(\varepsilon^{\frac{1}{2}\nu-\frac{1}{2}} + \varepsilon + \varepsilon^{\nu-1}|\xi|^{2\nu}) \qquad (\varepsilon \to +0)$$

uniformly for all $\xi \in (-\varepsilon^{-\frac{1}{2}},\varepsilon^{-\frac{1}{2}})$, so it is majorized by the "square-well" potential V,

$$V_\varepsilon(\xi) = \begin{cases} k(\varepsilon + \varepsilon^{\frac{1}{4}\nu-\frac{1}{4}}) & \text{if } |\xi| \le \varepsilon^{\frac{1}{4}\nu-\frac{1}{4}}, \\ \infty & \text{if } |\xi| > \varepsilon^{\frac{1}{4}\nu-\frac{1}{4}}. \end{cases}$$

It is easily seen that the eigenvalues of $d^2/d\xi^2 - V_\varepsilon$ become dense on $(-\infty,0)$ for $\varepsilon \to +0$ and by well-known comparison theorems this is transferred to $\sigma({}_\nu\hat{T}_\varepsilon)$ and hence to $\sigma({}_\nu T_\varepsilon)$, q.e.d.

REMARK 1: The operator associated with (2^-) satisfies the same result as ${}_\nu T_\varepsilon$ does; the only difference in the proof is that ω has to be replaced by $-\omega$ in formula (4).

From here on we shall deal with the intermediate case $\nu = 1$ only, so we will drop the subscript ν of ${}_\nu T_\varepsilon$ for $\nu = 1$. This case is the most interesting, both parts, $\varepsilon\tau$ and xd/dx of the operator T_ε have an influence of equal strength on the spectrum of the operator and these influences are in balance, such that the spectrum neither vanishes nor tends to a dense set:

THEOREM 2: *Let* $\{\lambda_k(\varepsilon) \mid k \in \mathbb{N}\}$ *be the set of eigenvalues of* T_ε, *arranged in decreasing order, i.e.* $\lambda_{k+1} < \lambda_k$, *then all eigenvalues satisfy:*

$$(6)\qquad \lambda_k(\varepsilon) = -k + \mathcal{O}(\varepsilon^{\frac{1}{2}}k^{3/2}) \qquad (\varepsilon \to +0)$$

uniformly with respect to k.

We will merely sketch the proof; for a detailed version we refer to [3] or [4]. Again we transform T_ε into $\tilde{T}_\varepsilon$ as in (4) and we consider the special (Hermite-) operator $\Pi_\varepsilon := \varepsilon d^2/dx^2 - x^2/4\varepsilon$ on the same domain of definition. By computing the solutions of $\Pi_\varepsilon v = \lambda v$ we can verify directly that the eigenvalues of Π_ε satisfy (6) and that the function χ_n,

(7) $\chi_n(x,\varepsilon) := \exp(-\tfrac{1}{4}x^2/\varepsilon)H_{k-1}(x/\sqrt{2\varepsilon})$, ($H_k$ is the k-th Hermite polynomial)

approximates the eigenfunction of Π_ε at λ_k up to $\mathcal{O}(\varepsilon^{-n}e^{-1/2\varepsilon})$ for $\varepsilon \to +0$. In order to compare $\tilde{T}_\varepsilon$ and Π_ε we connect them by the continuous chain $R_{\varepsilon,t} := (1-t)\Pi_\varepsilon + t\tilde{T}_\varepsilon$, which satisfies

(8) $$||R_{\varepsilon,t}\,u - R_{\varepsilon,s}\,u|| = \mathcal{O}((s-t)\{||R_{\varepsilon,t}\,u|| + ||u||\})$$

uniformly for all $s,t \in [0,1]$ and all u in the domain of definition. By spectral perturbation theorems, cf. [5] ch. 5, this implies that the eigenvalues of $R_{\varepsilon,t}$ depend continuously on t and that their variation is of order $\mathcal{O}((s-t)/|\lambda|)$, since $R_{\varepsilon,t}$ is (nearly) selfadjoint.

Next we prove that the approximate eigenfunctions of Π_ε satisfy

(9) $$||(R_{\varepsilon,t} - k)\chi_k|| = \mathcal{O}(\varepsilon^{\frac{1}{2}}k^{3/2}||\chi_k||), \qquad (\varepsilon \to +0),$$

uniformly for all $k \in \mathbb{N}$ and $t \in [0,1]$. If $\lambda_k(\varepsilon,t)$ is the k-th eigenvalue of $R_{\varepsilon,t}$, then $\lambda_k(\varepsilon,0)$ satisfies (6). By (8) we can find numbers $\varepsilon_n > 0$ and $t_n \geq d/n$ (with $d > 0$ and independent of ε) such that the circle $C(-j,\tfrac{1}{2})$ around $-j$ with radius $\tfrac{1}{2}$ is contained in the resolvent set of $R_{\varepsilon,t}$ for all $j \leq n$ ($j,n \in \mathbb{N}$), all $\varepsilon \in [0,\varepsilon_n]$ and $t \in [0,t_n]$ and that this circle contains only the eigenvalue $\lambda_j(\varepsilon,t)$ and no other. With the aid of the projection onto the eigenfunction of $\lambda_j(\varepsilon,t)$ it follows from (9) that $\lambda_j(\varepsilon,t)$ satisfies (6) for all $t \in [0,t_n]$ and $j \leq n$. Now we can repeat the argument, starting from the point $t = t_n$ instead of $t = 0$; however, since we did not prove anyting about $\lambda_{n+1}(\varepsilon,t)$, this possibly can enter $C(-n,\tfrac{1}{2})$ for some $t > t_1$. So we prove in the next step that $\lambda_j(\varepsilon,t)$ satisfies (6) for all $j \leq n-1$ and $t \in [t_n, t_n + t_{n-1}]$ and so on. Since we can choose n so large that $\sum_{j=k+1}^{n} t_j \geq d \sum_{j=k+1}^{n} 1/j \geq 1$, we have proved formula (6) after n-k steps, q.e.d.

REMARK 2: From the proof of theorem 2 we obtain the inequality

$$(10) \qquad ||\tilde{T}_\varepsilon u - \lambda u|| \geq ||u||\{\mathrm{dist}(\lambda,\{-n \mid n \in \mathbb{N}\}) - D\varepsilon^{\frac{1}{2}}\}$$

for some constant $D > 0$.

REMARK 3: By taking the adjoint of T_ε we find that the eigenvalues of the operator connected with (1^-) converge to the nonpositive integers for $\varepsilon \to +0$ with the same asymptotic estimate as in (6).

REMARK 4: Theorem 2 can be generalized to elliptic problems in a higher dimensional space which degenerate to a first order operator with a (simple) critical point for $\varepsilon \to +0$, cf. [4].

REMARK 5: The spectrum of the limit operator T_0 in $L^2(-1,1)$ is the set $\sigma(T_0) = \{\lambda \in \mathbb{C} \mid \mathrm{Re}\lambda \leq -\frac{1}{2}\}$; we see that there is an apparent lack of spectral continuity in L^2-sense. However, in distributional sense there is spectral continuity: the only (Schwartz-) distributions whose support is contained in $(-1,1)$ and which satisfy the equation $xu' = \lambda u$ (in distributional sense), are Dirac's δ-distribution and its derivatives. They satisfy $x \frac{d}{dx} \delta^{(n-1)} = -n\delta^{(n-1)}$, $n \in \mathbb{N}$; moreover, the (approximate) eigenfunctions of T_ε converge to them in distributional sense.

3. CONVERGENCE OF THE SOLUTIONS

In order to be able to prove convergence of a formal approximation of the solution of (1^+) we can use the following lemma:

Lemma 3: Constants $C > 0$ and $K > 0$ exist, such that $T_\varepsilon - \lambda$ is invertible and satisfies

$$||T_\varepsilon u - \lambda u|| \geq ||u||(\mathrm{Re}\lambda + \tfrac{1}{2} - K\varepsilon) + \varepsilon C||u'||$$

for all $u \in \mathcal{D}(T_\varepsilon)$, $\lambda \in \mathbb{C}$ with $\mathrm{Re}\lambda > -\frac{1}{2} + K\varepsilon$ and for all $\varepsilon \in [0,1]$.

For the proof we refer to [3] or [4].

Let u_ε be the solution of the full problem (1^+) and u_0 the solution of the reduced equation $xu' - \lambda u = f$ of (1^+), which satisfies both boundary conditions, i.e.:

(11) $$u_o(x,\lambda) := \int_{x/|x|}^{x} f(t)\left(\frac{x}{t}\right)^{\lambda}\frac{dt}{t} + \tfrac{1}{2}(A + B)|x|^{\lambda} + \tfrac{1}{2}(A - B)x|x|^{\lambda-1}.$$

If $f'' \in L^2(-1,1)$ and if $\mathrm{Re}\lambda > .3/2$, then $u_o'' \in L^2(-1,1)$ and satisfies

$$||(\varepsilon\tau + xd/dx - \lambda)u_o - f|| = ||\varepsilon\tau f|| = O(\varepsilon||f|| + \varepsilon||f''||), \qquad (\varepsilon \to +0);$$

since $u_\varepsilon(\pm 1,\lambda) - u_o(\pm 1,\lambda) = 0$ this implies by lemma 3:

(12) $$||u_\varepsilon - u_o|| \le (\mathrm{Re}\lambda + \tfrac{1}{2} - K\varepsilon)^{-1}||(T_\varepsilon - \lambda)(u_\varepsilon - u_o)|| =$$

$$= O(\varepsilon||f|| + \varepsilon||f''||), \qquad (\varepsilon \to +0).$$

By Sobolev's inequality $[u]^2 \le 2||u||^2 + 2||u||\ ||u'||$ we infer from (12) and lemma 3:

(13) $$[u_\varepsilon - u_o] = O(\varepsilon^{\frac{1}{2}}(||f|| + ||f''||)), \qquad (\varepsilon \to +0),$$

provided $\mathrm{Re}\lambda > 3/2$.

From (13) we see that (in first approximation) there are no boundary layers in the approximation, if $\mathrm{Re}\lambda > 3/2$. From (11) however, we see that a non-uniformity can be expected near the point $x = 0$ and that it will grow larger as $-\mathrm{Re}\lambda$ grows larger.

For $\mathrm{Re}\lambda \le 3/2$ a proof of convergence becomes more difficult, since $u_o''(\cdot,\lambda) \notin L^2(-1,1)$ in that case. Therefore we extend the boundary value problem (1^+) to the larger space $H_o^{-n}(-1,1)$, equiped with the norm $u \to |u|_{-n}$, with $n \in \mathbb{N}$, cf. [8] ch. 1.13. In this space we can prove the analogue of (12) provided $\mathrm{Re}\lambda > -n + 3/2$. By interpolation we can obtain convergence in stronger norms:

LEMMA 4: *A positive function* $C_k(\lambda)$ *exists such that*

(14) $$||x^k u|| + ||x^{k+1}u'|| + \varepsilon||x^k u''|| \le 8||x^k(\varepsilon\tau + xd/dx - \lambda)u|| + C_k(\lambda)|u|_{-k}$$

for all functions u, *for which* $||x^k u''|| < \infty$, *and for every* $k \in \mathbb{N}$.

PROOF: Cf. [4] lemma 3.13.

We observe that $u_o''(\cdot,\lambda) \in H_o^{-n}(-1,1)$ and $x^n u_o''(\cdot,\lambda) \in L^2(-1,1)$ provided $\mathrm{Re}\lambda > -n + 3/2$. With the aid of lemma 4 and the inequality

$$[x^{k+\frac{1}{2}}w]^2 \le (2k+2)||x^k w||^2 + 2||x^k w||\ ||x^{k+1}w'||, \qquad (k \in \mathbb{N}),$$

we obtain the final result:

THEOREM 5: *If* $n \in \mathbb{N}$ *and if* $\lambda \in \mathbb{C} \setminus (-\mathbb{N})$ *satisfies* $\mathrm{Re}\lambda > -n + 3/2$, *then*

$$\left.\begin{array}{ll}(15a) & \|x^n(u_\varepsilon - u_o)\| \\ (15b) & [x^{n+\frac{1}{2}}(u_\varepsilon - u_o)]\end{array}\right\} = O(\varepsilon\|f\| + \varepsilon\|f''\|), \qquad (\varepsilon \to +0).$$

We point out that the weight factors x^n and $x^{n+\frac{1}{2}}$ in the norms of the estimates (15a) and (15b) smoothe down the non-uniformity of u_ε at $x = 0$. If λ is a negative integer, we cannot expect convergence because of a neighbouring eigenvalue.

The solution of problem (1^-) converges also to a solution of the reduced equation $-xu' - \lambda u = f$ in the major part of the interval, as the solution of (1^+) does. However, in this case we cannot impose boundary conditions at $x = \pm 1$ on the set of solutions of the reduced equation. We have to select the right solution from this set by a smoothness condition at $x = 0$; this smoothness condition arises in a very natural way from the choice of suitable domains for the operator connected with (1^-) and for its limit (for $\varepsilon \to +0$). At the points $x = \pm 1$ (ordinary) boundary layers of width $O(\varepsilon)$ arise.

We will merely state the final result; for a proof we refer to [4]. For any $f \in C^n(-1,1)$ we define the function w_n ($n \in \mathbb{N} \cup \{0\}$) by

$$w_n(x;\lambda) := -\sum_{j=0}^{n-1} \frac{x^j f^{(j)}(0)}{(\lambda+j)j!} - \int_0^x \{f(t) - \sum_{j=0}^{n-1} t^j f^{(j)}(0)/j!\}(\tfrac{t}{x})^\lambda \frac{dt}{t},$$

provided $\lambda \in -\mathbb{N} \cup \{0\}$ and $\mathrm{Re}\lambda > -n$. Clearly this function is a solution of the reduced equation and is smooth at $x = 0$. It satisfies:

THEOREM 6: *If* $n \in \mathbb{N}$, $f \in L^2(-1,1)$ *such that* $\|f^{(n+2)}\| < \infty$ *and if* $\lambda \in \mathbb{C} \setminus (-\mathbb{N} \cup \{0\})$ *and* $\mathrm{Re}\lambda > -n + \frac{1}{2}$, *then the solution* v_ε *of* (1^-) *satisfies*

$$\text{(16)} \qquad v_\varepsilon(x;\lambda) = w_n(x;\lambda) + (A - w_n(1,\lambda))\exp\frac{x-1}{\varepsilon a(1)} + $$
$$+ (B - w_n(-1,\lambda))\exp\frac{-x-1}{\varepsilon a(-1)} + $$
$$+ O(\varepsilon(|A| + |B| + \|f\| + \|f^{(n+2)}\|))$$

for $\varepsilon \to +0$ *and uniformly for all* $x \in [-1,1]$.

REMARK 6: The method by which theorem 6 is proved, is in some sense dual to the one of theorem 5. In this proof we have to restrict the boundary value problem (1^-) to the smaller spaces $\dot{H}^{+n}(-1,1)$ ($n \in \mathbb{N}$) in order to be able to enlarge the part of the λ-plane in which an analogue of lemma 3 is true and in which we hence can prove validity of the asymptotic formula (16).

REFERENCES

1. Ackerberg, R.C. & R.E. O'Malley, *Boundary layer problems exhibiting resonance*, Studies in Appl. Math., 49 (1970), p.277-295.

2. Cook, L. Pamela & W. Eckhaus, *Resonance in a boundary value problem of singular perturbation type*, Studies in Appl. Math., 52 (1973), p.129-139.

3. Groen, P.P.N. de, *Spectral properties of second order singularly perturbed boundary value problems with turning points*, preprint: report 39 (May 1975) of the "Wiskundig Seminarium der Vrije Universiteit", Amsterdam, to appear in the Journal of Math. Anal. and Applications.

4. Groen, P.P.N. de, *Singularly perturbed differential operators of second order*, Mathematisch Centrum Amsterdam, tract 68 (to appear 1976).

5. Kato, T., *Perturbation theory of linear operators*, Springer Verlag, Berlin etc., 1966.

6. Rubenfeld, L.A. and B. Willner, *The general second order turning point problem and the question of resonance for a singularly perturbed second order ordinary differential equation*, to appear.

7. Lions, J.L., *Perturbations singulières dans les problèmes aux limites et en Contrôle optimal*, Lecture notes in Math. 323, Springer Verlag, Berlin etc., 1973.

8. Lions, J.L. & E. Magenes, *Problèmes aux limites non homogènes*, Dunod, Paris, 1968.

W. Eckhaus (ed.), New Developments in Differential Equations

ASYMPTOTICS FOR A CLASS OF PERTURBED INITIAL VALUE PROBLEMS

Bob Kaper

Department of Mathematics, University of Groningen, Groningen, the Netherlands

INTRODUCTION

In this paper we are dealing with initial value problems containing a small nonnegative perturbation parameter ε. On time intervals initiating the origin we will approximate the exact solution (provided it exists) asymptotically with respect to ε as $\varepsilon \downarrow 0$. The asymptotic solutions could be derived from so-called formal asymptotic solutions, i.e., functions which satisfy the differential equation and the initial conditions up to an asymptotic accuracy of certain order. These formal asymptotic solutions should then be compared asymptotically with the exact solution. The question arises whether such a function approximates the exact solution up to an asymptotic accuracy of the same order as it approximates the equation and the initial conditions. Or at least whether there exists a relation between these two orders. This question will be answered in connection with the type of the interval to be considered.

As an application we consider a class of perturbed oscillations described by the nonlinear second order ordinary differential equation with slowly varying coefficients

$$w'' + F(w,\varepsilon t) + \varepsilon f(w,w',\varepsilon t,\varepsilon) = 0, \qquad t \geq 0, \tag{1.a}$$

$$w(0,\varepsilon) = \alpha_1(\varepsilon),\ w'(0,\varepsilon) = \alpha_2(\varepsilon). \tag{1.b}$$

The force term F, on which εf is to be considered as a small perturbation, is assumed to be the derivative of a potential function which has an absolute minimum at the origin.

Let us briefly recall the concepts of (formal) asymptotic solution in dealing with the vector differential equation in $\mathbb{R}^n$:

$$x' = f(x,\varepsilon t,\varepsilon), \qquad t \in I, \tag{2.a}$$

$$x(0,\varepsilon) = \alpha(\varepsilon) \tag{2.b}$$

- A function $\tilde{x}$ is called *an asymptotic solution of order* κ ($\kappa(\varepsilon) = o(1)$ as $\varepsilon \downarrow 0$) if

$$x - \tilde{x} = O(\kappa) \text{ on } I$$

(x represents the exact solution of problem (2)).

- A function u is called *a formal asymptotic solution of order* η ($\eta(\varepsilon) = o(1)$ as $\varepsilon \downarrow 0$) if g and β,

$$g(t,\varepsilon) = u(t,\varepsilon) - f(u(t,\varepsilon), \varepsilon t,\varepsilon),\ \beta(\varepsilon) = \alpha(\varepsilon) - u(0,\varepsilon),$$

are $O(\eta)$ on I,

(call g and β *the residuals of u* for problem (2)).

An obvious modification leads to similar concepts for second order problems. Note that without loss of generality a slowly varying dependence of f on t is assumed in view of the application.
The order symbols O and o are understood to be related to the limitprocess $\varepsilon \downarrow 0$ uniformly in t on I.
In section 1 we will give a brief summary of the form of an Nth order formal assymptotic solution ϕ_N (whose residuals are $O(\varepsilon^{N+1})$) of problem (1). For a complete description I refer to [1].
In section 2 we will treat the remainder problem x - u in dealing with the vector problem (2). It includes the existence and uniqueness of the exact solution. Connected to this section we will construct an improved N^{th} order formal asymptotic solution ϕ_N in section 3.
In a final section 4 we will state some results that hold on the infinite interval $[0,\infty)$.

§1. ON FORMAL ASYMPTOTIC SOLUTIONS OF OSCILLATION PROBLEMS

We may expect the solutions of problem (1) to be oscillating functions with slowly varying amplitude and frequency. In order to include these large-scale variations in the approximation we consider for the moment intervals of order ε^{-1}, i.e., intervals of the type $[0, \varepsilon^{-1}L]$ where L is independent of ε. Asymptotically we may distinguish two different time scales, *a local- (or fast-) time scale* on which the solution is periodic with a period of order one and *a slow- (or stretched-) time scale*, characterized by the slow variable $\tau = \varepsilon t$, which accounts for the slow modulation of the oscillations. Both scales are made explicitely in the form of the formal asymptotic solution which technique is known as the two variable method. An Nth order formal asymptotic solution ϕ_N of problem (1) is given by

$$\phi_N(t,\varepsilon) = \eta(\tau) + A_o(\tau)\Phi_o(p,\tau) + \sum_{\nu=1}^{N}\varepsilon^{\nu}[A_\nu(\tau)z_2^*(p,\tau) + \Phi_\nu(p,\tau)],\ p = \varepsilon^{-1} S(\tau,\varepsilon;N), \tau = \varepsilon t,$$

$$S(\tau,\varepsilon;N) = \int_0^\tau \omega(\sigma)d\sigma + \varepsilon\sum_{j=o}^{N} \varepsilon^j s_j(\tau) \tag{1.1}$$

with residuals $\varepsilon^{N+1}g_{N+1}(t,\varepsilon)$, $g_{N+1} = O(1)$, and $\varepsilon^{N+1}\beta_i(\varepsilon)$, $\beta_i = O(1)$, i = 1,2.
The O-(1) contribution U_o, $U_o = \eta + A_o\Phi_o$, to the expansion of ϕ_N is the even, 2π-periodic solution of the nonlinear conservative system

$$\omega^2(\tau)\ U_{o,pp} + F(U_o,\tau) = 0, \quad p \geq 0, \tag{1.2}$$

in which τ is to be considered as a fixed parameter. The function η is the algebraic average of the extreme values of U_o, which makes it possible to introduce an amplitude function $A_o(\tau)$. The function ω follows from the normalization of the period of U_o to 2π and will therefore depend on A_o in the case of a nonlinear system (1.2). The higher order contributions U_ν, $U_\nu = A_\nu z_2^* + \Phi_\nu$, to the expansion of ϕ_N are determined by linear, second order equations whose homogeneous part is the first variational equation of (1.2) with respect to U_o,

$$\omega^2(\tau)U_{\nu,pp} + F_y\{U_o(p,\tau),\tau\}U_\nu = \gamma_\nu(p,\tau), \quad \nu = 1,\dots,N.$$

One homogeneous solution, $z_1^*(p,\tau)$, follows direcly by differentiating U_o with respect to p. A second solution, $z_2^*(p,\tau)$ could be found by the variation of constants method. Equations for A_j and S_j, $j = 0,\dots, N-1$, follow from boundedness requirements of U_{j+1}. They are of the type

$$\int_o^{2\pi} \gamma_{j+1}(p,\tau)z_i^*(p,\tau)dp = 0,\ i = 1,\ 2,\ j = 0,\dots,\ N-1.$$

(known as suppression of secular terms in γ_{j+1}). As an immediate consequence of the determination of ϕ_N we have for the residual function g_N

$$g_N(t,\varepsilon) = \gamma_{N+1}(p,\tau) + O(\varepsilon), \quad p = \varepsilon^{-1}S(\tau,\varepsilon,N), \tau = \varepsilon t.$$

At this stage we may draw the following conclusions.

- From the expansions (1.1) we see that

(i) $\phi_N(t,\varepsilon) = \phi_N^*(p,\tau,\varepsilon), \quad p = \varepsilon^{-1}S(\tau,\varepsilon;N),\ \tau = \varepsilon t,$

where

(ii) ϕ_N^* is defined on $\mathbb{R}^+ \times [0,L] \times [0,\varepsilon_o]$,

(iii) $\phi_N^*(p + 2\pi,\tau,\varepsilon) = \phi_N^*(p,\tau,\varepsilon)$.

Let us call such functions satisfying (i), (ii) and (iii) functions of the periodic two variable type.

- In consequence g_N is of the periodic two variable type.
- No equations for A_N and S_N have been determined yet.

We introduced the quantities anticipating a question on the order of asymptotic accuracy of ϕ_N conceiving it as an asymptotic solution on $[0,\varepsilon^{-1}L]$. We will treat this problem in the next section when dealing with the vector differential problem (2) on arbitrary intervals I.

§2. PROOF OF ASYMPTOTIC CORRECTNESS

In this section we consider the vector differential problem (2) in $\mathbb{R}^n$:

$$x' = f(x,\varepsilon t,\varepsilon), \quad t \in I, \tag{2.1a}$$

$$x(0,\varepsilon) = \alpha(\varepsilon), \tag{2.1b}$$

where I be some finite or infinite interval, possibly depending on ε. Let $|.|$ denote the vector- and matrix norm and $||x|| = \sup |x(t)|$, $t \in I$. Let u be a formal asymptotic solution of (2.1) of order η, i.e., the residuals g and β are $O(\eta)$. In order to compare the formal asymptotic solution u with the exact solution x of (2.1), whose existence and uniqueness should be established, we apply the change of variables

$$x = u + \rho.$$

Then the remainder function ρ should satisfy the nonlinear vector differential problem

$$\rho' = A(t,\varepsilon)\rho + h(\rho,\varepsilon t,\varepsilon) - g(t,\varepsilon), \tag{2.2a}$$

$$\rho(0,\varepsilon) = \beta(\varepsilon), \tag{2.2b}$$

where

$$A(t,\varepsilon) = f_x(u(t,\varepsilon),\varepsilon t,\varepsilon),$$
$$h(\rho,t,\varepsilon) = f(u(t,\varepsilon) + \rho,\varepsilon t,\varepsilon) - f(u(t,\varepsilon),\varepsilon t,\varepsilon) - A(t,\varepsilon)\rho.$$

Let $\Psi(t,s;\varepsilon)$, $\Psi(t,t;\varepsilon) = E$, be a fundamental matrix solution of the linear equation

$$z' = A(t,\varepsilon)z, \tag{2.3}$$

$\Psi(t,s;\varepsilon) = \Psi_0(t,\varepsilon)\, \Psi_0^{-1}(s,\varepsilon)$, $\Psi_0(t,\varepsilon) = \Psi(t,0;\varepsilon)$.
The initial value problem (2.2) for ρ may be transformed into the nonlinear Volterra integral equation

$$\rho = k + T\rho, \tag{2.4}$$

where

$$T\rho(t,\varepsilon) = \int_0^t \Psi(t,s;\varepsilon)h(\rho(s,\varepsilon),\varepsilon s,\varepsilon)ds,$$

$$k(t,\varepsilon) = \Psi_0(t,\varepsilon)\ \beta(\varepsilon) + \int_0^t \Psi(t,s;\varepsilon)g(s,\varepsilon)ds. \tag{2.5}$$

Provided f is sufficiently smooth we can show by means of a contraction mapping principle the existence and uniqueness of the solution ρ of (2.4) within a ball B(R) with radius R if

$$\gamma(\varepsilon)\kappa(\varepsilon) \leq \tfrac{1}{4} \text{ with } R = 2\kappa(\varepsilon),$$

where

$$\kappa(\varepsilon) = ||\kappa(t,\varepsilon)||,$$
$$\gamma(\varepsilon) = \left|\left|\int_0^t |\Psi(t,s;\varepsilon)|ds\right|\right|.$$

Let κ be an asymptotic order function, i.e., $\kappa = o(1)$ as $\varepsilon \downarrow 0$. Then we have the following

Theorem 1.

Let u be a formal asymptotic solution of (2.1) with residuals g and β of order η. If

$$\gamma(\varepsilon) = o(\kappa^{-1})$$

then for sufficiently small $\varepsilon \geq 0$ problem (2.1) has an exact solution $x = u + \rho$ with

$$x - u = O(\kappa) \quad \text{(uniformly in } t \text{ on } I)$$

(Hence u is an asymptotic solution of order κ).

The relation between κ and η, given by the function $k(t,\varepsilon)$ defined in (2.5), depends on the order of magnitude of the interval I and the behaviour of the fundamental matrix Ψ. Let us assume throughout this section that $I = [0,\varepsilon^{-m}L]$ for some integer m and that $|\Psi(t,s;\varepsilon)| \leq K$, $0 \leq s \leq t$, $t \in I$.

In this case $\gamma = O(\varepsilon^{-m})$ and the condition on $\gamma.\kappa$ of Theorem 1 imposes a minimum order of asymptotic accuracy of ϕ_N as an asymptotic solution and hence also as a formal asymptotic solution.

The relation between κ and η is simply $\kappa = \varepsilon^{-m}\eta$, which means a reduction of the order of asymptotic accuracy of u as an asymptotic solution when the order of magnitude of I(m) increases. On intervals of order ε^{-1} this means a loss of one ε in the order of u as a asymptotic solution compared to u as a formal asymptotic solution. For a subclass of initial value problems related to the oscillation problem (1) we may improve the order with one ε by the application of a partial integration rule and by imposing a condition on the residual function g (hence on u). Therefore we need the concept of

- *first order formal asymptotic matrix solution* Φ of (2.3), i.e., a matrix valued function for which $A(t,\varepsilon)\,\Phi(t,\varepsilon) - \Phi'(t,\varepsilon) = \varepsilon\, G(t,\varepsilon)$, $G = O(1)$, $\det \Phi > 0$ (hence Φ^{-1} exists).

Apply the change $\Psi(t,s;\varepsilon) = \Phi(t,\varepsilon)\, Y(t,s;\varepsilon)\, \Phi^{-1}(s,\varepsilon)$, then Y is a fundamental matrix of

$$\dot{z} = -\varepsilon\, \Phi^{-1}(t,\varepsilon) G(t,\varepsilon) z.$$

In view of the oscillation problems we have the following

Theorem 2.

Let u be a formal asymptotic solution of (2.1) of order η, which is of the periodic two variable type, $u(t,\varepsilon) = u^*(p,\tau,\varepsilon)$, $p = \varepsilon^{-1} S(\tau,\varepsilon)$, $\tau = \varepsilon t$.

Let Φ be a first order formal asymptotic matrix solution of (2.3) of the periodic two variable type, $\Phi(t,\varepsilon) = \Phi^*(p,\tau,\varepsilon)$, $p = \varepsilon^{-1} S(\tau,\varepsilon)$, $\tau = \varepsilon t$. If

$$\begin{aligned} &(i)\ \gamma(\varepsilon) = o(\kappa^{-1}) \\ &(ii)\ \Phi = O(1) \\ &(iii)\ \int_0^{2\pi} \Phi^{*-1}(p,\tau) g^*(p,\tau,0)\,dp = 0, \end{aligned} \tag{2.6}$$

where g^* is the two variable counterpart of g, then problem (2.1) has an exact solution $x = u + \rho$ with

$$x - u = O(\kappa) \text{ on } I \text{ with } \kappa = \varepsilon^{-m+1}\eta.$$

Proof.

The theorem is almost an immediate consequence of Theorem I except for the improved relation between κ and η. Substitute the above assumptions on u, Ψ and Φ in k (defined in (2.5))

$$k(t,\varepsilon) = \Psi_0(t,\varepsilon)\beta(\varepsilon) + \Phi(t,\varepsilon)\int_0^t Y(t,s;\varepsilon)\Phi^{*-1}(p(s),\varepsilon s,\varepsilon)g^*(p(s),\varepsilon s,\varepsilon)ds,$$
$$p(s) = \varepsilon^{-1}S(\varepsilon s,\varepsilon).$$

The variable s of integration appears in the integrand in two different ways: via the periodicity variable p and elsewhere characterized by a derivative of order $O(\varepsilon)$ ($\dot{Y}$ is of $O(\varepsilon)$ since Φ and G are $O(1)$). A partial integration rule gives a gain of one ε in the integral which means $\kappa = \varepsilon.\varepsilon^{-m}\eta$.

Till now we made the assumption of uniform boundedness of Ψ. In a corrolory we will replace the assymption by a condition on the matrix A in the special case of intervals of order ε^{-1}.

Corollary 1.

Let U and Φ be formal asymptotic solutions in the sense of Theorem 2. If

$$\begin{aligned} &(i) && \gamma(\varepsilon) = o(\eta^{-1}),\\ &(ii) && \lim_{\varepsilon\downarrow o}\int_0^t \operatorname{tr} A(s,\varepsilon)ds = O(1) \text{ on } [0, \varepsilon^{-1}L],\\ &(iii) && \Phi = O(1),\\ &(iv) && \int_0^{2\pi}\Phi^{*-1}(p,\tau)g^*(p,\tau,0)dp = 0 \end{aligned} \qquad (2.7)$$

then u is an asymptotic solution of (2.1) of order η on intervals of order ε^{-1}.

Proof.

Once Ψ_0 is bounded, condition (2.7) (ii) assures the boundedness of the inverse Ψ_0^{-1} which means $|\Psi(t,s;\varepsilon)| \leq K$, $0 \leq s \leq t \leq \varepsilon^{-1}L$. Since $\Phi = O(1)$ the boundedness of Y_0 (and hence of Ψ_0) follows from an application of Gronwall's lemma to the equation for Y_0 on intervals of order ε^{-1}.

§3. IMPROVED FORMAL ASYMPTOTIC SOLUTION ϕ_N OF THE OSCILLATION PROBLEM (1)

With respect to the formal asymptotic solution ϕ_N constructed in Section 1 for the class of oscillation problems we already know from Theorem 1 that ϕ_N is an $(N-1)^{st}$ order asymptotic solution of (1) on $[0,\varepsilon^{-1}L]$ provided $N \geq 2$. In order that ϕ_N be an Nth order asymptotic solution of (1) on $[0, \varepsilon^{-1}L]$ the conditions (2.7) should be verified if we apply the change of variables to a first order system in $\mathbb{R}^2$:

$$x(t,\varepsilon) = \begin{pmatrix} w(t,\varepsilon) \\ w'(t,\varepsilon) \end{pmatrix}.$$

On intervals of order ε^{-1} we have $\gamma(\varepsilon) = O(\varepsilon^{-1})$. Since $\eta = \varepsilon^{N+1}$ for an Nth order formal asymptotic solution ϕ_N condition (2.7) (i) implies $N \geq 1$. Once ϕ_1 is a first order asymptotic solution the asymptotic correctness of ϕ_0 as a zeroth order asymptotic solution follows from a comparison of ϕ_0 and ϕ_1.
It is a straightforward check to see that z_i, $i = 1, 2$,

$$z_i(t,\varepsilon) = z_i^*(p,\tau), \quad p = \varepsilon^{-1} S(\tau,\varepsilon;N), \quad \tau = \varepsilon t,$$

are first order formal asymptotic solutions of the linear variational equation of (1) with respect to ϕ_N:

$$y'' + F_w\{\phi_N(t,\varepsilon),\varepsilon t\}y + \varepsilon q_1(t,\varepsilon)y + \varepsilon q_2(t,\varepsilon)y' = 0 \tag{3.1}$$

where $q_1(t,\varepsilon) = \frac{\partial f}{\partial w}$ and $q_2(t,\varepsilon) = \frac{\partial f}{\partial w'}$, at $\phi_N(t,\varepsilon)$.
Conditions (2.7) (ii) and (iii) follow by a direct verification (note that the trace of the matrix $A(t,\varepsilon)$ is formed by the coefficient of y' in (3.1)).
For the second order problem condition (2.7) (iv) reads as

$$\int_0^{2\pi} \gamma_{N+1}(p,\tau)z_i^*(p,\tau)dp = 0, \quad i = 1, 2.$$

(c.f. also the conclusions at the end of section 1). These conditions (suppression of secular terms in γ_{N+1}) are satisfied if A_N and S_N are solutions of linear first order differential equations (initial conditions follow in the same way as the values for A_j and S_j). In consequence of Corollary 1 we now have that ϕ_N is an Nth order asymptotic solution of (1) on intervals of order ε^{-1}.

The above described form of an Nth order asymptotic solution ϕ_N gives rise to a procedure generating asymptotic solutions of all order. This is due to the independence of the truncating index N of the equations for all quantities involved by the expansions for ϕ_N (c.f. [1]).

§4. OSCILLATION PROBLEMS WITH SMALL DECAY

Throughout the examination of the relation between κ and η in section 2 we assumed boundedness of the fundamental matrix Ψ of (2.3). In case of the second order oscillation problems this equation corresponds with the linear equation (3.1). A much better result in the relation between κ, η and the order of magnitude of the interval I could be obtained in cases of an absolute integrable fundamental matrix on $[0,\infty)$. For simplicity let us restrict ourselves to the class of weakly nonlinear oscillation problems. Its variational equation with respect to ϕ_N is

$$y'' + y + \varepsilon q_1(t,\varepsilon)y + \varepsilon q_2(t,\varepsilon)y' = 0 \tag{4.1}$$

where $q_1(t,\varepsilon) = \frac{\partial f}{\partial w}(\phi_N(t,\varepsilon),\phi_N'(t,\varepsilon),\varepsilon t,\varepsilon)$, $q_2(t,\varepsilon) = \frac{\partial f}{\partial w'}(\phi_N(t,\varepsilon),\phi_N'(t,\varepsilon),\varepsilon t,\varepsilon)$.

The coefficient functions q_1 and q_2 are determined by the form of the original equation as well as in consequence by the formal asymptotic solution ϕ_N.
Let us start with two subclasses, the class of weakly nonlinear oscillation problems with an exponential decay (class I)

$$w'' + w + 2\varepsilon\gamma(\varepsilon t)w' + \varepsilon f(w,w',\varepsilon t,\varepsilon) = 0, \tag{4.2}$$

where γ is some positive function, and the class with an algebraic decay and constant coefficients (class II)

$$w'' + w + (w')^{2m+1} + \varepsilon f(w,w',\varepsilon) = 0, \qquad m \geq 1 \tag{4.3}$$

In both cases positivity conditions should be imposed on the function f. In a forthcoming paper we will go into details about estimates on ϕ_n, g_N and the fundamental matrix Ψ of (4.1). Here we will restrict ourselves to a brief survey of the results.

In case I the formal asymptotic solution ϕ_N could be estimated as follows

$$|\phi_N(t,\varepsilon)| \leq M_o\, p_N(\varepsilon t,\varepsilon) \exp[-\int_0^{\varepsilon t} \gamma(\sigma)d\sigma],$$

where p_N is a polynomial in εt of a degree depending on N and less than N. A similar estimate holds for the residual function g_N of ϕ_N. The linear variational equation is given by

$$y'' + y + 2\varepsilon\gamma(\varepsilon t)y' + \varepsilon q_1(t,\varepsilon)y + \varepsilon q_2(t,\varepsilon)y' = 0.$$

There exists a constant M_o such that for the fundamental matrix Ψ holds

$$|\Psi(t,s;\varepsilon)| \leq M_o \exp[-\int_{\varepsilon s}^{\varepsilon t} \gamma(\sigma)d\sigma], \quad 0 \leq s \leq t < \infty.$$

A combination of the above results and the partial integration rule gives the following

Theorem 3.

Let ϕ_N be an Nth order formal asymptotic solution of (4.2). For $N \geq 1$ problem (4.2) has an exact solution w of the form $w = \phi_N + \rho$ with

$$|\rho| \text{ and } |\rho'| \leq M_o\, \varepsilon^{N+1}\, p_{N+1}(\varepsilon t,\varepsilon)\, \exp[-\int_0^{\varepsilon t} \gamma(\sigma)d\sigma], \quad 0 \leq t < \infty$$

for some constant M_o.

A similar result on $[0,\infty)$ could be derived in case II. The exponential damping should be replaced by a algebraically decreasing function of order $(2m)^{-1}$, the polynomial in εt multiplying the exponentially decreasing function in case II by a polynomial in logarithmically increasing functions.

REFERENCE

[1] Kaper, B., (1975). Perturbed Nonlinear Osciallations, SIAM Jrnl. on Appl. Maths.

W. Eckhaus (ed.), New Developments in Differential Equations

ON THE SOLUTIONS OF PERTURBED DIFFERENTIAL EQUATIONS

H.- D. Niessen
Department of Mathematics
University of Essen
Essen, Germany

INTRODUCTION

In the last years a lot of results have been obtained which guarantee that all solutions of some perturbed ordinary differential equation are in some sense integrable, e.g. belong to L^p, if this is true for the unperturbed equation. (Compare e.g. Bradley [5], Halvorsen [10], Patula and Wong [13,p.24], Zettl [18]).

Here a very elementary theorem on perturbed systems of differential equations is presented which contains the above mentioned theorems and some others (Bellman [2],[3],[4 ,p.43], Walker [16]) as special cases. Although the assumptions of this theorem are weaker and the assertion is stronger than those of the earlier theorems, its proof is much more elementary (and nearly obvious).

In sections 1,2 and 3 various applications are made to linearly perturbed systems, to ordinary differential equations and to limit circle criteria. Section 4 contains generalizations of the theorem which e.g. imply results of Cesari [6], Halvorsen [10], Levinson [11] and Wong [17].

1.SOLUTIONS OF PERTURBED SYSTEMS

All functions occuring in this and the following sections are supposed to be measurable and complex vector- or matrix-valued of some suitable size. Let I be any real interval and A a locally integrable (n,n) - matrix-valued function on I. Consider on I the perturbed system

$$(1.1)\qquad z' = A(t)z+f(t,z)$$

arrising from the unperturbed system

$$(1.2)\qquad x' = A(t)x$$

by the perturbation

$$(1.3)\qquad f(t,z)=B(t)O(1)+C(t)O(|D(t)z|)$$

(with matrix-valued functions B,C,D,O). f is assumed to be defined on $I\times\mathbb{C}^n$. Then we have

(1.4) THEOREM. *Let X be a fundamental matrix of (1.2) and suppose that*

$$(1.5)\qquad |X^{-1}B|\ ,\ |X^{-1}C||DX|\in L^1(I).$$

Then, for any solution z *of (1.1),*

$$z(t)=X(t)O(1) \qquad (t\in I).$$

Proof. Defining u by $z=:Xu$ we get by (1.1),(1.2) and (1.3)

$$u'=X^{-1}f(t,Xu)=X^{-1}BO(1)+X^{-1}CO(|DXu|).$$

By the first part of (1.5) integration yields for fixed $t_0\in I$

$$u(t)=O(1)+O(\left|\int_{t_0}^{t} |X^{-1}C||DX||u|\right|).$$

Applying the Gronwall inequality and using the second part of (1.5) we get

$$u(t)=O(1)\ \exp\ O(\left|\int_{t_0}^{t} |X^{-1}C||DX|\right|)=O(1).$$

For applications of this theorem the following remarks

are useful:

(1.6) REMARK. *For a linear perturbation* $f(t,z)=\tilde{A}(t)z$ *(1.5) may be replaced by*

$$X^{-1}\tilde{A}X\in L^1(I). \tag{1.7}$$

To see this, choose $B:=o$, $C:=\tilde{A}X$, $D:=X^{-1}$. Then f is of the form (1.3) since $Dz=O(|Dz|)$. Furthermore, (1.5) holds:

$$|X^{-1}C||DX|=|X^{-1}\tilde{A}X|\in L^1(I).$$

(1.8) REMARK. *Consider*

$$y'=-A^*(t)y \tag{1.9}$$

(where * *denotes the adjoining or the transposing operator). Then (1.5) is equivalent to*

$$|y^*B|,\ |y^*C||Dx|\in L^1(I) \text{ for all solutions } x \text{ of (1.2)}, \tag{1.10}$$

y *of (1.9), and (1.7) is equivalent to*

$$y^*\tilde{A}x\in L^1(I) \text{ for all solutions } x \text{ of (1.2)}, y \text{ of (1.9)}. \tag{1.11}$$

This follows from the fact that if Y denotes a fundamental matrix of (1.9), then

$$X^{-1}(t)=KY^*(t)$$

with some constant- and obviously nonsingular-matrix K.

By the preceding remarks theorem 1.4 implies immediately the first assertion of the following

(1.12) THEOREM. *Let X be a fundamental matrix of (1.2), and let* Z[1] *denote a fundamental matrix of*

$$z'=(A(t)+\tilde{A}(t))z. \tag{1.13}$$

[1] The assumption of the theorem implies that $\tilde{A}$ is locally integrable. Therefore Z exists.

If for all solutions x *of (1.2) and for all solutions* y *of (1.9)*

$$y^*\tilde{A}x \in L^1(I),$$

then

$$Z(t)=X(t)O(1) \quad \textit{and} \quad X(t)=Z(t)O(1).$$

To prove the second assertion, we apply the first part of the theorem to the unperturbed system (1.9) and to the perturbed system

$$(1.14) \qquad w'=(-A^*(t)-\tilde{A}^*(t))w$$

obtaining

$$(1.15) \qquad W(t)=Y(t)O(1),$$

where W denotes a fundamental matrix of (1.14). Since (1.2),(1.9) and (1.13),(1.14) are adjoint systems, we get with nonsingular constant matrices K_1, K_2:

$$X^{-1}(t)=K_1Y^*(t), \quad Z^{-1}(t)=K_2W^*(t).$$

Together with (1.15) this implies

$$Z^{-1}(t)=O(1)X^{-1}(t)$$

proving the second assertion.

Since $X^{-1}=(\det X)^{-1}X^t_{ad}$, where X_{ad} denotes the matrix of the algebraic complements of X, and since

$$\det X(t)= \det X(t_0) \cdot \exp\left(\int_{t_0}^{t} trA(\tau)d\tau\right),$$

e.g. (1.7) may be formulated as

$$\exp\left(-\int_{t_0}^{t} trA(\tau)d\tau\right) X^t_{ad}\,\tilde{A}x \in L^1(I).$$

Then theorem 1.4 and remark 1.6 especially yield

(1.16) THEOREM. Let $\int_{t_0}^{t} \Re e\, trA(\tau)d\tau$ *be bounded below, let* $\tilde{A}$ *be integrable and let any solution* x *of (1.2) be bounded. Then*

$$z(t)=X(t)O(1)$$

for any solution z *of (1.13). Especially,* z *is bounded.*

The last assertion is theorem 6 of [4], Ch.2.

2. APPLICATIONS TO DIFFERENTIAL EQUATIONS

Consider on $I \subset \mathbb{R}$ the differential operator

$$(2.1) \qquad L\xi := \sum_{i=0}^{n} p_i \xi^{(i)},$$

its formal adjoint

$$(2.2) \qquad L^+\eta := \sum_{i=0}^{n} (-1)^i (p_i^* \eta)^{(i)}$$

and the perturbed equation

$$(2.3) \qquad L\zeta = \varphi(t,\zeta,\zeta',\ldots,\zeta^{(n-1)}).$$

Here p_i^* denotes throughout either p_i or its complex conjugate. In the first case for any matrix M, M^* denotes the transposed matrix, in the second case it denotes the adjoint matrix. We suppose that p_n is never zero, that

$$\frac{p_i}{p_n} \quad (i=0,\ldots,n-1)$$

is locally integrable on I[2] and that

$$(2.4) \qquad |\varphi(t,z_1,\ldots,z_n)| \le k_0(t) + \sum_{j=1}^{n} k_j(t)|z_j| \quad (t \in I, z_j \in \mathbb{C})$$

with nonnegative (measurable) functions k_j.

Furthermore, denote by

$$S := \{\xi \mid L\xi = 0\}, \quad S^+ := \{\eta \mid L^+\eta = 0\}.$$

Then theorem 1.4 and remark 1.8 imply

(2.5) THEOREM. *If for all* $\xi \in S$, $\eta \in S^+$

$$(2.6) \qquad k_0\eta, k_j\xi^{(j-1)}\eta \in L^1(I) \qquad (j=1,\ldots,n),$$

then for any solution ζ *of* (2.3) *and for any basis* $\xi_1,\ldots,\xi_n$ *of* S

$$(2.7) \qquad \zeta^{(i)}(t) = (\xi_1^{(i)}(t),\ldots,\xi_n^{(i)}(t))\,O(1) \quad (i=0,\ldots,n-1).$$

[2] If the p_i's are not smooth enough, (2.2) has to be interpreted as a quasi-differential operator.

Proof. If

$$A := \begin{pmatrix} 0 & 1 & & & \\ & \cdot & \cdot & & \\ & & \cdot & \cdot & \\ & & & 0 & 1 \\ -\frac{p_0}{p_n} & \cdot & \cdot & \cdot & -\frac{p_{n-1}}{p_n} \end{pmatrix}$$

and if for $t \in I$, $z \in \mathbb{C}^n$

$$f(t,z) := \frac{1}{p_n} \varphi(t,z) e_n$$

(with $e_n = (\delta_{jn})_{j=1}^n$), then $L\xi = 0$ is equivalent to

(2.8) $$x' = Ax \text{ with } x = (\xi^{(i-1)})_{i=1}^n,$$

and (2.3) is equivalent to

(2.9) $$z' = Az + f(t,z) \text{ with } z = (\zeta^{(i-1)})_{i=1}^n.$$

Analogously, $L^+\eta = 0$ is equivalent to

(2.10) $$y' = -A^* y \text{ with } y = (\sum_{j=i}^n (-1)^{j-i} (p_j^* \eta)^{(j-i)})_{i=1}^n.$$

Defining

$$D := \text{diag}(k_1, \ldots, k_n)$$

and using the l^1-norm for vectors we get by (2.4)

$$\varphi(t,z) = k_0(t) O(1) + O(|D(t)z|).$$

Therefore,

$$f(t,z) = B(t) O(1) + C(t) O(|D(t)z|)$$

with

$$B := \frac{k_0}{p_n} e_n, \quad C := \frac{1}{p_n} e_n.$$

To verify (1.10) let x be a solution of (2.8), y a solution of (2.10). Then

$$x = (\xi^{(i-1)}) \text{ with } \xi \in S,$$

$$y = (\sum_{j=i}^n (-1)^{j-i} (p_j^* \eta)^{(j-i)}) \text{ with } \eta \in S^+.$$

Thus by (2.6)

$$y^* B = \frac{k_0}{p_n} y^* e_n = \frac{k_0}{p_n} (p_n \eta^*) = k_0 \eta^* \in L^1(I),$$

$$|y^*q||Dx| = |\eta^*| \sum_{j=1}^{n} |k_j\xi^{(j-1)}| \in L^1(I).$$

Therefore, if ζ is a solution of (2.3), i.e., if $z=(\zeta^{(i-1)})$ is a solution of (2.9), theorem 1.4 and remark 1.8 imply the assertion:

$$(\zeta^{(i-1)}(t))=(\xi_j^{(i-1)}(t))O(1).$$

This theorem implies a result of Zettl:

<u>(2.11) THEOREM (Zettl [18]). Consider on</u> $I:=[a,\infty)$

$$L\xi := \xi^{(n)} + \sum_{i=0}^{n-1} p_i\xi^{(i)} = 0$$

and

(2.12) $$L\zeta=\Psi(t,\zeta)$$

<u>with</u> p_i <u>realvalued, continuous and</u>

(2.13) $$|\Psi(t,\zeta)| \le k_o(t)+k_1(t)|\zeta|,$$

<u>where</u> k_o <u>and</u> k_1 <u>are nonnegative. Suppose</u> $S\cup S^+\subset L^2(I), k_o\eta\in L^1(I)$ <u>for</u> $\eta\in S^+$ <u>and either</u>

(i) $k_1^{1/2}\xi\in L^2(I)$ <u>for all</u> $\xi\in S\cup S^+$ <u>or</u>

(ii) $k_1\eta\in L^2(I)$ <u>for all</u> $\eta\in S^+$.

<u>Then all solutions of (2.12) are in</u> $L^2(I)$.

<u>Proof</u>. Since in either case

(2.14) $$k_o\eta, k_1\xi\eta\in L^1(I) \text{ for } \xi\in S, \eta\in S^+,$$

theorem 2.5 especially implies

(2.15) $$\zeta=(\xi_1,\ldots,\xi_n)O(1)$$

for any solution ζ of (2.12). Since $S\subset L^2(I)$, (2.15) yields $\zeta\in L^2(I)$.

Obviously, in this proof neither the assumption $S^+\subset L^2(I)$ nor the assumption that the p_i are real is needed.

Furthermore, this proof shows that (2.14) (and more general (2.6)) is the essential assumption of the theorem. It may be interpreted as a smallness condition on the perturbation rather than an integrability condition on the solutions of the unperturbed equation and its adjoint. Then the square-integrability of all solutions is only transferred by (2.15) from the unperturbed to the perturbed equation. It is clear that (2.15) (and more general (2.7)) gives more information on the structure of the solutions of the perturbed equation than does the assertion of Zettl's theorem.

Similarly, (1.5) and (1.7) or (1.10), (1.11) may be interpreted as smallness conditions on the perturbation.

As an application of theorem 1.12 we get analogously to theorem 2.5 for linear perturbations the following

(2.16) THEOREM. Suppose that q_n is never zero and that for all $\xi\in S, \eta\in S^+$

$$\sum_{j=0}^{n-1} \left(q_j \frac{p_n}{q_n} - p_j\right)\xi^{(j)}\eta\in L^1(I). \tag{2.17}$$

Let $\xi_1,\dots,\xi_n$ denote a basis of S and $\zeta_1,\dots,\zeta_n$ a fundamental system[3)] of solutions of

$$\sum_{j=0}^{n} q_j\zeta^{(j)}=0. \tag{2.18}$$

Then for any solution ζ of (2.18) and for any $\xi\in S$

$$\left.\begin{aligned}\zeta^{(i)}(t)&=(\xi_1^{(i)}(t),\dots,\xi_n^{(i)}(t))O(1)\\ \xi^{(i)}(t)&=(\zeta_1^{(i)}(t),\dots,\zeta_n^{(i)}(t))O(1)\end{aligned}\right\}\text{for } i=0,\dots,n-1.$$

Proof. In the terminology of theorem 1.12 and of the proof of theorem 2.5

$$\tilde{A}= -e_n\left(\frac{q_0}{q_n} - \frac{p_0}{p_n},\dots,\frac{q_{n-1}}{q_n} - \frac{p_{n-1}}{p_n}\right).$$

3) Since by (2.17) $\frac{q_j}{q_n} - \frac{p_j}{p_n}$ $(j=0,\dots,n-1)$ are locally integrable (compare footnote 1)), such a fundamental system exists.

Therefore (2.17) implies

$$y^{*}\tilde{A}x = -p_n \eta^{*} \sum_{j=0}^{n-1} \left(\frac{q_j}{q_n} - \frac{p_j}{p_n} \right) \xi^{(j)} \in L^1(I).$$

Now theorem 1.12 yields the assertion.

Theorem 2.16 generalizes and strengthens some recent (and earlier) results on linearly perturbed differential equations. As an immediate consequence (compare the proof of theorem 2.11) we get

(2.19) THEOREM (Zettl [18]). *Consider on* $I:=[a,\infty)$

$$L\xi := \xi^{(n)} + \sum_{i=0}^{n-1} p_i \xi^{(i)} = 0$$

and

(2.20) $$L\zeta = h\zeta$$

with p_i, h *realvalued and continuous. Suppose* $S \cup S^{+} \subset L^2(I)$ *and either*

(i) $|h|^{1/2} \xi \in L^2(I)$ *for all* $\xi \in S \cup S^{+}$ *or*

(ii) $h\eta \in L^2(I)$ *for all* $\eta \in S^{+}$.

Then all solutions of (2.20) belong to $L^2(I)$.

Another corollary is

(2.21) THEOREM (Halvorsen [10]). *Consider on a half-open interval* I *the equations*

(2.22) $$\xi'' + p_0(t)\xi = 0,$$

(2.23) $$\zeta'' + q_0(t)\zeta = 0$$

with p_0, q_0 *real-valued and locally integrable. If*

(2.24) $$|q_0 - p_0|^{1/2} \xi \in L^2(I)$$

for every solution ξ *of (2.22), then (2.23) is of limit circle type or of limit point type according as which is the case for equation (2.22). In addition, (2.24) will be satisfied with any solution* ξ *of (2.23).*

Halvorsen's theorem partly generalizes the following theorem, which is also contained in theorem 2.16:

(2.25) THEOREM (Bellman [3]). *Consider on* $I:=[0,\infty)$ *the equations*

(2.22) $$\xi''+p_0(t)\xi=0,$$

(2.23) $$\zeta''+q_0(t)\zeta=0.$$

If

(2.26) $$q_0-p_0\in L^\infty(I)$$

and if for some p,p' *with*

$$1\le p\le 2\le p'\le\infty,\ \frac{1}{p}+\frac{1}{p'}=1$$

all solutions of (2.22) belong to $L^p(I)\cap L^{p'}(I)$, *then all solutions of (2.23) belong to* $L^p(I)\cap L^{p'}(I)$.

Proof. Choosing $p_0^*:=p_0$, we have $L=\frac{d^2}{dt^2}+p_0=L^+$. Therefore (2.17) reduces to

(2.27) $$(q_0-p_0)\xi\eta\in L^1(I)$$

for all solutions ξ,η of (2.22). But this is implied by (2.26) since

$$\xi,\eta\in L^p(I)\cap L^{p'}(I)\subset L^2(I)$$

(which is true even without the assumption $\frac{1}{p}+\frac{1}{p'}=1$). Thus by theorem 2.16

(2.28) $$\zeta=(\xi_1,\xi_2)O(1)\in L^p(I)\cap L^{p'}(I)$$

for any solution ζ of (2.23).

Patula and Wong recently generalized Bellman's theorem partially to L^p-perturbations under the additional assumption that all solutions of the unperturbed equation are bounded:

(2.29) THEOREM (Patula and Wong [13]). *Consider on* $I:=[0,\infty)$

(2.22) $$\xi''+p_0(t)\xi=0,$$

(2.23) $$\zeta''+q_0(t)\zeta=0$$

with

(2.30) $$q_0-p_0\in L^p(I)$$

for some p with $1\leq p\leq\infty$ [4]. Suppose that all solutions of (2.22) belong to $L^2(I)\cap L^\infty(I)$. Then all solutions of (2.23) are in $L^2(I)\cap L^\infty(I)$.

Proof. Since $2\leq\frac{2p}{p-1}\leq\infty$ we have

$$L^2\cap L^\infty\subset L^{\frac{2p}{p-1}}.$$

Furthermore,

$$\frac{1}{p}+\frac{1}{\frac{2p}{p-1}}+\frac{1}{\frac{2p}{p-1}}=1.$$

Therefore (2.30) and $\xi,\eta\in L^2\cap L^\infty$ imply (2.27). Then the first part of (2.28) holds and yields the assertion.

The same argument shows that the following modification of theorem 2.29 is true:

(2.31) THEOREM. If (2.30) holds and if all solutions of (2.22) belong to $L^r(I)\cap L^s(I)$, where

$$0<r\leq\frac{2p}{p-1}\leq s\leq\infty,$$

then all solutions of (2.23) are in $L^r(I)\cap L^s(I)$.

Especially to prove that (2.23) is of limit circle type, it is sufficient to suppose that (2.22) is of limit circle type and that every solution of (2.22) belongs to $L^{\frac{2p}{p-1}}(I)$.

If p_2 is locally absolutely continuous, the operator $L=\frac{d}{dt}(p_2\frac{d}{dt})+p_0$ may be written in the form (2.1) and it is self-

---

[4] for $p=\infty$ put $\frac{p}{p-1}:=1$

adjoint (if $p_2^* := p_2, p_0^* := p_0$). Therefore theorem 2.5 and theorem 2.16 in this case especially imply the following theorems:

(2.32) THEOREM (Bradley [5]). Consider on $I:=(a,\infty)$

(2.33) $$(p_2\xi')' + p_0\xi = 0,$$

(2.34) $$(p_2\zeta')' + p_0\zeta = \Psi(t,\zeta)$$

with $p_2>0, p_0, \Psi$ continuous and

(2.13) $$|\Psi(t,\zeta)| \leq k_0(t) + k_1(t)|\zeta|,$$

where k_0 and k_1 are nonnegative. If for every solution ξ of (2.33)

$$\xi, k_1^{1/2}\xi \in L^2(I) \text{ and } k_0\xi \in L^1(I),$$

then all solutions of (2.34) are in $L^2(I)$.

(2.35) THEOREM (Bradley [5]). If $|q_0 - p_0|^{1/2}\xi$ belongs to $L^2(I)$ for all solutions of (2.33), then $|q_0 - p_0|^{1/2}\zeta$ belongs to $L^2(I)$ for all solutions of

(2.36) $$(p_2\zeta')' + q_0\zeta = 0$$

and (2.33) is of limit circle type iff (2.36) is of limit circle type.

If p_2 is not locally absolutely continuous, (2.33) has to be considered as a quasi-differential equation. Similar to the operators (2.1) and (2.2) every quasi-differential operator (for definition compare [15]) and its adjoint may be transformed to a differential system and its adjoint system, resp.. Therefore, theorem 1.4 and theorem 1.12 imply e.g. results on quasi-differential equations similar to those stated in theorem 2.5 and theorem 2.16. Of the various possible results we mention only the following two, which generalize Bradley's theorems:

(2.37) THEOREM. *Consider on* $I\subset\mathbb{R}$

$$L\xi := \sum_{i=0}^{n} (p_i\xi^{(i)})^{(i)}=0, \tag{2.38}$$

$$L\zeta=\varphi(t,\zeta,\zeta',\ldots,\zeta^{(n)}) \tag{2.39}$$

where $p_0,\ldots,p_{n-1},\frac{1}{p_n}$ *are locally integrable and*

$$|\varphi(t,z_1,\ldots,z_{n+1})|\le k_0(t)+\sum_{i=1}^{n+1} k_i(t)|z_i|$$

with $k_i\,(i=0,\ldots,n+1)$ *nonnegative. Suppose that*

$$k_0\xi, k_j\xi^{(j-1)}\eta, k_{n+1}p_n\xi^{(n)}\eta\in L^1(I) \quad (j=1,\ldots,n)$$

for all solutions ξ,η *of (2.38). Then for any fundamental system* $\xi_1,\ldots,\xi_{2n}$ *of solutions of (2.38) and for any solution* ζ *of (2.39)*

$$\zeta^{(i)}(t)=(\xi_1^{(i)}(t),\ldots,\xi_{2n}^{(i)}(t))O(1) \quad (i=0,\ldots,n).$$

(2.40) THEOREM. *Suppose that for all solutions* ξ,η *of (2.38)*

$$\sum_{j=0}^{n-1}(-1)^{n+j}(p_j-q_j)\xi^{(j)}\eta^{(j)}+\left(\frac{p_n^2}{q_n}-p_n\right)\xi^{(n)}\eta^{(n)}\in L^1(I).$$

Let $\xi_1,\ldots,\xi_{2n}$ *and* $\zeta_1,\ldots,\zeta_{2n}$ *be fundamental systems of solutions of (2.38) and*

$$\sum_{i=0}^{n}(q_i\zeta^{(i)})^{(i)}=0, \tag{2.41}$$

resp.. Then for any solution ζ *of (2.41) and* ξ *of (2.38)*

$$\left.\begin{aligned}\zeta^{(i)}(t)&=(\xi_1^{(i)}(t),\ldots,\xi_{2n}^{(i)}(t))O(1)\\ \xi^{(i)}(t)&=(\zeta_1^{(i)}(t),\ldots,\zeta_{2n}^{(i)}(t))O(1)\end{aligned}\right\}\ \textit{for}\ i=0,\ldots,n-1,$$

$$\zeta^{(n)}(t)=(\xi_1^{(n)}(t),\ldots,\xi_{2n}^{(n)}(t))O\left(\frac{p_n(t)}{q_n(t)}\right),$$

$$\xi^{(n)}(t)=(\zeta_1^{(n)}(t),\ldots,\zeta_{2n}^{(n)}(t))O\left(\frac{q_n(t)}{p_n(t)}\right).$$

The proofs proceed as those of theorems 2.5 and 2.16 except for the transformations:

(2.38) is transformed by

$$(2.42)\qquad x:=\begin{pmatrix}x_1\\ x_2\end{pmatrix}, x_1:=(\xi^{(j-1)})_{j=1}^{n}, x_2=\Big(\sum_{i=n-j+1}^{n}(p_i\xi^{(i)})^{(i-n+j-1)}\Big)_{j=1}^{n}$$

to

$$x'=Ax$$

with

$$A=\left(\begin{array}{cccc|cccc} 0 & 1 & & & & & & \\ & 0 & \ddots & & & & \bigcirc & \\ & & \ddots & 1 & & & & \\ & & & 0 & \frac{1}{p_n} & & & \\ \hline & & & -p_{n-1} & 0 & 1 & & \\ & & \cdot^{\cdot^{\cdot}} & & & 0 & \ddots & \\ & & & & & & \ddots & 1 \\ -p_o & & & & & & & 0 \end{array}\right)$$

and – as its (real) adjoint – by

$$y:=\begin{pmatrix}y_1\\ y_2\end{pmatrix}, y_1:=\Big((-1)^j\sum_{i=j}^{n}(p_i\xi^{(i)})^{(i-j)}\Big)_{j=1}^{n}, y_2:=((-1)^{n-j}\xi^{(n-j)})_{j=1}^{n}$$

to

$$y'=-A^*y.$$

In the case of theorem 2.37, (2.39) is transformed by (2.42) to

$$z'=Az+\varphi e_{2n};$$

in the case of theorem 2.40 the differential equation (2.41) is transformed by

$$z:=\begin{pmatrix}z_1\\ z_2\end{pmatrix}, z_1:=(\zeta^{(j-1)})_{j=1}^{n}, z_2:=\Big(\sum_{i=n-j+1}^{n}(q_i\zeta^{(i)})^{(i-n+j-1)}\Big)_{j=1}^{n}$$

to

$$z'=A_1z$$

with

$$A_1:=\left(\begin{array}{cccc|cccc} 0 & 1 & & & & & & \\ & 0 & \ddots & & & & \bigcirc & \\ & & \ddots & 1 & & & & \\ & & & 0 & \frac{1}{q_n} & & & \\ \hline & & & -q_{n-1} & 0 & 1 & & \\ & & \cdot^{\cdot^{\cdot}} & & & \ddots & \ddots & \\ & & & & & & \ddots & 1 \\ -q_o & & & & & & & 0 \end{array}\right).$$

The rest of the proofs is then easily carried out.

Finally, applying theorem 1.16 to the systems arrising from (2.1) and (2.18), we get the following theorem which is a generalized and strenghtened version of a theorem due to Bellman [2]

(2.43) THEOREM. *Let the real part of* $\int_{t_0}^{t} \frac{p_{n-1}(\tau)}{p_n(\tau)}\, d\tau$ *be bounded above and let*

$$\frac{q_j}{q_n} - \frac{p_j}{p_n} \in L^1(I) \qquad (j=0,\ldots,n-1).$$

If $\xi^{(i)}$ *is bounded for all solutions* ξ *of*

$$\sum_{j=0}^{n} p_j \xi^{(j)} = 0 \tag{2.44}$$

and for any $i=0,\ldots,n-1$, *then for any solution* ζ *of*

$$\sum_{j=0}^{n} q_j \zeta^{(j)} = 0 \tag{2.45}$$

and for any fundamental system $\xi_1,\ldots,\xi_n$ *of solutions of (2.44)*

$$\zeta^{(i)}(t) = (\xi_1^{(i)}(t),\ldots,\xi_n^{(i)}(t))O(1) \qquad (i=0,\ldots,n-1).$$

Especially, $\zeta^{(i)}(i=0,\ldots,n-1)$ *is bounded for any solution of (2.45).*

3. ON THE LIMIT CIRCLE CASE FOR WEIGHTED EIGENVALUE PROBLEMS.

Let the differential operator

$$l(\xi) := \sum_{i=0}^{n} a_i \xi^{(i)} \tag{3.1}$$

be formally selfadjoint on $I \subset \mathbb{R}$ with $a_n(t) \neq 0$ $(t \in I)$, $a_i \in C^i(I)$ and let w be a nonnegative continuous function on I. Then the (possibly singular) eigenvalue problem

$$l(\xi) = \lambda w \xi, \tag{3.2}$$

considered in the space

$$L^2(w,I):=\{\xi/w^{1/2}\ \xi\in L^2(I)\},$$

is said to be in the <u>limit circle case</u> if for every $\lambda\in\mathbb{C}$ every solution of (3.2) belongs to $L^2(w,I)$. Recently Walker proved the following

<u>(3.3) THEOREM (Walker [16]). If for some $\lambda_o\in\mathbb{C}$ all solutions of</u>

(3.4) $$l(\xi)=\lambda_o w\xi$$

<u>and</u>

(3.5) $$l(\xi)=\overline{\lambda}_o w\xi$$

<u>belong to $L^2(w,I)$, then (3.2) is in the limit circle case.</u>

For $w\equiv 1$ and n even Everitt [9] proved this theorem assuming only that for some $\lambda_o\in\mathbb{C}$ all solutions of (3.4) are square-integrable. The same result follows for arbitrary $n\geq 2$ and for an arbitrary weight function w from theorem 9.11.2 of [1] by transforming (3.2) to an n-th order system using the transformations given in [14]. As stated above, the theorem is a simple consequence of theorem 2.16 applied to

$$L:=l-\lambda_o w,\quad L^+:=l-\overline{\lambda}_o w$$

(where L^+ denotes the complex adjoint)

$$p_o:=a_o-\lambda_o w, q_o:=a_o-\lambda w, p_j:=q_j:=a_j(j=1,\ldots,n).$$

Theorem 3.3 may easily be generalized to eigenvalue problems of the form

(3.6) $$m(\xi)=\lambda n(\xi)$$

with formally selfadjoint differential operators m,n such that n is positive on a suitable function space.

More generally, we get an analogous result for singular right-definite $\mathfrak{S}$-hermitian eigenvalue problems. Such eigenvalue problems (for definition and properties compare e.g. [12]) may be reduced to systems of the form

$$(3.7) \qquad F_1x'+F_2x=\lambda Gx$$

with (n,n)-matrix-valued functions F_1, F_2, G, defined and continuous on some interval $I\subset\mathbb{R}$ such that $F_1(t)$ is nonsingular for $t\in I$. A right-definite S-hermitian eigenvalue problem (3.7) especially has the following properties:

There exists a continuously differentiable (n,n)-matrix-valued function H and a positive-semidefinite- valued continuous function W such that for every $\lambda\in\mathbb{C}$

$$(3.8) \qquad H'=HF_1^{-1}(F_2-\lambda G)+[F_1^{-1}(F_2-\bar{\lambda}G)]^*H,$$

$$(3.9) \qquad G=WF_1^{*-1}H^*.$$

Let K denote the continuous and positive-semidefinite square-root of W and define

$$U:=KF_1^{*-1}H^*,\quad L^2(U,I):=\{x\,/\,Ux\in L^2(I)\}.$$

Then the eigenvalue problem (3.7) considered in the space $L^2(U,I)$ is said to be in the <u>limit circle case</u> if for every $\lambda\in\mathbb{C}$ every solution of (3.7) belongs to $L^2(U,I)$.

Then the following theorem holds:

<u>(3.10) THEOREM. If for some</u> $\lambda_0\in\mathbb{C}$ <u>all solutions of</u>

$$(3.11) \qquad F_1x'+F_2x=\lambda_0Gx$$

<u>and of</u>

$$(3.12) \qquad F_1x'+F_2x=\bar{\lambda}_0Gx$$

<u>belong to</u> $L^2(U,I)$, <u>then (3.7) is in the limit circle case</u>.

<u>Proof.</u> Let $\lambda\in\mathbb{C}$ be fixed. Defining

$$A:=-F_1^{-1}(F_2-\lambda_0G),\quad \tilde{A}:=(\lambda-\lambda_0)F_1^{-1}G$$

we have

$$A+\tilde{A}= -F_1^{-1}(F_2-\lambda G).$$

Therefore, (3.11) and (3.7) are equivalent to

(3.13) $$x'=Ax$$

and

(3.14) $$x'=(A+\tilde{A})x,$$

resp.. Furthermore, by (3.9) and the definition of K and U we obtain

$$\tilde{A}=(\lambda-\lambda_o)F_1^{-1}KKF_1^{*-1}H^*=(\lambda-\lambda_o)H^{-1}U^*U.$$

Now let x and y be solutions of (3.13) and

$$y'= -A^*y$$

resp.. Then $x\in L^2(U,I)$ by assumption, and using (3.8) it is easily seen that $H^{*-1}y$ is a solution of (3.12) and therefore belongs to $L^2(U,I)$, too. Thus

$$y^*\tilde{A}x=(\lambda-\lambda_o)(UH^{*-1}y)^*Ux\in L^1(I).$$

Now theorem 1.12 implies that all solutions of (3.14), i.e., all solutions of (3.7) belong to $L^2(U,I)$.

Since the eigenvalue problem (3.6) may be transformed to a right-definite S-hermitian eigenvalue problem, theorem 3.10 implies an analogous result for (3.6).

4. GENERALIZATIONS

The preceding results may be generalized in various directions. E.g., if $I=[t_o,b)$, in the case of a linear perturbation we may replace the integrability condition (1.7) by the assumption that the integral of the largest eigenvalue of the real part of $X^{-1}\tilde{A}X$ is bounded above. More generally, we get

(4.1) THEOREM. *Let X denote a fundamental matrix of (1.2) and z any solution of (1.13). Then*

$$z(t)=X(t)O\left(e^{\operatorname{sign}(t-t_0)\int_{t_0}^{t}\mu(\operatorname{sign}(t-t_0)X^{-1}(\tau)\tilde{A}(\tau)X(\tau))d\tau}\right)\quad (t\in I).$$

Here

$$\mu(M):=\lim_{h\to+0}\frac{|E+hM|-1}{h}\,.$$

For properties of μ compare [8],p.41. Especially $|\mu(M)|\leq|M|$, and if the vector norm is taken to be the Euclidean norm, then $\mu(M)$ is the largest eigenvalue of the real part of M. Theorem 4.1 is an easy application of a theorem due to Lozinskiĭ (compare [8],p.58) to

$$u'=X^{-1}\tilde{A}Xu.$$

As an example we consider for $a\neq 0$

$$z'=\begin{pmatrix}0 & \sqrt{a}\\ -\frac{q}{\sqrt{a}} & 0\end{pmatrix}z,$$

which arrises from

(4.2) $$\zeta''+q\zeta=0$$

by the transformation

$$z:=\begin{pmatrix}\sqrt{a}\ \zeta\\ \zeta'\end{pmatrix}.$$

Choosing $A:=0$, $X:=E$,

$$\tilde{A}:=\begin{pmatrix}0 & \sqrt{a}\\ -\frac{q}{\sqrt{a}} & 0\end{pmatrix}$$

and the Euclidean norm, theorem 4.1 implies

$$\zeta^{(i)}(t)=O(\exp[\tfrac{1}{2}|a|^{-\frac{1}{2}}|\int_{t_0}^{t}||a|-q(\tau)|\,d\tau|])\qquad (i=0,1)$$

for all solutions ζ of (4.2). For $a>0$ and $i=0$ this result is theorem I of [11].

Another application to differential equations gives the following theorem, which is an improved version of [10], theorem 2:

(4.3) THEOREM. Let p,q be realvalued and locally integrable, and denote by ξ_1, ξ_2 a real fundamental system of

$$\xi''+p\xi=0 \tag{4.4}$$

such that $\xi_1\xi_2' - \xi_1'\xi_2=1$. Then (4.2) (and (4.4)) are of limit circle type provided

$$\sqrt{\xi_1^2+\xi_2^2}\; e^{\frac{1}{2}\left|\int_{t_0}^{t}|p-q|\,(\xi_1^2+\xi_2^2)\right|} \in L^2(I).$$

Proof. Let

$$A:=\begin{pmatrix}0 & 1\\ -p & 0\end{pmatrix},\quad \tilde{A}:=(p-q)\begin{pmatrix}0 & 0\\ 1 & 0\end{pmatrix}.$$

Then (4.4) is equivalent to

$$x'=Ax \text{ with } x=\begin{pmatrix}\xi\\ \xi'\end{pmatrix}, \tag{4.5}$$

(4.2) is equivalent to

$$z'=(A+\tilde{A})z \text{ with } z=\begin{pmatrix}\zeta\\ \zeta'\end{pmatrix}. \tag{4.6}$$

Furthermore,

$$X:=\begin{pmatrix}\xi_1 & \xi_2\\ \xi_1' & \xi_2'\end{pmatrix}$$

is a fundamental matrix of (4.5) and the real part of $X^{-1}\tilde{A}X$ has eigenvalues

$$\pm\frac{1}{2}|p-q|\,(\xi_1^2+\xi_2^2).$$

Thus, by theorem 4.1, for any solution $z=\begin{pmatrix}\zeta\\ \zeta'\end{pmatrix}$ of (4.6)

$$z(t)=X(t)O\left(e^{\frac{1}{2}\left|\int_{t_0}^{t}|p-q|\,(\xi_1^2+\xi_2^2)\right|}\right)$$

implying $\zeta\in L^2(I)$.

Theorem 1.4 may be generalized to

(4.7) THEOREM. Let X be a fundamental matrix of $x'=Ax$ and let X_1, X_2 be locally absolutely continuous and such that $X=X_1X_2$. For fixed $\sigma \in [0,1]$ let

$$f(t,z)=B(t)g(t,z)+C(t)|D(t)z|^{\sigma}h(t,z),$$

where B is locally integrable and $|g(t,z)|, |h(t,z)| \leq 1$. If

$$|X_2(t)X^{-1}(\tau)C(\tau)||D(\tau)X_1(\tau)|^{\sigma} \leq k_1(t)k_2(\tau) \quad (\tau \in [t_0,t]^{5)}), \tag{4.8}$$

where k_1, k_2 are locally integrable nonnegative functions, then for any solution z of

$$z'=A(t)z+f(t,z) \tag{4.9}$$

we have

a)

$$z(t)=X_1(t)O(|X_2(t)|+|\int_{t_0}^{t} |X_2(t)X^{-1}(\tau)B(\tau)|d\tau| + +(1-\sigma)|\int_{t_0}^{t} k_1(t)k_2(\tau)d\tau|+\sigma|\int_{t_0}^{t} k_1(t)k_2(\tau)[|X_2(\tau)|+ +|\int_{t_0}^{\tau} |X_2(\tau)X^{-1}(s)B(s)|ds|+(1-\sigma)|\int_{t_0}^{\tau} k_1(\tau)k_2(s)ds|] \times e^{|\int_{\tau}^{t} k_1(s)k_2(s)ds|} d\tau|). \tag{4.10}$$

b) If for some $\gamma \geq 0$

$$|X_2(t)X^{-1}(\tau)| \leq \gamma \quad (\tau \in [t_0,t]), \tag{4.11}$$

$$|X_1^{-1}C||DX_1|^{\sigma} \in L^1(I), \tag{4.12}$$

$$\int_{t_0}^{t} |X_2(t)X^{-1}(\tau)B(\tau)|d\tau \in L^{\infty}(I), \tag{4.13}$$

5) for $t < t_0$, $[t_0,t]:=[t,t_0]$

then

(4.14) $$z(t)=X_1(t)O(1) \qquad (t\in I).$$

Proof. Using $a^\sigma \leq 1-\sigma+\sigma a$ for $a\geq 0$, we obtain with $u:=X^{-1}z$ similarly to the proof of theorem 1.4

$$|X_2(t)u(t)|\leq|X_2(t)u(t_0)|+\left|\int_{t_0}^{t}|X_2(t)X^{-1}(\tau)B(\tau)|\,d\tau\right|+$$

$$+(1-\sigma)\left|\int_{t_0}^{t}k_1(t)k_2(\tau)d\tau\right|+\sigma\left|\int_{t_0}^{t}k_1(t)k_2(\tau)|X_2(\tau)u(\tau)|\,d\tau\right|$$

and by a modification of the Gronwall inequality (e.g. [7] with a product kernel K) this yields (4.10).

To prove b) choose $k_1:=\gamma, k_2:=|X_1^{-1}C||DX_1|^\sigma$. Then (4.8) holds by (4.11) and (4.12), X_2 is bounded by (4.11), k_1 is bounded and k_2 is integrable by assumption. Thus (4.14) follows from (4.10).

As special cases we mention

1) Choosing $X_1:=X, X_2:=E$ and $\sigma=1$, theorem 4.7 b) reduces to theorem 1.4. In this case, partial integration of (4.10) with $k_1:=1$, $k_2:=|X^{-1}C||DX|$ yields

$$z(t)=X(t)O\Big(e^{\left|\int_{t_0}^{t}|X^{-1}(\tau)C(\tau)||D(\tau)X(\tau)|\,d\tau\right|}+$$

$$+\left|\int_{t_0}^{t}|X^{-1}(\tau)B(\tau)|\,e^{\left|\int_{\tau}^{t}|X^{-1}(s)C(s)||D(s)X(s)|\,ds\right|}d\tau\right|\Big).$$

2) Choosing $X_2:=X, X_1:=E$ theorem 4.7 b) implies

(4.15) THEOREM. If for some $\gamma\geq 0$, $\sigma\in[0,1]$

(4.16) $$|X(t)X^{-1}(\tau)|\leq\gamma \qquad (\tau\in[t_0,t]),$$

(4.17) $$|C||D|^\sigma, |B|\in L^1(I),$$

then all solutions of (4.9) are bounded.

(4.16) implies that X is bounded. By Floquet's theory the converse is true if A is periodic:

(4.18) REMARK. *Let A be periodic and X bounded. Then for some* $\gamma \geq 0$ *(4.16) is valid.*

Theorem 4.15 may be considered as a generalization of the following

(4.19) THEOREM (Cesari [6]). *Suppose that*

$$f_i - a_i \in L^1[t_o,\infty),\, f_i(t) \to a_i\ (t\to\infty),\ (i=0,\dots,n-1)$$

and that all solutions of

$$\xi^{(n)} + \sum_{i=0}^{n-1} a_i \xi^{(i)} = 0 \tag{4.20}$$

together with their first n-1 derivatives are bounded. Then all solutions of

$$\zeta^{(n)} + \sum_{i=0}^{n-1} f_i(t)\zeta^{(i)} = 0 \tag{4.21}$$

together with their first n-1 derivatives are bounded.

Proof. Choosing $B:=0, C:=e_n(a_0-f_0,\dots,a_{n-1}-f_{n-1}), D:=E, h:=|z|^{-1}z, \sigma:=1,$

$$A := \begin{pmatrix} 0 & 1 & & \\ & \ddots & \ddots & \\ & & \ddots & 1 \\ -a_0, & \dots & , & -a_{n-1} \end{pmatrix},$$

(4.20), (4.21) are equivalent to

$$x' = Ax,\quad x = (\xi^{(i-1)}) \tag{4.22}$$

and

$$z' = Az + f(t,z),\quad z = (\zeta^{(i-1)}),$$

resp.. By assumption any fundamental matrix X of (4.22) is bounded. Since A is constant, remark 4.18 implies (4.16). Finally the integrability of C shows that (4.17) holds. Thus the boundedness of $\zeta^{(i)}(i=0,\dots,n-1)$ follows from theorem 4.15.

Obviously this proof does not need the convergence assumption on the f_i. Furthermore, the a_i may be allowed to be periodic functions. It may be mentioned that the original proof of theorem 4.19 took about 16 pages. The proof given here also seems to be more simple than that indicated in [2]. Finally we remark that theorem 4.15 implies the second part of theorem 1.16 and the theorem of [2].

Recently Wong announced the following

(4.23) THEOREM (Wong [17]). *Suppose that all solutions of*

$$L\xi := \sum_{i=0}^{n} (p_i \xi^{(i)})^{(i)} = 0 \tag{4.24}$$

belong to $L^2[0,\infty)\cap L^\infty[0,\infty)$ *and that for some* $\sigma\in[0,1]$

$$|\varphi(t,\zeta)| \leq \lambda(t)|\zeta|^\sigma$$

with $\lambda\in L^p[0,\infty)$ *for some* $p, 1 \leq p \leq \frac{2}{1-\sigma}$. *Then, all solutions of*

$$L\zeta = \varphi(t,\zeta) \tag{4.25}$$

belong to $L^2[0,\infty)\cap L^\infty[0,\infty)$.

This theorem is also a corollary to theorem 4.7 b): Transforming (4.24) and (4.25) to systems as in the proof of theorem 2.37, choosing $X_1 := X, X_2 := E$ and using a similar argument as remark 1.9 the assumptions of theorem 4.7 b) are seen to be fulfilled if

$$\lambda|\xi|^\sigma\eta \in L^1[0,\infty) \tag{4.26}$$

for all solutions ξ,η of (4.24). To prove (4.26) we remark that by

$$2 \leq \frac{\sigma+1}{1-\frac{1}{p}} \leq \infty$$

$$L^2\cap L^\infty \subset L^{\frac{\sigma+1}{1-1/p}}.$$

Thus

$$\lambda \in L^p, |\xi|^\sigma \in L^{\frac{1+1/\sigma}{1-1/p}}, \eta \in L^{\frac{\sigma+1}{1-1/p}}.$$

Therefore

$$\frac{1}{p} + \frac{1-1/p}{1+1/\sigma} + \frac{1-1/p}{1+\sigma} = 1$$

implies (4.26). Now the assertion follows from theorem 4.7 b).

This proof shows that (4.26) holds if we only suppose that all solutions of (4.24) belong to $L^{\frac{\sigma+1}{1-1/p}}$ (with $\sigma \in [0,1]$ and arbitrary $p \geq 1$). Then by theorem 4.7 b)

$$\zeta^{(i)}(t) = (\xi_1^{(i)}(t), \ldots, \xi_{2n}^{(i)}(t))o(1) \qquad (i=0,\ldots,n)$$

for any solution ζ of (4.25) and any fundamental system $\xi_1, \ldots, \xi_{2n}$ of (4.24).

REFERENCES

[1] Atkinson, F.V.: Discrete and continuous boundary problems. Academic Press, New York 1964.

[2] Bellman, R.: The stability of solutions of linear differential equations. Duke Math.J. 10 (1943), 643-647.

[3] Bellman, R.: A stability property of solutions of linear differential equations. Duke Math.J. 11 (1944), 513-516.

[4] Bellman, R.: Stability theory of differential equations. Dover, New York 1953.

[5] Bradley, J.S.: Comparison theorems for the square integrability of solutions of $(r(t)y')' + q(t)y = f(t,y)$. Glasgow Math.J. 13 (1972), 75-79.

[6] Cesari, L.: Sulla stabilità delle soluzioni delle equazioni differenziali lineari. Ann. Scuola Norm. Sup. Pisa (2) 8 (1939), 131-148.

[7] Chu, S.C. and F.T. Metcalf: On Gronwall's inequality. Proc. AMS 18 (1967), 439-440.

[8] Coppel, W.A.: Stability and asymptotic behavior of differen-

tial equations. Heath, Boston 1965.

[9] Everitt, W.N.: Singular differential equations I: The even order case. Math. Ann. 156 (1964), 9-24.

[10] Halvorsen,S.: On the quadratic integrability of solutions of $d^2x/dt^2+f(t)x=0$. Math. Scand. 14 (1964), 111-119.

[11] Levinson, N.: The growth of the solutions of a differential equation. Duke Math. J. 8 (1941), 1-10.

[12] Niessen, H.D.: Singuläre S-hermitesche Rand-Eigenwertprobleme. manuscripta math. 3 (1970), 35-68.

[13] Patula, W.T. and J.S.W. Wong: An L^p-analogue of the Weyl alternative. Math. Ann. 197 (1972), 9-28.

[14] Schneider, A.: Zur Einordnung selbstadjungierter Rand- Eigenwertprobleme bei gewöhnlichen Differentialgleichungen in die Theorie S-hermitescher Rand-Eigenwertprobleme. Math. Ann. 178 (1968), 277-294.

[15] Shin, D.: Existence theorems for the quasi-differential equation of the n-th order. C.R. Acad. Sci. URSS 18 (1938), 515-518.

[16] Walker, Ph. W.: Weighted singular differential operators in the limit circle case. J. London Math. Soc.(2), 4 (1972), 741-744.

[17] Wong, J.S.W.: Square integrable solutions of L^p perturbations of second order linear differential equations. Springer lecture notes no. 145 (1974), 282-292.

[18] Zettl, A.: Square integrable solutions of Ly=f(t,y). Proc. AMS 26 (1970), 635-639.

W. Eckhaus (ed.), New Developments in Differential Equations

ON CERTAIN ORDINARY DIFFERENTIAL EXPRESSIONS AND ASSOCIATED INTEGRAL INEQUALITIES

W N Everitt and M Giertz

1. This paper is concerned with inequalities of the form

$$\alpha\|(pf')'\|^2 + \beta\|(pq)^{1/2}f'\|^2 + \gamma\|qf\|^2 \leq \|M[f]\|^2 + A\|f\|^2 \tag{1.1}$$

$$\|p^{1/2}f'\|^2 + \|q^{1/2}f\|^2 \leq \epsilon\|M[f]\|^2 + B\|f\|^2 \tag{1.2}$$

for functions f in certain linear manifolds of the integrable-square function space $L^2(a,\infty)$ with the usual norm $\|\cdot\|$. Here α,β,γ, A,B and ϵ (with ϵ to be taken as 'small') are non-negative real numbers, and the differential expression M is defined, in terms of the positive valued coefficients p and q, by

$$M[f] = -(pf')' + qf \quad \text{on} \quad [a,\infty) \qquad (' \equiv d/dx). \tag{1.3}$$

In [7, Chapter VI, Sections 6 and 8] Goldberg has given certain _a priori_ estimates of the form (1.2) for differential expressions of arbitrary order with bounded coefficients. The inequalities considered in this paper give an extension of this type of estimate to second-order expressions with, in general, unbounded coefficients.

The inequality (1.1) is considered by Everitt and Giertz in [6]. The results given here avoid one of the difficulties met in determining conditions on the coefficients p and q for [6; Theorem 3] to hold.

As in [6] _separation_ results for the differential expression M follow from inequalities of the form (1.1); for the definition of separation see [4; Section 1] or [6; Section 6]. Additionally a complete separation result is obtained here; for this concept see the definition in [5; Section 1]

Inequalities of the form (1.1) and (1.2) also yield results in the theory of relatively bounded perturbations of the differential operators generated by M in $L^2(a,\infty)$. For results in this direction see [6; Section 9], which also depend on the work of Kato, see [8; Chapter V, Section 4].

There are also connections with inequalities considered by Everitt in [3].

In section 2 of this paper there is a statement of the results to be proved; following sections contain the proof of these results together with some comments on their consequences. There is a list of references.

2. R and C denote the real and complex number fields respectively. For $a \in R$ the half-line $[a,\infty)$ is closed at a and open at ∞. AC denotes absolute continuity and L Lebesgue integration; 'loc' for local, i.e. a property satisfied on all compact subintervals of $[a,\infty)$. $L^2(a,\infty)$ denotes the classical Lebesgue complex function space, which is also indentified with the Hilbert function space of equivalence classes.

Let the coefficients p and q satisfy the following basic conditions : $p, q : [a,\infty) \to R$ and

$$p \in AC_{loc}[a,\infty),\ p' \in L^2_{loc}[a,\infty)\ \underline{\text{and}}\ p(x) > 0 \quad (x \in [a,\infty)) \tag{2.1}$$

$$q \in AC_{loc}[a,\infty)\ \underline{\text{and}}\ q(x) > 0 \quad (x \in [a,\infty)). \tag{2.2}$$

These conditions on p and q imply that the differential expression M, given by (1.3), is regular at all points of $[a,\infty)$ but has a singular point at ∞; see [9; Section 15.1]. Since q is bounded below on $[a,\infty)$ it is known that M is in the limit-point condition at the singular point ∞; see [9; Sections 17.5 and 23.6] and [2].

Following the notation in [6; Section 3] we define the linear manifold $D_1 \equiv D_1(p,q)$ of $L^2(a,\infty)$ by

$$D_1 = \{f \in L^2(a,\infty) : f' \in AC_{loc}[a,\infty) \ \underline{\text{and}} \ M[f] \in L^2(a,\infty)\}. \tag{2.3}$$

D_1 is the domain of the maximal operator T_1 generated by M in $L^2(a,\infty)$ and defined by $T_1 f = M[f]$ $(f \in D_1)$; it is known that T_1 is closed but is not symmetric in $L^2(a,\infty)$; for these results see [9; Sections 17.4 and 5], also the remarks in [6; Section 3]. The basic operator theoretic definitions are given in [1; Sections 39 and 41].

We say that M is <u>separated in</u> $L^2(a,\infty)$ if (see [4; Section 1] and [6; Section 6])

$$qf \in L^2(a,\infty) \ \underline{\text{for all}} \ f \in D_1; \tag{2.4}$$

and <u>completely separated in</u> $L^2(a,\infty)$ if (see [5; Section 1])

$$pf'', p'f', qf \ \underline{\text{are all in}} \ L^2(a,\infty) \ \underline{\text{for all}} \ f \in D_1. \tag{2.5}$$

Note that both (2.4) and (2.5) are conditions to be satisfied only at ∞ since the basic conditions (2.1) and (2.2) on the coefficients p and q imply that all terms in (2.4) and (2.5) are in $L^2_{loc}[a,\infty)$.

We may now state (recall that $\|\cdot\|$ denotes the usual norm in $L^2(a,\infty)$)

<u>Theorem 1</u> <u>Let the coefficients p and q satisfy the basic conditions</u> (2.1) <u>and</u> (2.2); <u>let additionally</u> p <u>and</u> q <u>satisfy the conditions</u>;

let $K \in (0,\infty)$ *be given; let* η *be chosen so that* $0 < \eta < \min(1, 4C_1^2 q(a))$ *where* $C_1 = \max(7, 3K)$

(a) *let*

$$\{p(x)\}^{1/2}|q'(x)| \le (1 - \eta)\{q(x)\}^{3/2} \quad (\textit{almost all } x \in [a,\infty)) \tag{2.6}$$

(b) let

$$|p'(x)| \le 2K^2\{p(x)\}^{1/2}q(x) \quad (\textit{almost all } x \in [a,\infty)) \tag{2.7}$$

then

$$(1) \quad (pf')', \{pq\}^{1/2}f', qf \ \textit{are all in} \ L^2(a,\infty) \ \textit{for all} \ f \in D_1 \tag{2.8}$$

and (2) *for all* $\epsilon \in (0,\eta)$ *the following inequality is valid*

$$(1 - \epsilon)||(pf')'||^2 + (1 + \eta - \epsilon)||\{pq\}^{1/2}f||^2 + (\eta - \epsilon)||qf||^2$$
$$\le ||M[f]||^2 + A\epsilon^{-5}||f||^2 \quad (f \in D_1) \tag{2.9}$$

where the positive number A *depends only on the coefficients* p *and* q; A *is given explicity in* (3.17) *below*.

Proof This is given in the sections which follow.

Notes The explicit dependence of the number A is given in the sections devoted to the proof of (2.9).

The inequality (2.9) has some similarity with the inequality in [6; (8.3) of Theorem 3] but the condition (2.7) above avoids the unsatisfactory nature of the condition [6; (8.2)] depending as it does on the verification of another inequality.

As in [6; Theorem 3] the control condition (2.6) is necessary to the proof of (2.9); it prevents too much oscillation in the coefficients p and q in the neighbourhood of ∞.

Corollary 1 Under the conditions of Theorem 1 the differential expression M is separated in $L^2(a,\infty)$.

Proof It follows at once from the inequality (2.9) that condition (2.4) is satisfied.

Corollary 2 If in addition to all the conditions of Theorem 1 the coefficients p and q satisfy

(3) for some $L \in (0,\infty)$

$$|p'(x)| \leq L\{p(x)q(x)\}^{1/2} \quad (\text{almost all } x \in [a,\infty)) \tag{2.10}$$

then M is completely separated in $L^2(a,\infty)$).

Proof It follows from the inequality (2.9) that $(pf')' \in L^2(a,\infty)$ $(f \in D_1)$. Now $(pf')' = pf'' + p'f'$ and from (2.10) we obtain $|p'f'| \leq L\{pq\}^{1/2}f' \in L^2(a,\infty)$ on using (2.9) again and this for all $f \in D_1$. It now follows that condition (2.5) is satisfied.

Note that in general the conditions (2.7) and (2.10) are independent of each other; however if the coefficient q has a positive lower bound on $[a,\infty)$ then (2.10) implies (2.7).

Finally we have

Theorem 2 Under the basic conditions (2.1) and (2.2) on the coefficients p and q

$$p^{1/2}f' \text{ and } q^{1/2}f \text{ are in } L^2(a,\infty) \text{ for all } f \in D_1; \tag{2.11}$$

under the additional conditions (2.6) and (2.7) of Theorem 1 the following inequality is valid for all $\epsilon \in (0,1)$

$$\| p^{1/2}f' \|^2 + \| q^{1/2}f \|^2 \le \epsilon \| M[f] \|^2 + B\epsilon^{-5} \| f \|^2 \quad (f \in D_1) \tag{2.12}$$

where the positive number B depends only on the coefficients p and q.

Proof This is given in the sections which follow.

Note that the inequality (2.12) should be compared with the a priori estimates given by Goldberg in [7; Chapter VI, Sections 6 and 8]; see in particular Theorem VI 8.1 However in (2.12) the coefficients are, in general, unbounded on $[a,\infty)$.

3. In this section we give the proof of Theorem 1.

The linear manifold D_1 is defined in (2.3). Now define $D_{1,0} \equiv D_{1,0}(p,q)$ as the collection of all f in D_1 which vanish in some neighbourhood, which may change with f, of ∞, i.e.

$$D_{1,0} = \{f \in D_1 : \text{for some } X \equiv X(f) > a,\ f(x) = 0 \quad (x \in [X,\infty)). \tag{3.1}$$

Note that in defining $D_{1,0}$ no restriction is placed on the values taken by f at the regular end-point a.

The reason for the introduction of $D_{1,0}$ is as follows. The inequality (2.9) is first established on $D_{1,0}$. From known results in the theory of differential operators it may be claimed that given any $f \in D_1$ there is a sequence $\{f_n : f_n \in D_{1,0}$ for $n = 1,2,3,\ldots\}$ such that the sequences $\{f_n\}$ and $\{M[f_n]\}$ are convergent in $L^2(a,\infty)$ to f and M[f] respectively. Furthermore the terms on the left-hand side of (2.9), with f replaced by f_n, also converge in $L^2(a,\infty)$ to the corresponding terms with f in D_1. In this way the inequality is established in D_1.

This closure argument leans heavily on the fact that the maximal operator T_1 introduced in the previous section, following (2.3), is closed. In fact if $S : D_{1,0} \to L^2(a,\infty)$ is defined by $Sf = M[f]$ $(f \in D_{1,0})$ then S is closeable in $L^2(a,\infty)$ and the closure $\tilde{S} = T_1$. This result depends on no restriction being placed on the values of $f \in D_{1,0}$ at the regular end-point a (see above) and the property of M being limit-point at ∞ (see the previous section following (2.2)). Some additional analytical details may be found in [6; Sections 4 and 8].

We now prove that (2.9) holds for all $f \in D_{1,0}$. It is clearly sufficient to prove the result for real-valued f in $D_{1,0}$ since then there is an immediate extension to complex-valued f in $D_{1,0}$.

For the remainder of this section we take f to be real-valued and in $D_{1,0}$; since then f vanishes in some neighbourhood of ∞ there are no convergence problems in the integrals concerned.

As in [6; Section 8] we have the following identity obtained on integration by parts (we recall from (2.1) and (2.2) that p and q are positive on $[a,\infty)$)

$$\int_a^\infty qf(pf')' = -(qfpf')(a) - \int_a^\infty pqf'^2 - \int_a^\infty pq'ff'$$

$$\leq (pq|ff'|)(a) - \int_a^\infty pqf'^2 + \int_a^\infty p|q'ff'|. \tag{3.2}$$

From (2.6)

$$2\int_a^\infty p|q'ff'| \le 2(1-\eta)\int_a^\infty p^{1/2}q^{3/2}|ff'|$$

$$\le 2(1-\eta)\left\{\int_a^\infty pqf'^2 \int_a^\infty p^2f^2\right\}^{1/2}$$

$$\le (1-\eta)\int_a^\infty pqf'^2 + (1-\eta)\int_a^\infty q^2f^2.$$

Thus (3.2) may be rewritten in the form

$$(1+\eta)\int_a^\infty pqf'^2 + \eta\int_a^\infty q^2f^2 \le \int_a^\infty q^2f^2 - 2\int_a^\infty qf(pf')'$$

$$+ 2(pq|ff'|)(a). \qquad (3.3)$$

We now find an estimate for the expression $(pq|ff'|)(a)$ in terms of $\|M[f]\|^2$ and $\|f\|^2$; it is for this reason that we have to have available the condition (2.7) on p and q.

Note that we cannot take $q(a) = 0$ to avoid this estimate for the last term of (3.3), unless we take $q(x) = 0$ $(x \in [a,\infty))$. It was established in [6; Section 10] that whenever a condition of the form of (2.6) holds then either q is positive or identically zero on $[a,\infty)$. It is for this reason that the condition $q(x) > 0$ $(x \in [a,\infty))$ forms part of (2.2).

The main difficulty in establishing inequalities of the kind considered in this paper is to allow for the effect of having no boundary condition on the elements of D_1 at the regular end-point a. This difficulty also occurs, for example, in consideration of an inequality of the form

$$\int_a^\infty \{p|f'|^2 + q|f|^2\} \le K\left\{\int_a^\infty |M[f]|^2 \int_a^\infty |f|^2\right\}^{1/2} \quad (f \in D_1) \tag{3.4}$$

which is closely related to the inequality (2.12) of this paper. This last inequality is discussed in detail in [3]; see in particular the remarks in [3; Sections 2, 3 and 4].

In the analysis which follows we make use of the following inequalities

$$2ab \le (ta)^2 + (b/t)^2 \tag{3.5}$$

and

$$\sum_{k=1}^n a_k^2 \le \left(\sum_{k=1}^n a_k\right)^2 \le n\sum_{k=1}^n a_k^2 \tag{3.6}$$

valid for all positive a, b, t and a_k $(k = 1,2,\ldots,n)$.

On integration by parts, use of the condition (2.7), i.e. $|p'| \le 2K^2p^{1/2}q$, and then (3.5) we obtain for $f \in D_{1,0}$

$$\begin{aligned}(p^{1/2}f^2)(a) &= -2\int_a^\infty p^{1/2}ff' - \frac{1}{2}\int_a^\infty p'p^{-1/2}f^2\\ &\le 2\|p^{1/2}f'\|\,\|f\| + K^2\|q^{1/2}f\|^2\\ &\le t^2\|p^{1/2}f'\|^2 + t^{-2}\|f\|^2 + K^2\|q^{1/2}f\|^2\end{aligned} \tag{3.7}$$

$$\begin{aligned}(p^{3/2}f'^2)(a) &= -2\int_a^\infty (pf')'pf'p^{-1/2} + \frac{1}{2}\int_a^\infty p'p^{-1/2}pf'^2\\ &\le 2\|(pf')'\|\,\|p^{1/2}f'\| + K^2\|(pq)^{1/2}f'\|^2\\ &\le (2t)^2\|(pf')'\|^2 + (2t)^{-2}\|p^{1/2}f'\|^2 + K^2\|(pq)^{1/2}f'\|^2\end{aligned} \tag{3.8}$$

where t is an arbitrary positive number.

Multiplying (3.7) by (3.8) we obtain, after applying the first inequality in (3.6) with n = 3, and then taking square roots on both sides

$$|pff'|(a) \leq (t\|p^{1/2}f'\| + R)((2t)^{-1}\|p^{1/2}f'\| + S) \tag{3.9}$$

where we have put

$$R = t^{-1}\|f\| + K\|qf\| \ \underline{\text{and}}\ S = 2t\|(pf')'\| + K\|(pq)^{1/2}f'\| . \tag{3.10}$$

From the identity

$$\int_a^\infty pf'^2 = -(pff')(a) - \int_a^\infty (pf')'f \tag{3.11}$$

we obtain, using (3.9) and with $T = t\,S + R/2t$

$$\begin{aligned}\|p^{1/2}f'\|^2 &\leq |pff'|(a) + \|(pf')'\|\,\|f\| \\ &\leq \tfrac{1}{2}\|p^{1/2}f'\|^2 + T\|p^{1/2}f'\| + RS + \|(pf')'\|\,\|f\| .\end{aligned} \tag{3.12}$$

After multiplying both sides by the factor 2 and applying the inequality (3.5) to the last two terms we obtain, for any real positive numbers h and k,

$$(\|p^{1/2}f'\| - T)^2 \leq T^2 + (hR)^2 + (S/h)^2 + k^2\|(pf')'\|^2 + k^{-2}\|f\|^2.$$

Again using the first inequality in (3.6), taking square roots on both sides, substituting firstly $T = t\,S + R/2t$ and secondly for R and S from (3.10),

$$\begin{aligned}\|p^{1/2}f'\| &\leq (2t + h^{-1})S + (t^{-1} + h)R + k\|(pf')'\| \\ &= (4t^2 + 2th^{-1} + k)\|(pf')'\| + (2t + h^{-1})K\|(pq)^{1/2}f'\| \\ &\quad + (t^{-2} + ht^{-1} + k^{-1})\|f\| + (t^{-1} + h)K\|q^{1/2}f\|.\end{aligned} \tag{3.13}$$

In this last result t, k and h are, thus far, arbitrary positive numbers. We now choose $t \in (0,1)$ and put $h^{-1} = k = t^{1/2} = s$ (say) and estimate the last term in (3.13) by, using again (3.5),

$$\|q^{1/2}f\|^2 \leq \|qf\|\,\|f\| \leq \frac{1}{2}\ell^2\|qf\|^2 + \frac{1}{2}\ell^{-2}\|f\|^2$$

$$\leq (\ell\|qf\| + \ell^{-1}\|f\|)^2.$$

Since $s \in (0,1)$ it follows from (3.13) that

$$\|p^{1/2}f'\| \leq 7s\|(pf')'\| + 3sK\|(pq)^{1/2}f'\| + 2Ks^{-2}\ell\|qf\|$$

$$+ (3s^{-4} + 2Ks^{-2}\ell^{-1})\|f\|$$

and then, with $\ell = s^3$ and defining

$$C_1 = \max\{7,\ 3K\} \quad C_2 = 3 + 2K,$$

we obtain

$$\|p^{1/2}f'\| \leq C_1 s\left(\|(pf')'\| + \|(pq)^{1/2}f'\| + \|qf\|\right) + C_2 s^{-5}\|f\|.$$

This gives on using the second inequality in (3.6), with $n = 4$, and replacing $4C_1^2 s^2$ by $\epsilon[4q(a)]^{-1}$, with $\epsilon > 0$ given,

$$\|p^{1/2}f'\|^2 \leq \epsilon[4q(a)]^{-1}\left(\|(pf')'\|^2 + \|(pq)^{1/2}f'\|^2 + \|qf\|^2\right)$$

$$+ C_3\,\epsilon^{-5}\|f\|^2 \qquad (3.14)$$

where

$$C_3 = 4C_2^2[16C_1^2 q(a)]^5.$$

This last inequality (3.14) is valid for all ϵ satisfying

$0 < \epsilon\,[4q(a)]^{-1} < 4C_1^2$, that is

$$\epsilon \in (0,\ 16C_1^2\,q(a)) \tag{3.15}$$

and for all real-valued $f \in D_{1,0}$.

Returning now to (3.11) we obtain, on using (3.14) and then (3.5) again, for all real-valued $f \in D_{1,0}$

$$\begin{aligned} 2|pqff'|(a) &\le 2q(a)\left(\|p^{1/2}f'\|^2 + \|(pf')\|\,\|f\|\right) \\ &\le \frac{1}{2}\epsilon\left(\|(pf')'\|^2 + \|(pq)^{1/2}f'\|^2 + \|qf\|^2\right) + 2C_3q(a)\epsilon^{-5}\|f\|^2 \\ &\quad + \frac{1}{2}\epsilon\,\|(pf')'\|^2 + 2\epsilon^{-1}(q(a))^2\|f\|^2 \\ &\le \epsilon\left(\|(pf')'\|^2 + \|(pq)^{1/2}f'\|^2 + \|qf\|^2\right) + A\epsilon^{-5}\|f\|^2 \end{aligned} \tag{3.16}$$

where the positive number A is defined by

$$A = q(a)[2C_3 + 2(4C_1)^8 q(a)^5] \tag{3.17}$$

on taking into account the upperbound for ϵ given in (3.15).

Returning now to (3.3), with η chosen as in the statement of Theorem 1, we substitute the estimate (3.17) to obtain the required inequality (2.9) but valid for all real-valued $f \in D_{1,0}$. To complete the proof of Theorem 1 it only remains to recall the extension to complex-valued $f \in D_{1,0}$ and then to invoke the closure argument given at the start of this section to extend to the maximal domain D_1.

A detailed examination of the above analysis shows that the order of the term ϵ^{-5}, for small ϵ, is best possible and cannot be reduced.

4. In this section we give the proof of Theorem 2.

We outline the proof of the Theorem only and omit the details which determine an estimate for the positive number B, since this follows the pattern of the analysis in section 3.

The proof of (2.11) follows from known results for the differential expression M; see, for example, [2; Section 5].

To prove (2.12) we start with (3.14) but now extended by the closure argument to the maximal domain D_1; this extension follows from the argument given at the start of section 3 above.

From (2.9) and (3.14) it follows that we may establish a result of the form

$$\|p^{1/2}f'\|^2 \leq \frac{1}{2}\epsilon\|M[f]\|^2 + B_1\epsilon^{-5}\|f\|^2 \qquad (f \in D_1) \tag{4.1}$$

valid for all $\epsilon \in (0,1)$, where B_1 is positive and depends only on the coefficients p and q.

Also we have for any positive k and $f \in D_1$

$$\|q^{1/2}f\|^2 \leq \|qf\|\,\|f\| \leq \frac{1}{2}k\|qf\|^2 + \frac{1}{2}k^{-1}\|f\|^2. \tag{4.2}$$

Turning again to (2.9), using only the third term on the left-hand side we obtain an inequality of the form, putting $\epsilon = \frac{1}{2}\eta$,

$$\|qf\|^2 \leq 2\eta^{-1}\|M[f]\|^2 + A(2\eta^{-1})^4\|f\|^2 \tag{4.3}$$

Combining (4.2) and (4.3) we obtain, on choosing k to be small,

$$\|q^{1/2}f\|^2 \leq \frac{1}{2}\epsilon\,\|M[f]\|^2 + B_2\epsilon^{-5}\|f\|^2 \quad (f \in D_1) \tag{4.4}$$

valid for all $\epsilon \in (0,1)$, where B_2 is positive and depends only on the coefficients p and q.

Taken together (4.1) and (4.4) give the required inequality (2.12) with $B = B_1 + B_2$.

This completes the proof of Theorem 2.

References

1. N. I. Akhiezer and I. M. Glazman, Theory of linear operators in Hilbert space (Ungar, New York, 1961; translated from the Russian edition).

2. W. N. Everitt, 'On the limit-point classification of second-order differential operators', J. London Math. Soc. 41 (1966), 531-534.

3. W. N. Everitt, 'On an extension to an integro-differential inequality of Hardy, Littlewood and Polya', Proc. Royal Soc. Edinburgh (A) 69 (1971/72), 295-333.

4. W. N. Everitt and M. Giertz, 'Some properties of the domains of certain differential operators', Proc. London Math. Soc. (3) (1971), 301-24.

5. W. N. Everitt and M. Giertz, 'On some properties of the domains of powers of certain differential operators', Proc. London Math. Soc. (3) 24 (1972), 756-768.

6. W. N. Everitt and M. Giertz, 'Some inequalities associated with certain ordinary differential operators', Math. Zeit. 126 (1972), 308-326.

7. S. Goldberg, Unbounded linear operators (McGraw-Hill New York, 1966).

8. T. Kato, Perturbation theory for linear operators (Springer-Verlag, Heidelberg, 1966).

9. M. A. Naimark, Linear differential operators; Part II (Ungar, New York, 1968; translated from the Russian edition).

W N Everitt
Department of Mathematics
The University
Dundee, Scotland

M Giertz
Department of Mathematics
The Royal Institute of Technology
Stockholm, Sweden

W. Eckhaus (ed.), New Developments in Differential Equations

ON LEGENDRE'S POLYNOMIALS

by

Åke Pleijel.

1. With $Su = DpDu + qu$, $D = id/dx$, $Tu = ru$, the equation $Su = \lambda Tu$ is considered on $I = \{x: a < x < b\}$. It is assumed that $p(x) > 0$, $0 < r(x) \leq C\, q(x)$ with a constant C, and observed that these conditions are fulfilled for Legendre's differential equation. This equation is obtained when $p(x) = 1 - x^2$, $q(x) = r(x) = 1$, $a = -1$, $b = +1$. Green's or Lagrange's formula tells that $\int_\alpha^\beta (Su\cdot\bar{v} - u\cdot\overline{Sv})dx$ = out-integrated part. If $Su = r\dot{u}$, $Sv = r\dot{v}$, $J = [\alpha,\beta]$, $(u,v)^T_J = \int_J r\, u\, \bar{v}\, dx$, Green's formula reads $i^{-1}((\dot{u},v)^T_J - (u,\dot{v})^T_J) = Q^T_J$, where $Q^T_J = q^T_\beta - q^T_\alpha$, $q^T = p\, Du\cdot\bar{v} + u\cdot\overline{p\, Dv}$. It is valid on the linear relation $E(I)$ of ordered pairs $U = (u,\dot{u})$, where the functions $u,\dot{u}$ satisfy $Su = T\dot{u}$ and belong to $AC^1_{loc}(I) = A(I)$ i.e. have locally absolutely continuous first order derivatives on I.

2. $A[I]^T$ is the subspace of $A(I)$ containing only those u for which $\int_I r|u|^2 dx < \infty$. By this condition for u and the same for $\dot{u}$, the linear space $E[I]^T$ is cut out from $E(I)$. For U,V in $E[I]^T$, $Q^T_I(U,V) = q^T_b(U,V) - q^T_a(U,V)$, where the terms are the finite limits of Q^T_J, q^T_β, q^T_α when $J \to I$, $\beta \to b$, $\alpha \to a$.

3. $A(\hat{I})$, $E(\hat{I})$ are defined as $A(I)$, $E(I)$ only that I is replaced by the union $\hat{I}$ of $I_a = \{x: a < x \leq \xi\}$ and $I_b = \{x: \eta \leq x < b\}$, $\xi < \eta$. The interval $J_0 = \{x: \xi < x < \eta\}$ reduces to c if $\xi = c - 0$, $\eta = c + 0$. For $U,V \in E(\hat{I})$, $\alpha \in I_a$, $\beta \in I_b$ one defines $Q^T_J(U,V) = q^T_\beta(U,V) - q^T_\alpha(U,V)$. From $E(\hat{I})$ the space $E[\hat{I}]$ is cut out by the conditions that $(u,u)^T_{I-J_0} < \infty$, $(\dot{u},\dot{u})^T_{I-J_0} < \infty$. On $E[\hat{I}]$ the hermitean

form $Q_I^T = q_b^T - q_a^T$ exists. Q_I^T, q_b^T and q_a^T are defined as limits. The space $E[I]^T$ is a subspace of $E[\hat{I}]^T$.

4. Solutions of $Su = \lambda Tu$ on I determine the elements $(u,\dot{u})$, $\dot{u} = \lambda u$, of the 2-dimensional space $E_\lambda(I)$. The request $(u,u)_I^T < \infty$ defines the subspace $E_\lambda[I]^T$ of $E_\lambda(I)$. Clearly $0 \leq \dim E_\lambda[I]^T \leq$ ≤ 2. Green's formula shows that $c(\lambda)\, Q_I^T$, $c(\lambda) = i^{-1}(\lambda - \bar{\lambda})$, is positive definite on $E_\lambda[I]^T$. Since $c(\bar{\lambda}) = -c(\lambda)$ the same form is negative definite on $E_{\bar{\lambda}}[I]^T$.

5. According to the preceding results the form Q_I^T is non-degenerate on the direct sum (λ non-real) $R_\lambda^T = E_\lambda[I]^T \dotplus E_{\bar{\lambda}}[I]^T$. It can be proved that if U belongs to $E[\hat{I}]$ but is not in R_λ^T, then Q_I^T is degenerate on the linear hull $\{U, R_\lambda^T\}$. Because of this, R_λ^T is called maximal regular for Q_I^T in $E[\hat{I}]^T$. In general there exist also other maximal regular subspaces R. Every U in $E[\hat{I}]^T$ has a unique Q_I^T-projection U' on a regular R so that $Q_I^T(U - U', R) = 0$. If R is maximal regular it can be seen that $Q_I^T(U,V) = Q_I^T(U',V')$ under Q_I^T-projections on R. The Sylvester signature (p,n) of Q_I^T on R is the same for all maximal regular R. The results of Section 4,5 show that $p = \dim E_\lambda[I]^T$, $n = \dim E_{\bar{\lambda}}[I]^T$ when $\mathrm{Im}(\lambda) > 0$. Clearly $(0,0) \leq (p,n) \leq (2,2)$.

6. By definition, the limit-point (ℓ-p) case over I occurs when $(0,0) = (p,n)$ or $E_\lambda[I]^T = \{0\}$ for all non-real λ. The limit-circle (ℓ-c) case over I occurs when $(p,n) = (2,2)$ i.e. when $E_\lambda[I]^T =$ $= E_\lambda(I)$ for λ non-real. For solutions of $Su = \lambda Tu$ on I, the integral $(u,u)_I^T = \int_I r|u|^2 dx$ is ∞ (but for the case u identically 0) in the ℓ-p case, and is finite in the ℓ-c case.

Because of $Q^T_I(U,V) = Q^T_I(U',V')$, Section 5, the ℓ-p case over I takes place iff $Q^T_I = 0$ on $E[\hat{I}]^T$.

7. $E_\lambda[I_a]^T$ and $E_\lambda[I_b]^T$ are defined as $E_\lambda[I]^T$ in Section 4, but the integrability conditions at $\xi \in I_a$ and at $\eta \in I_b$ are automatically fulfilled. According to Weyl's theory $1 \leq \dim E_\lambda[I_b]^T \leq 2$, and similarly $1 \leq \dim E_\lambda[I_a]^T \leq 2$. Also $\dim E_\lambda[\hat{I}]^T = \dim E_\lambda[I]^T + 2$. The last equality has been generalized by Kodaira and is therefore referred to as Kodaira's identity.

8. A linear subspace Z of $E[\hat{I}]^T$ is a Q^T_I-nullspace if $Q^T_I(Z,Z) = 0$. For U,V in $Z \cap E[I]^T$, Green's formula is valid and implies the symmetry $(\dot{u},v)^T_I = (u,\dot{v})^T_I$. It can be proved that the restriction of $r^{-1}S$ to $Z \cap E[I]^T$ has a unique maximal symmetric Hilbert space extension provided the projection Z'_R of Z into one (then into all) maximal regular R has the maximal dimension $\dim Z'_R = \min(p,n)$ according to Sylvester's inertia theorem (Z'_R is a Q^T_I-nullspace in R since $Q^T_I(U',V') = Q^T_I(U,V)$). Z is then a symmetric boundary condition. The maximal symmetric extension is selfadjoint in the ℓ-c and in the ℓ-p case since $p = n$ in these cases.

9. Parallel to the previous T-positive theory there is an S-positive theory based upon $(u,v)^S_J = \int_J (pu'\overline{v}' + qu\overline{v})dx$ instead of $(u,v)^T_J$. A Green's formula $i^{-1}((\dot{u},v)^S_J - (u,\dot{v})^S_J) = [q^S(U,V)]_J = Q^S_J$ is valid on $E(I)$ with $q^S(U,V) = p\,Du\cdot\overline{\dot{v}} + \dot{u}\cdot\overline{pDv}$. The space $E[I]^S \subset E(I)$ is determined by the conditions $(u,u)^S_I < \infty$, $(\dot{u},\dot{u})^S_I < \infty$. $E[\hat{I}]^S$ is defined similarly as $E[\hat{I}]^T$. On $E[\hat{I}]^S$ the formula $Q^S_I = q^S_b - q^S_a$ holds true in which the expressions are limits as in Section 3. In terms of the dimensions of solution spaces $E_\lambda[I]^S$, $E_{\overline{\lambda}}[I]^S$ the ℓ-p and ℓ-c

cases over I are defined. Symmetric boundary conditions refer to Q^S_I instead of Q^T_I.

10. It is assumed that $Su = \lambda Tu$ is in the ℓ-c case in the T-positive theory, and in the ℓ-p case in the S-positive theory, as the situation is for Legendre's equation. Because of $r(x) \leq C\, q(x)$, the inequality $(u,u)^T_J \leq C(u,u)^S_J$ holds true and shows that $E[\hat{I}]^S \subset E[\hat{I}]^T$.

11. To see that $Z = E[\hat{I}]^S$ is a Q^T_I-nullspace in $E[\hat{I}]^T$, consider $S^\alpha = S + \alpha T$, $\alpha > 0$. It follows that $(u,v)^{S^\alpha}_J = (u,v)^S_J + \alpha(u,v)^T_J$. Hence, $(u,u)^{S^\alpha}_J$ is positive. The ℓ-p case over I in the S-positive theory implies the ℓ-p case over I in the S^α-positive theory. For, $Su = \lambda Tu$ can be written $S^\alpha u = (\lambda + \alpha)Tu$, and $\lambda, \lambda + \alpha$ are simultaneously non-real.

From $(u,u)^T_J \leq C(u,u)^S_J$ it follows that if $U = (u,\dot{u}) \in E[\hat{I}]^S$, then $U^\alpha = (u, \dot{u} + \alpha u)$ belongs to $E[\hat{I}]^{S^\alpha}$. If also $V = (v,\dot{v}) \in E[\hat{I}]^S$, $V^\alpha = (v, \dot{v} + \alpha v) \in E[\hat{I}]^{S^\alpha}$, a simple computation shows that $q^{S^\alpha}(U^\alpha, V^\alpha) = q^S(U,V) + \alpha q^T(U,V)$, and $Q^{S^\alpha}_J(U^\alpha,V^\alpha) = Q^S_J(U,V) + \alpha\, Q^T_J(U,V)$. According to the generalization to an S-positive theory of the statement at the end of Section 6 it follows, when J tends to I, that $Q^T_I(U,V) = 0$ on $Z = E[\hat{I}]^S$.

12. To show that $Z = E[\hat{I}]^S$ is a symmetric boundary condition in the T-positive theory it remains to prove that Z'_R in Section 8, for one R which is maximal regular (Q^T_I) in $E[\hat{I}]^T$, has the right maximal dimension. Since $Su = \lambda Tu$ is in the ℓ-c case in the T-positive theory $(p,n) = (2,2)$, and the right dimension must be 2. According to Weyl's theory as presented in Titchmarsh's textbook, the

4-dimensional space $R = E_\lambda[\hat{I}]^T = E_\lambda(\hat{I})$ is maximal regular. Clearly $Z'_R \supset Z \cap R = E[\hat{I}]^S \cap E_\lambda(\hat{I}) = E_\lambda[\hat{I}]^S$. The last space is 2-dimensional according to Kodaira's identity in the ℓ-p case of the S-positive theory, see Section 7. This suffices to conclude that $\dim Z'_R = 2$. Hence, $Z = E[\hat{I}]^S$ is a symmetric boundary condition in the T-positive theory.

13. An element $U = (u,\dot{u})$ satisfies $Z = E[\hat{I}]^S$ iff the integrals $(u,u)^S$, $(\dot{u},\dot{u})^S$ are finite when extended over intervals I_a and I_b. According to a well known theorem, the spectrum is discrete in the ℓ-c case. A solution of $Su = \rho Tu$ determines an element U with $\dot{u} = \rho u$, and Z is satisfied iff the integral

$$\int (p|u'|^2 + q|u|^2)dx$$

is finite over neighbourhoods of a and b. For the Legendre equation the Hilbert space of the theory is $L^2(-1,1)$, and $U = (u,\rho u)$, $Su = \rho Tu$, satisfies $Z = E[\hat{I}]^S$ when

$$\int (1-x^2)\ |u'|^2\ dx$$

is finite over neighbourhoods of -1 and 1. This condition is known (Akhiezer-Glazman). It has here been deduced from the S-positive theory of Legendre's equation.

A more detailed account of the subject treated above will be published in Annales Academiae Scientiarum Fennicae, in the volume dedicated to Rolf Nevanlinna, under the title "On the boundary condition for the Legendre polynomials".

Uppsala University, Sweden.

REFERENCES

Akhiezer, N.I., and Glazman, I.M., Theory of Linear Operators in Hilbert Space, Volume II, Frederick Ungar Publishing Co, New York 1963, 218 p.

Titchmarsh, E.C., Eigenfunction Expansions Associated with Second-order Differential Equations, Part I, Oxford University Press 1962, 203 p.

W. Eckhaus (ed.), New Developments in Differential Equations

INVARIANTS AND CANONICAL FORMS FOR MEROMORPHIC SECOND ORDER DIFFERENTIAL EQUATIONS

W.Jurkat,[1] Syracuse University
D.A.Lutz,[2] University of Wisconsin-Milwaukee
A.Peyerimhoff, Universität Ulm

ABSTRACT

Given a differential equation $x''+a(z)x'+b(z)x=0$, where $a(z)$ and $b(z)$ are analytic functions in a neighborhood of ∞ (generally an irregular singular point of the solutions), a complete system of invariants is computed which characterizes the solutions at ∞ to within linear combinations of meromorphic functions. A "classical" differential equation having these same invariants can be named and a transformation from the given equation to this classical one can be contructed. Thus the solutions of the given equation can be effectively calculated in terms of certain classical functions and computable meromorphic functions.

STATEMENT OF THE PROBLEM

We consider (scalar) differential equations of the form

(1) $$x''+ a(z)x'+ b(z)x = 0 \quad ,$$

where $a(z) = \sum_0^\infty a_\nu z^{-\nu}$, $b(z) = \sum_0^\infty b_\nu z^{-\nu}$

converge for $|z|$ large. The point at ∞ is "generally" an irregular singular point of the solutions and the differential equation is said to have Poincaré rank r=1 at ∞ .

A linear transformation

(2) $$y = t_1(z)x + t_2(z)x' \quad ,$$

where $t_1(z)$ and $t_2(z)$ are assumed to be at least meromorphic in a neighborhood of ∞ takes (1) into a second order linear differential equation for y whose coefficients are meromorphic at ∞ provided the matrix $T(z)$, defined by

---

1 Supported in part by a grant from the National Science Foundation.

2 Supported in part by grants from the National Science Foundation and the Alexander von Humboldt-Stiftung.

$$T(z) = \begin{bmatrix} t_1 & t_2 \\ t_1'-t_2b & t_1+t_2'-t_1a \end{bmatrix}$$

has not identically vanishing determinant in a neighborhood of ∞. Moreover, the transformation can be reverted to yield $x=m_1(z)y+m_2(z)y'$, which takes the differential equation for y back into (1).

We are concerned with the question of what simplification can be made in the differential equation through the use of such transformations.

G.D.Birkhoff had the idea of isolating the singularity at ∞ by transforming the differential equation into a new one which has only two singular points. Only one singular point would be, of course, not generally possible since that would imply that all solutions would be single-valued in a neighborhood of ∞. Birkhoff treated, however, only a "main" case and his statements are not correct without the introduction of some exceptional cases. In order to identify a certain exceptional case which arises, we say that the solution space of (1) _totally splits_ if the general solution of (1) can be expressed as a linear combination of solutions of analytic linear homogeneous first order differential equations.

STATEMENT OF THE RESULTS

By completing arguments of Birkhoff [1] and Turrittin [6], we have the following

Theorem. (Birkhoff, Turrittin, Jurkat-Lutz-Peyerimhoff)

Let $x''+a(z)x'+b(z)x=0$, _where_ $a(z)$ _and_ $b(z)$ _are analytic at_ ∞ _and assume that the solution space does not totally split. Then there exists a meromorphic transformation_ $x=m_1(z)y+m_2(z)y'$ _which takes the differential equation into a "classical" differential equation of the form_

$$y''+(\hat{a}_0+\frac{\hat{a}_1}{z})y'+(\hat{b}_0+\frac{\hat{b}_1}{z}+\frac{\hat{b}_2}{z^2})y=0. \tag{3}$$

Note that the differential equation (3) has only two singular points, 0 and ∞, and 0 is a _regular singular point_ in the sense of Frobenius-Fuchs. The differential equation is called "classical" because it can be solved explicitly by classical functions, e.g., Kummer functions, Barnes functions, and certain elementary functions.

This theorem may be viewed as a _qualitative_ statement in that it describes the solutions of the original differential equation as a meromorphic linear combination of certain classical functions. We are mainly interested in _quantitative_ versions of this theorem, namely, we wish to identify which of these classical functions are associated with a given differential equation and how the solution of the original differential equation may be expressed in terms of them.

To do this, we must determine the coefficients in the classical equation (3). It turns out that they are almost unique in a sense which will be described below. In order to make this discussion precise, it is proper to introduce at this point the concept of invariants, that is, quantities which are uniquely associated with the differential equation and which remain unchanged with respect to certain classes of linear transformations of the differential equation. A complete system of invariants characterizes the differential equation up to linear transformations from the specified class and the calculation of such a complete system leads to the determination of the coefficients in (3) as well as a linear transformation which reduces (1) to a classical equation.

In discussing systems of invariants, it is convenient to specialize the class of transformations under consideration so that the invariants become numbers instead of equivalence classes. Invariants corresponding to more general classes of transformations are then built from these.

If $t_1(z)$ and $t_2(z)$ are analytic at ∞ and if

$$[t_1(\infty)]^2 - a_0\, t_1(\infty)\, t_2(\infty) + b_0[t_2(\infty)]^2 \neq 0\,, \text{ i.e. } \det T(\infty) \neq 0\,,$$

then $y = t_1(z)x + t_2(z)x'$ is called an analytic transformation of the differential equation. Such a transformation takes (1) into another second order linear differential equation with analytic coefficients at ∞. If, in particular, $t_1(\infty) = 1$ and $t_2(\infty) = 0$, then $y = t_1(z)x + t_2(z)x'$ is called a strict transformation. It is for the class of strict transformations that the invariants appear in their simplest form and such invariants are called Birkhoff invariants in honor of G.D.Birkhoff who introduced quantities related to them [1] in connection with the problem we discuss here. Invariants corresponding to the classes of analytic and meromorphic transformations are called, simply, analytic and meromorphic invariants, respectively.

We now come to the problem of calculating a complete system of invariants and in order to do this we will work with the formal solutions of the differential equation. In computing formal solutions, there are two naturally disjoint cases to consider, depending upon whether $a_0^2 \neq 4b_0$ or not, i.e., whether the characteristic polynomial $\lambda^2 + a_0\lambda + b_0$ has distinct roots or equal roots. Since a_0 and b_0 are meromorphic invariants, this distinction is an invariant concept.

Case I : $(a_0^2 \neq 4b_0)$.

Let the distinct roots of $\lambda^2 + a_0\lambda + b_0$ be denoted λ_1, λ_2 . Then there exist two linearly independent formal solutions of the form

$$f_1(z)z^{\lambda_1'}e^{\lambda_1 z} \quad \text{and} \quad f_2(z)z^{\lambda_2'}e^{\lambda_2 z}\,,$$

where $\lambda_1' = (a_1\lambda_1 + b_1)/(\lambda_2 - \lambda_1)$, $\lambda_2' = (a_1\lambda_2 + b_1)/(\lambda_1 - \lambda_2)$,

$f_i(z) = \sum_0^\infty f_n^{(i)} z^{-n}$, $f_0^{(i)} = 1$, $i = 1,2$, and the coefficients $f_n^{(i)}$

are calculated recursively in a well-known manner. The series $f_i(z)$ generally diverge everywhere and the meaning usually attached to the formal solutions is that in sectors of sufficiently small angle they asmptotically represent actual solutions as $z \to \infty$. In this discussion we are not concerned with that interpretation.

If $a_0^2 = 4b_0$, then $\lambda^2 + a_0\lambda + b_0$ has equal roots $\lambda_1 = \lambda_2 \equiv \lambda = -a_0/2$. In this case the quantity $2a = a_0a_1 - 2b_1$ is a meromorphic invariant.

If $a=0$, this leads to convergent formal solutions and the problem of constructing a meromorphic transformation which takes (1) into a classical differential equation is a purely formal problem which is algebraic in nature and which we will not further consider here. (This case of $a=0$ is, however, treated in detail in [3], Section 6.)

<u>Case II</u>: $(a_0^2 = 4b_0 \ , \ a_0a_1 \neq 2b_1)$.

Define $\mu = 2\left(\frac{a_0a_1}{2} - b_1\right)^{1/2}$

by selecting a branch of the square root. Then there exist two linearly independent formal solutions of the form

$$f(t)t^{\lambda'}e^{\mu t+\lambda t^2} \quad \text{and} \quad f(-t)t^{\lambda'}e^{-\mu t+\lambda t^2}$$

where $t^2 = z$, $\lambda' = -a_1/2$, $f(t) = \sum_0^\infty f_n t^{-n}$, $f_0 = 1$, and the coefficients f_n are calculated recursively.

<u>Formal invariants</u> are defined with respect to formal transformations $y = t_1(z)x + t_2(z)x'$, i.e., $t_1(z)$ and $t_2(z)$ are just assumed to be formal series. The calculation of formal invariants is an algebraic problem. A complete system of formal Birkhoff invariants in Case I consists of $\{\lambda_1,\lambda_2,\lambda_1',\lambda_2'\}$ and in Case II consists of $\{\lambda,\mu,\lambda',\delta\}$, where δ is related to f_1 and is calculated in terms of a_0, a_1, b_0, b_1, and b_2.

Beyond these formal invariants, we now seek (non-trivial) invariants which will complete the system of (actual) invariants. They are related in a transcendental manner to the coefficients of the differential equation. The information containing these invariants comes from the asymptotic behavior of the coefficients in the formal series of the formal solutions in the following way:

In Case I, we have

$$\lim_{n\to\infty} \frac{f_n^{(1)}(\lambda_1-\lambda_2)^n}{\Gamma(n)n^{\lambda_2'-\lambda_1'}} = \gamma_1 \quad \text{and} \quad \lim_{n\to\infty} \frac{f_n^{(2)}(\lambda_2-\lambda_1)^n}{\Gamma(n)n^{\lambda_1'-\lambda_2'}} = \gamma_2 \ .$$

The numbers γ_1, γ_2 are Birkhoff invariants and complete the system, i.e., $\{\lambda_1,\lambda_2,\lambda_1',\lambda_2',\gamma_1,\gamma_2\}$ is a complete system of Birkhoff invariants. These invariants are free, subject only to the condition $\lambda_1 \neq \lambda_2$. The invariants γ_1 and γ_2 are related to the convergence-divergence properties of the formal series in that $\gamma_1 = 0$ iff $f_1(z)$ converges for $|z|$ large and $\gamma_2 = 0$ iff $f_2(z)$ converges for $|z|$ large. Moreover, total splitting of the differential equation occurs iff $\gamma_1 = \gamma_2 = 0$. A complete system of analytic invariants is given by $\{\lambda_1,\lambda_2,\lambda_1',\lambda_2',\gamma_1\gamma_2\}$ in case $\gamma_1\gamma_2 \neq 0$ and if $\gamma_1\gamma_2 = 0$, we must include the knowledge of which one (s) is zero. (See Theorem V and Section 7, [2].)

In Case II,

$$\lim_{n\to\infty} f_n(2\mu)^n/\Gamma(n) = \gamma \quad ,$$

where γ is a Birkhoff invariant and completes the system, i.e., $\{\lambda_1 = \lambda_2, \mu, \lambda', \delta, \gamma\}$ forms a complete system of Birkhoff invariants.

Moreover, $\gamma = 0$ iff the formal series $f(t)$ converges for $|t|$ large. A complete system of analytic invariants in this case is given by $\{\lambda_1 = \lambda_2, \mu, \lambda', \gamma\}$. (See Theorem II and Section 7, [3].)

In Case I when $\gamma_1\gamma_2 \neq 0$ (this corresponds to Birkhoff's "main" case [1]) and in Case II, it turns out that the differential equation (1) can be transformed analytically into a classical equation (3). The parameters $\hat{a}_0$, $\hat{a}_1$, $\hat{b}_0$ and $\hat{b}_1$ are in these cases equal to a_0, a_1, b_0 and b_1, respectively and it only remains to determine $\hat{b}_2$. The parameter $\hat{b}_2$ is not generally equal to b_2, but is influenced by the invariants $\gamma_1\gamma_2$, resp. γ and is calculated from them as follows.

We introduce an auxiliary parameter d which satisfies

$$\cos[2\pi d - \pi(\lambda_1' + \lambda_2')] = \cos \pi(\lambda_1' - \lambda_2') - 2\pi^2\gamma_1\gamma_2 \quad \text{in Case I, } \gamma_1\gamma_2 \neq 0,$$

respectively,

$$\cos[2\pi d + \pi a_1] = \pi\gamma \quad \text{in Case II },$$

and these determine d up to sign ($\pm$) and modulo one. Then $\hat{b}_2$ is given by

$$d^2 + (a_1 - 1)d + \hat{b}_2 = 0 .$$

Every solution d in the above equations is permitted and with each admissible d and corresponding value for $\hat{b}_2$, the resulting classical differential equations are related by explicit analytic transformations, in fact, they are polynomials in z^{-1}. This gives rise to a set of functional equations which relate the solutions of the equivalent classical differential equations and these functional equations are the only such ones possible. These discrete changes in the parameter $\hat{b}_2$ which do not change the invariants is what was meant when we said earlier that the parameters in the classical differential equations are "almost unique".

In the cases other than the ones treated above, i.e., $\gamma_1\gamma_2 = 0$ or $a = 0$, it is sometimes necessary to utilize meromorphic transformations in order to bring about the reduction of the original differential equation to a classical equation. The introduction of the invariants allows us to keep track of the deviations from the normal cases treated above. These cases are treated in detail in [2] and [3]. In Case I when $\gamma_1 = \gamma_2 = 0$ and $\lambda_1' \not\equiv \lambda_2'$ (mod 1), then (1) is not meromorphically equivalent to a classical differential equation of the form (3), but is equivalent to a second order differential equation with two regular singular points in the finite complex plane and the original singular point at ∞ .

Once the invariants have been calculated and a classical differential equation (3) having these same invariants has been named, the construction of an analytic or meromorphic transformation from the given differential equation to the classical differential equation can be accomplished using the formal series in the corresponding formal solutions. This procedure is more natural to discuss for two-dimensional differential systems than for second order differential equations. See Theorems V, VI of [2] and Theorems II, V of [3] for a statement of these results. The transformation is given as a certain quotient of formal series and the theory tells us that the quotient must, in fact, converge in a neighborhood of ∞ . Hence the calculation of a transformation (2) is purely a formal matter.

We conclude with an application of the Theorem in Case II to see how single-valued functions appear in the actual solutions, even though the formal solutions carry power series in $z^{-1/2}$.

Assume now that $a_0^2 = 4b_0$ and $a_0 a_1 \neq 2b_1$. Then according to our discussion above, the given differential equation (1) is analytically equivalent to a classical differential equation of the form (3). If we let

$$y = \exp[-(\hat{a}_0/2)z]\,\hat{y} ,$$

we obtain

$$\hat{y}'' + \frac{\hat{a}_1}{z}\hat{y}' + \left(\frac{2b_1 - \hat{a}_0\hat{a}_1}{2z} + \frac{\hat{b}_2}{z^2}\right)\hat{y} = 0 \tag{4}$$

and letting $z = t^2$, $\tilde{y}(t) = \hat{y}(t^2)$, we obtain

$$\frac{d^2\tilde{y}}{dt^2} + \frac{2\hat{a}_1 - 1}{t}\frac{d\tilde{y}}{dt} + \left[2(2\hat{b}_1 - \hat{a}_0\hat{a}_1) + \frac{4\hat{b}_2}{t^2}\right]\tilde{y} = 0 . \tag{5}$$

Note that in (5) the coefficient of $\tilde{y}$ is an even function of t , while the coefficient of $d\tilde{y}/dt$ is odd. This feature likewise holds if we would have applied these same transformations to the general differential equation. We now make two normalizations to further simplify (5). Letting $s = \sigma t$ we obtain

$$\frac{d^2\tilde{y}}{ds^2} + \frac{2\hat{a}_1 - 1}{s}\frac{d\tilde{y}}{ds} + \left[\frac{2(2\hat{b}_1 - \hat{a}_0\hat{a}_1)}{\sigma^2} + \frac{4\hat{b}_2}{s^2}\right]\tilde{y} = 0$$

and letting $\tilde{y} = s^{\rho}u$, we obtain

$$\frac{d^2u}{ds^2} + \frac{2\hat{a}_1 - 1 + 2\rho}{s}\frac{du}{ds} + \left[\frac{2(2\hat{b}_1 - \hat{a}_0\hat{a}_1)}{\sigma^2} + \frac{4\hat{b}_2 + \rho(\rho - 1)}{s^2}\right]u = 0 .$$

Hence we may select σ and ρ to obtain the normalized differential equation

$$\frac{d^2u}{ds^2} + \frac{2c}{s}\frac{du}{ds} - 4u = 0 ,$$

which is solved explicitly by the (linearly independent) functions

$${}_0F_1(c + \tfrac{1}{2}\,;\, s^2) \qquad \text{and} \qquad s^{1-2c}\,{}_0F_1(\tfrac{3}{2} - c\,;\, s^2)$$

in case 2c is not an integer. Here ${}_0F_1(a;x)$ denotes the <u>Barnes function</u>

$${}_0F_1(a;x) = \sum_{n=0}^{\infty} \frac{\Gamma(a)x^n}{\Gamma(n+a)n!} , \quad a \neq 0,-1,\ldots .$$

If 2c is an integer, modifications must be made in these functions to obtain two linearly independent solutions. We do not include that discussion here since the procedure is well-known for these classical differential equations. (See, for example, [5].)

Hence we obtain (for $2c \neq$ integer) two linearly independent solutions for the classical differential equation (3) of the form

$$\exp[-(a_0/2)z]\, z^{\alpha/2}\; {}_0F_1(c+\tfrac{1}{2};\, \sigma^2 z) \qquad \text{and}$$

$$\exp[-(a_0/2)z]\, z^{\alpha/2+1/2-c}\; {}_0F_1(\tfrac{3}{2}-c;\, \sigma^2 z)\;.$$

We remark that ${}_0F_1(\cdot\,;z)$ is an entire function of order $1/2$ and this explains how the singularity of the solutions is built in this case of equal roots of the characteristic polynomial.

REFERENCES

[1] G.D.Birkhoff, On a simple type of irregular singular point, Trans.Amer.Math.Soc. 14(1913), 462-476.

[2] W.Jurkat, D.Lutz, and A.Peyerimhoff, Birkhoff invariants and effective calculations for meromorphic linear differential equations, I, J.Math.Anal.Appl. (to appear).

[3] W.Jurkat, D.Lutz, and A.Peyerimhoff, Birkhoff invariants and effective calculations for meromorphic linear differential equations, II, Houston J.Math. (to appear).

[4] W.Jurkat, D.Lutz, and A.Peyerimhoff, Effective solutions for meromorphic second order differential equations, Symposium on Ordinary Diff.Equs., Lect. Notes in Math.#312, Springer-Verlag New York, 1973, 100-107.

[5] W.Magnus and F.Oberhettinger, Formulas and Theorems for the Special Functions of Mathematical Physics, Chelsea, New York, 1949.

[6] H.L.Turrittin, Reduction of ordinary differential equations to the Birkhoff canonical form, Trans. A.M.S. 107(1963), 485-507.

W. Eckhaus (ed.), New Developments in Differential Equations

ON GENERALIZED EIGENFUNCTIONS AND LINEAR TRANSPORT THEORY

C.G. Lekkerkerker,
Mathematisch Instituut,
Universiteit van Amsterdam,
The Netherlands

ABSTRACT

The main purpose of this paper is to formulate and prove a theorem on generalized eigenfunctions. This theorem can be applied in the theory of the linear transport equation. Particular attention will be given to the case that the parameter c occurring in that equation is equal to 1.

1. INTRODUCTION

The equation meant in the abstract reads

$$(1)\quad \mu \frac{\partial\psi(x,\mu)}{\partial x} + \psi(x,\mu) = \frac{c}{2}\int_{-1}^{1}\psi(x,\mu')d\mu' + g(x,\mu).$$

Here, ψ and g are real functions of two real variables x and μ ranging over some interval J in $\mathbb{R}$ and the segment $I=[-1,1]$, respectively. Furthermore, c is a positive constant. Equation (1) can be written in a more concise form, as follows. Consider the Hilbert space $L^2(I)$. Define two operators A and T by putting

$$(Af)(\mu) = f(\mu) - \frac{c}{2}\int_{-1}^{1}f(\mu')d\mu'$$

$$(Tf)(\mu) = \mu f(\mu) \qquad\qquad (f\in L^2(I)\ ,\ \mu\in I).$$

Then, if ψ and g are conceived as functions of the single variable x, with values in $L^2(I)$, and differentation of such a function with respect to x is defined by taking limits in $L^2(I)$, (1) takes the form

$$(1')\quad T\frac{d\psi}{dx} = -A\psi + g \quad \text{ or also } \frac{d\psi}{dx} = -T^{-1}A\psi + T^{-1}g.$$

[By modifying this procedure, more precisely by admitting only functions ψ and g with values in the space Lip(I) of Hölder continuous functions on I, one is led to classical solutions of (1). See [5, chapter 8]].

Equation (1) combined with various types of boundary conditions was treated amongst others by Case (see [2]). Also, more general instances of the linear transport equation were investigated (see [9]). A functional analytic treatment of (1), in the case $0 < c < 1$, was given by Hangelbroek [5]. The case $c = 1$ was dealt with in [7]. In the present paper we bring a general theorem on eigenfunction expansions and show how it can be applied in the discussion of (1).

This means that we restore the rôle of the eigenfunctions of $T^{-1}A$ the use of which was suppressed in [5] and [7].

The basis for our considerations is the spectral theorem. It will be applied in the following form. Let B be a bounded self-adjoint operator in a Hilbert space H, with spectrum N. Suppose that the spectrum is simple, i.e., there exists a cyclic element $e_0 \in H$ for the operator B. Then there exist a finite positive regular Borel measure σ on N and a unitary map F from H onto the Hilbert space $L^2(N,\sigma)$ such that

(i) $Fe_0 = 1_N$ (the function on N that is identically equal to 1)

(ii) F diagonalizes B, i.e., FBF^{-1} is the multiplication in $L^2(N,\sigma)$ with the function $h(\nu) = \nu$.

If we wish to apply this theorem to a particular problem, e.g., a boundary value problem connected with (1), we do this by performing the transformation F associated with a suitably chosen operator B. Then our task consists of two parts, viz.

a) to determine F explicitly

b) to solve the transformed problem.

This program will be carried out in sections 3 and 4. Some details will be left out; for them the reader is referred to [5] and [7].

2. GENERALIZED EIGENFUNCTIONS

Let H,B,N,e_0,σ and F be as in the introduction. There is a close connection between F and the eigenfunctions of B. Results describing this connection were given by Berezanskii [1], Foiaş [4] and Maurin [8, chapter 2]. However, these theorems are of no help to us, since in our case the conditions are not fulfilled. [These conditions are that a certain subspace Φ of H having the property that the dual space contains the eigenfunctions of B is either a nuclear space or the embedding of Φ in H is nuclear]. Here, we choose a different approach in which we presuppose the existence of eigenfunctions. We shall not enter upon the general question when this assumption is fulfilled.

Let L be a locally convex space fulfilling the following conditions (P is the set of all polynomials)

1°. L is a subspace of H which is densely and continuously embedded in H

2°. the set $\{p(B)e_0 : p \in P\}$ is sequentially dense in L, with respect to the topology of L

3°. B maps L continuously into itself.

Let us consider the anti-dual space L', i.e. the space of all continuous conjugate linear functionals on L (we prefer to consider the anti-dual rather than the dual space). The space H can be embedded in L' in a very natural way; on account of 1°, such an embedding is furnished by the map $f \mapsto u_f$, where $u_f(g) = (f,g)(g \in L)$. Furthermore, the operator B can be extended to L' by putting

(2) $(B\psi,g) = (\psi,Bg) \qquad (\psi \in L'\ ,\ g \in L).$

Here, we have used round brackets in indicating the values of a functional on L. For each $\psi \in L'$, the functional $B\psi$ defined by (2) again belongs to L', on account of 3°. We now state one further condition, viz.

4°. for each $\nu \in N$, L' contains an element ψ_ν (eigenfunction of B) such that $B\psi_\nu = \nu\psi_\nu$ and $(\psi_\nu,e_0) = 1$.

We observe that the normalization condition $(\psi_\nu,e_0) = 1$ is not a real restriction. For, if $(\psi_\nu,e_0) = 0$, then also $(\psi_\nu,Be_0) = (B\psi_\nu,e_0) = \nu(\psi_\nu,e_0) = 0$; more generally, $(\psi_\nu,p(B)e_0)= 0$ for all $p \in P$, and so $\psi_\nu = 0$.

Theorem 1. Suppose that the conditions stated above are fulfilled. Then, on the subspace L, the transformation F is given by

(3) $(Ff)(\nu) = \overline{(\psi_\nu,f)} \qquad (\nu \in N).$

Proof. The right-hand member of (3) is well defined for each $\nu \in N$ and each $f \in L$; let it be denoted by $\tilde{f}(\nu)$. On account of 4° and (2), $(\psi_\nu,Bf) = \nu(\psi_\nu,f)$. Thus, since ν is a real number, $\widetilde{Bf}(\nu) = \nu\tilde{f}(\nu)$. We also have $\tilde{e}_0(\nu) = 1 \quad (\nu \in N)$. Thus the map $f \mapsto \tilde{f}$ satisfies the properties (i) and (ii) of the transformation F. It follows that it coincides with F, at least on the set of elements $f = p(B)e_0$. We observe that $F\,p(B)e_0 = p$.

Now let $f \in L$ be arbitrary. By 2°, there is a sequence of polynomials p_n such that $p_n(B)e_0 \to f$ in the space L, as $n \to \infty$. Since ψ_ν is a continuous functional on L, it follows that $(\psi_\nu,p_n(B)e_0) \to (\psi_\nu,f)$ as $n \to \infty$. Thus

$$p_n(\nu) \to \tilde{f}(\nu) \quad \text{as } n \to \infty, \text{ for each } \nu \in N.$$

On the other hand, it is also true that $p_n(B)e_0 \to f$ in the coarser topology of H. Therefore, $p_n = Fp_n(B)e_0 \to Ff$ in the space $L^2(N,\sigma)$, as $n \to \infty$. Then there is a subsequence p_{n_k} tending point-wise to Ff a.e.. It follows that, as elements of $L^2(N,\sigma)$, Ff and $\tilde{f}$ coincide.

Remark. The spectral theorem and theorem 1 remain true if we allow B to be an unbounded operator, but retain the other conditions, in particular condition 3°. Consider the case that $H = L^2(-\infty,\infty)$ and that B is the operator $(1/i)d/dx$. Take L to be the Schwartz space S. Then the conditions of theorem 1 are fulfilled, with some choice of e_0; apart from numerical factors, the normalized eigenfunctions of B are the locally integrable functions $\psi_\xi(x) = e^{i\xi x}$ $(\xi,x \in \mathbb{R})$.

We remark that it has some advantage to take $e_0(x) = e^{-x^2/2}$, to choose

F in such a way that $(Fe_0)(\nu) = e^{-\nu^2/2}$ and to normalize the ψ_ξ by requiring that $(\psi_\xi, e_0) = e^{-\nu^2/2}$. Then the conclusion of theorem 1 again holds true, whereas $\psi_\xi(x) = (2\pi)^{-\frac{1}{2}} e^{i\xi x}$. At any rate, the result is that the transformation F diagonalizing B is the Fourier transformation. This is a first and very simple application of theorem 1; it serves to stress the analogy between the Fourier transformation and the transformation F appearing in our discussion of the linear transport equation.

3. THE OPERATOR $T^{-1}A$. DETERMINATION OF F

First let $0 < c < 1$. Then A is invertible, and so we may consider the operator $A^{-1}T$ as well. This operator is bounded. It is self-adjoint provided the space $L^2(I)$ (henceforth denoted by $L^2(I,A)$) is endowed with the inner product $(f,g)_A = (f,Ag)$. We say a few words on how to determine the spectrum N of $A^{-1}T$ (see [5]).

Let E denote the identity operator in $L^2(I,A)$ and let C be defined by $Cf = \frac{1}{2}\int_{-1}^{1} f(\mu')d\mu' . 1_I$. A simple calculation yields

$$(A^{-1}T - \lambda E)(T - \lambda E)^{-1} = E + \frac{c}{1-c} CT(T - \lambda E)^{-1} \quad (\lambda \notin sp(T) = I).$$

The operator on the right differs from the identity by an operator of rank 1. We can introduce the determinant of this operator; it is called the perturbation determinant of the pair $A^{-1}T$, T. We denote this determinant by $(1-c)^{-1}\,\Lambda(\lambda)$. Then

$$\Lambda(\lambda) = 1 - c + \frac{c}{2}\int_{-1}^{1} \frac{\mu}{\mu-\lambda}\, d\lambda = 1 + \frac{\lambda c}{2}\int_{-1}^{1}\frac{d\mu}{\mu-\lambda} \quad (\lambda \notin I). \tag{4}$$

We also define

$$\lambda(\nu) = 1 + \frac{\nu c}{2} ⨍_{-1}^{1} \frac{d\mu}{\mu-\nu} \quad (-1 < \nu < 1), \tag{5}$$

where we have taken the Cauchy principal value of the integral. Using function theoretic arguments one can prove that the function $\Lambda(\lambda)$ has two simple zeros $\pm\nu_0$, with $\nu_0 > 1$, and no other zeros in the extended complex plane cut along the segment I. It is then easy to show that

$$N = I \cup \{\text{zeros of } \Lambda(\lambda)\} = I \cup \{\pm \nu_0\}. \tag{6}$$

The spectrum N is simple; as a cyclic element for the operator $A^{-1}T$, one may take $e_0 = (1-c)1_I$.

By the foregoing, the spectral theorem may be applied to $A^{-1}T$ considered as an operator in $L^2(I,A)$. This gives rise to a unitary map F from $L^2(I,A)$ onto some space $L^2(N,\sigma)$ such that F diagonalizes $A^{-1}T$, while $Fe_0 = 1_N$. The conditions of theorem 1 are fulfilled if one takes $B = A^{-1}T$ and $L = Lip(I)$ (confer [5, chapter 6]). Here, the space Lip(I) is defined as follows. For $0 < \alpha \le 1$, let $Lip_\alpha(I)$ denote the linear space of functions f on I that are (uniformly) Hölder continuous with exponent α. It is a Banach space in the norm

$$||f||_\alpha = \sup|f(x)| + \sup|f(x) - f(x')| \Big/ |x - x'|^\alpha .$$

We put $\mathrm{Lip}(I) = \cup_{0<\alpha\leq 1}\mathrm{Lip}_\alpha(I)$ and give this space the inductive limit topology [observe that $\mathrm{Lip}_\alpha(I) \supset \mathrm{Lip}_\beta(I)$ if $\alpha < \beta$].

We have to take some care if we want to extend $B = A^{-1}T$ to the space $\mathrm{Lip}(I)'$, since $A^{-1}T$ is self-adjoint in $L^2(I,A)$, but not in $L^2(I)$. For any $\psi \in \mathrm{Lip}(I)'$, let us define a new functional $(\psi,\cdot)_A$ by putting

$$(\psi,f)_A = (\psi,Af) \qquad (f \in \mathrm{Lip}(I)).$$

Then the extension of $A^{-1}T$ to $\mathrm{Lip}(I)'$ is given by

$$(A^{-1}T\psi,f)_A = (\psi,A^{-1}Tf)_A .$$

[We would have obtained the same extension if we had taken the second adjoint $\left((A^{-1}T)^*\right)'$, where $(A^{-1}T)^*$ is the adjoint of $A^{-1}T$ in $L^2(I)$ and the second adjoint is taken with respect to (,)].

Applying theorem 1 we find that, on the subspace $\mathrm{Lip}(I)$, the transformation F is given by $(Ff)(\nu) = \tilde{f}(\nu) = \overline{(\psi_\nu,f)_A}$, where ψ_ν satisfies the relations

$$(\psi_\nu,Bf)_A = \nu(\psi_\nu,f)_A \quad , \quad (\psi_\nu,e_0)_A = 1 \qquad (\nu \in N).$$

If we define ϕ_ν by $(\phi_\nu,f) = (\psi_\nu,f)_A = (\psi_\nu,Af)$, then we get $\tilde{f}(\nu) = \overline{(\phi_\nu,f)}$, where the ϕ_ν are determined by

$$(7) \qquad (\phi_\nu,Bf) = \nu(\phi_\nu,f) \quad , \quad (\phi_\nu,e_0) = 1 .$$

From (7) we deduce $(\phi_\nu,(T-\nu)f) = -\frac{c}{2}\cdot\overline{(Tf,1_I)}$. This relation is satisfied by $\phi_\nu = \alpha\delta_\nu - \frac{c}{2}\frac{\mu}{\mu-\nu}$, where α is a constant, δ_ν denotes the delta-function centered at the point ν and $\frac{\mu}{\mu-\nu}$ stands for the functional with values $⨏_{-1}^{1}\frac{\mu}{\mu-\nu}f(\mu)d\mu$. Taking into account the second relation (7) we get $\alpha = \lambda(\nu)$. The result obtained may be written as

$$(8) \qquad \tilde{f}(\nu) = \overline{(\phi_\nu,f)} = \langle\lambda(\nu)\delta_\nu - \frac{c}{2}\frac{\mu}{\mu-\nu} , f\rangle \quad (\nu \in N; \nu \neq \pm 1).$$

This is an expansion of $\tilde{f} = Ff$ in terms of f and the eigenfunctions of $B = A^{-1}T$. For $\nu = \pm \nu_0$, $\lambda(\nu) = \Lambda(\nu) = 0$.

Conversely, one may express f in $\tilde{f}$ by considering the operator FTF^{-1} in $L^2(N,\sigma)$; this operator is diagonalized by F^{-1} and serves as the analogue of $A^{-1}T$. As in [5] one finds

$$(9) \qquad f(\mu) = \left(\lambda(\mu)\delta_\mu + \frac{c}{2}\frac{\nu}{\nu-\mu} ,\tilde{f}\right)_\sigma \quad \left(\tilde{f} \in \mathrm{Lip}(N),\ -1 < \mu < 1\right),$$

where the subscript σ means that the values are taken with respect to the measure σ. Thus, more explicitly,

$$(9') \qquad f(\mu) = \lambda(\mu)\tilde{f}(\mu) + \frac{c}{2}⨏_N \frac{\nu}{\nu-\mu}\tilde{f}(\nu)d\sigma(\nu).$$

The measure σ is absolutely continuous with respect to Lebesgue measure on I, and its Radon-Nikodym derivative on I is Hölder continuous. [To compute σ it is necessary to write (8) and (9') as contour integrals]. Another observation is that F maps Lip(I) one-to-one onto Lip($\dot{N}$).

Next, let c = 1. Then the operator A is no longer invertible. So we work with the unbounded operator $T^{-1}A$. It has spectrum $I^{-1} \cup \{0\}$, where we have written $I^{-1} = (-\infty,-1] \cup [1,\infty)$ (observe that $1/\nu_0 \to 0$ as $c \uparrow 1$). There are two difficulties which are closely related to each other, viz.

(i) $T^{-1}A$ has a double eigenvalue 0

(ii) $T^{-1}A$ no longer is hermitian.

To overcome these difficulties we proceed as follows (see [7]). The space $L^2(I)$ $\big($or $L^2(I,A)\big)$ can be decomposed as a direct sum $G_0 \oplus G_1$ such that G_0 and G_1 are invariant under $T^{-1}A$ and that

$$\sigma(T^{-1}A|G_0) = I^{-1} \quad , \quad \sigma(T^{-1}A|G_1) = \{0\}.$$

It is even possible to give an explicit formula for the projection onto G_1 along G_0 (see [4, p.178] or [6, section VII.3]). One finds

$$(10) \qquad \begin{cases} G_1 = \text{the set of linear functions} \\ G_0 = \{f \in L^2(I) : m_1(f) = m_2(f) = 0\}, \end{cases}$$

where $m_i(f) = \int_{-1}^{1} \mu^i f(\mu)d\mu$. It may be observed that G_0 and G_1 are not orthogonal and that G_1 has dimension 2, in accordance with the fact that 0 is an eigenvalue of multiplicity 2. On G_1, the operator $T^{-1}A$ behaves badly, in the sense that it is not hermitian; in fact, with respect to the basis $\{1_I, d\}$, where d denotes the function $d(\mu) = \mu$, it is represented by the matrix $\begin{pmatrix} 0 & 1 \\ 0 & 0 \end{pmatrix}$. On G_0, $T^{-1}A$ is invertible, since 0 does not belong to the spectrum. We put $B_0 = \left(T^{-1}A|G_0\right)^{-1}$; this is the analogue of the operator $A^{-1}T$ considered in the case $0 < c < 1$. Concerning B_0 the following can be said.

First of all, B_0 is hermitian with respect to the inner product $(\ , \ )_A$. Next, there is a cyclic element for B_0. It is obtained as follows. An elementary calculation yields

$$(11) \qquad B_0 g = Tg - \frac{3}{2} m_3(g) . 1_I \qquad (g \in G_0).$$

Therefore, B_0 sends polynomials into polynomials, the degree of a polynomial increasing by 1 under the map B_0. Now the coefficients of the polynomials in G_0 satisfy two independent homogeneous linear relations, on account of (10). So G_0 contains a polynomial d_0 of degree 2, and by applying **repeatedly** B_0 to this polynomial and taking linear combinations one gets all polynomials in G_0.

From this fact one may conclude that d_0 is a cyclic element. One has

(12) $$d_0(\mu) = -3\mu^2 + 9/5.$$

Of course, d_0 is only determined up to a nonzero constant factor. We have chosen this factor in such a way that $(\phi_\nu, d_0) = 1$, where the ϕ_ν are the eigenfunctions of B_0. These eigenfunctions are obtained from the functions ϕ_ν considered earlier by simply substituting $c = 1$. Finally, B_0 has spectrum I.

The conditions of theorem 1 are fulfilled with $H = G_0$, $B = B_0$, $L = \mathrm{Lip}(I) \cap G_0$. Hence, there exists a (unique) unitary map F from G_0 onto some space $L^2(I,\sigma)$ such that F diagonalizes B_0, while $Fd_0 = 1_I$. On the subspace $\mathrm{Lip}(I) \cap G_0$, the anti-dual of which contains the ϕ_ν, the transformation F is given by

$$(Ff)(\nu) = \tilde{f}(\nu) = \overline{(\phi_\nu, f)} = \langle \lambda(\nu)\delta_\nu - \frac{1}{2}\frac{\mu}{\mu-\nu}, f\rangle ,$$

where $\lambda(\nu) = 1 + \frac{\nu}{2} \oint_{-1}^{1} \frac{d\mu}{\mu-\nu} = \frac{1}{2}\int_{-1}^{1} \frac{\mu}{\mu-\nu}\, d\mu$. Thus

(13) $$(Ff)(\nu) = \tilde{f}(\nu) = -\frac{1}{2}\int_{-1}^{1} \frac{\mu}{\mu-\nu} \big(f(\mu) - f(\nu)\big)d\mu.$$

Conversely, f may be expressed in $\tilde{f}$.

4. APPLICATIONS

We indicate very briefly how the foregoing results can be applied to solve boundary value problems connected with the differential equation (1). We restrict ourselves to the discussion of the homogeneous equation for which $g = 0$. Moreover, we take $c = 1$. For simplicity, we suppose that $0 \in J$.

Let us recall that an arbitrary function $f \in L^2(I)$ can be decomposed as

(14) $$f(\mu) = a + b\mu + f_0(\mu) \quad , \text{ with } f_0 \in G_0.$$

The function f_0 satisfies the relations $m_1(f) = m_2(f) = 0$. Hence, since $\int_{-1}^{1} \mu d\mu = \int_{-1}^{1} \mu^3 d\mu = 0$ and $\int_{-1}^{1} \mu^2 d\mu = \frac{2}{3}$, the coefficients a and b in (14) are given by

(15) $$a = \frac{3}{2}\, m_2(f) \quad , \quad b = \frac{3}{2}\, m_1(f).$$

We further recall that $T^{-1}A$ acts on G_1 as the matrix $\begin{pmatrix} 01 \\ 00 \end{pmatrix}$ and that $B_0 = (T^{-1}A|G_0)^{-1}$ is diagonalized by F. Thus the homogeneous equation (1), or rather the equation $d\psi(x)dx = -T^{-1}A\psi$ is equivalent with the system

(16) $$\begin{cases} dm_2\big(\psi(x)\big)/dx = -m_1\big(\psi(x)\big)\, ,\ dm_1\big(\psi(x)\big)/dx = 0 \\ d\widetilde{\psi_0(x)}/dx = -T^{-1}\widetilde{\psi_0(x)}. \end{cases}$$

The relations (16) imply that $m_1\big(\psi(x)\big)$ is a constant, $m_2\big(\psi(x)\big)$ is a linear function of x, and $\widetilde{\psi_0(x)}$ has the form $\widetilde{\psi_0(x)}(\nu) = e^{-x/\nu}\widetilde{\psi_0(0)}$. Now suppose that one has imposed boundary and/or growth conditions on the function ψ in such a way that $\psi(0)$ is determined by these conditions. Then $m_1\big(\psi(0)\big)$, $m_2\big(\psi(0)\big)$ and

$\widetilde{\psi_0(0)}$ are known. Then also $m_1(\psi(x))$, $m_2(\psi(x))$ and $\widetilde{\psi_0(x)}$ are known. Applying F^{-1} we get $\psi_0(x)$. Thus $\psi(x)$ is determined.

In particular, let $J = [0,\infty)$. Write $I_+ = [0,1], I_- = [-1,0]$. Then $\psi_0(x)$, and so $\psi(x)$, is exponentially increasing if $\widetilde{\psi_0(0)}$ does not vanish on I_-, whereas $\psi_0(x)$ is exponentially decreasing in the opposite case. In realistic problems it is asked whether one can solve the differential equation (1) under conditions of the following type

a) $\widetilde{\psi_0(0)} \in L^2(I_+,\sigma)$

b) $\psi(0)|I_+$ is a prescribed function in $L^2(I_+)$ (half-range boundary condition).

The answer to this question is given in the so-called half-range theory (see [7]). It turns out that there is a unique solution ψ if one imposes one extra condition of the form $m_1(\psi(x)) = \alpha$, where α is a given complex number. We give some details.

Put $G_p = F^{-1}L^2(I_+,\sigma)$ and denote by P_+ the projection in $L^2(I)$ that projects onto $L^2(I_+)$ along $L^2(I_-)$. One can prove that P_+ maps G_p one-to-one onto a subspace $H_{+,0}$ of $H_+ = L^2(I_+)$ of codimension 1, and one can derive an explicit formula for $P = (P_+|G_p)^{-1}$. The subspace $H_{+,0}$ does not contain the constants $\neq 0$. Now, by a), $\psi(0)$ must have the form $a + bd + g$, where a and b are constants, d is the function $d(\mu) = \mu$ and $g \in G_p$. Since G_p is a subspace of G_0, $m_1(\psi(0)) = \frac{2}{3}b$. So if we require that $m_1(\psi(0))$ has a given value α, this fixes the constant b. For this value $b = \frac{3}{2}\alpha$, there exists exactly one number a such that the function $f_+ = (\psi(0) - a - bd)|I_+$ belongs to $H_{+,0}$. Having found a and f_+ we obtain $\psi(0)$ in the form $\psi(0) = a + bd + Pf_+$.

The Milne problem concerning the density of light emanating from a star is the particular case $\psi(0)|I_+ = 0$. In [7], also the inhomogeneous equation (1) is considered. In [5], the case $0 < c < 1$ is treated.

REFERENCES

[1] BEREZANSKII, JU.M., Expansions in eigenfunctions of selfadjoint operators, Translations Math. Monographs XVII (1968).

[2] CASE, K.M. - ZWEIFEL, P.F., Linear Transport Theory, Addison-Wesley Publ. Co, Reading 1967.

[3] DUNFORD, N. - SCHWARTZ, J.T., Linear Operators I, Interscience Publ., New York 1958.

[4] FOIAŞ, C., Décompositions en opérateurs et vecteurs propres I et II, Rev. Roumaine Math. Pures et Appl. 7, 241-282 et 571-602 (1962).

[5] HANGELBROEK, R.J., A functional analytic approach to the linear transport equation, Thesis, Groningen 1973.

[6] KATO, T., Perturbation theory for linear operators, Springer, Berlin 1967.

[7] LEKKERKERKER, C.G., The linear transport equation. The degenerate case c = 1. I. Full range theory; II. Half-range theory. To appear in Proc. Edinburgh Math. Soc.

[8] MAURIN, K., General eigenfunction expansions and unitary representations of topological groups, Warszawa 1968.

[9] Mc CORMICK, N.J. - KUŠČER, I., Singular eigenfunction expansions in neutron transport theory, Advances in Nuclear Science and Technology 7, 181-282 (1973).

W. Eckhaus (ed.), New Developments in Differential Equations

INTEGRAL-ORDINARY DIFFERENTIAL-BOUNDARY SUBSPACES AND SPECTRAL THEORY

Aalt Dijksma

1. Introduction. In recent years a number of papers have appeared that deal with a class of linear integral-ordinary differential-boundary (eigenvalue) problems. See A.M. Krall's paper [9] where a long list of references can be found. In vector notation such problems are associated with formal expressions and side conditions of the form

$$(1.1)\quad (Sy)(x) = P_1(x)y'(x) + P_0(x)y(x) + H(x)[M_1y(a) + N_1y(b)] + K(x)\int_a^b F(t)\, y(t)dt + L(x)c,$$

$$(1.2)\quad My(a) + Ny(b) + \int_a^b G(t)\, y(t)dt = 0.$$

Here y is a vector-valued function and P_1, P_0, H, K, F, L and G are matrix-valued functions on a compact interval $[a,b] \subset \mathbb{R}$, M_1, N_1, M and N are constant matrices and c is an arbitrary parameter vector (cf. [10]). Some of these papers are concerned with finding a suitable formulation of the problem adjoint to the given one. Such formulations should deal with formal expressions and side conditions similar to (1.1) and (1.2) except that in (1.1) $P_1D + P_0$ is to be replaced by $-DP_1^* + P_0^*$, $D = d/dx$. The adjoint is then used to determine the conditions under which the original problem is selfadjoint and to calculate the eigenvalues.

Restricting ourselves to problems in the Hilbertspace $H = L^2(\iota)$ of vector-valued functions on $\iota = (a,b) \subset \mathbb{R}$, we shall discuss a method of classifying the operators associated with such problems along with their adjoints, and eigenfunction expansion results for the case that the operators are self-adjoint. In H operators associated with side conditions of the form (1.2) are usually not densely defined and therefore the adjoints of such operators do not exist as operators. This is the reason why we shall use the notion of subspaces (= closed linear manifolds) in the Hilbertspace $H^2 = H \oplus H$, which is a generalization of the notion of closed operators by replacing them by their graphs in H^2. The subspace ideas have their beginning in a paper of R. Arens [1] and have been further developed by E.A. Coddington [2], [3], [4].

The report that follows is on work done in collaboration with Prof. Coddington. Part of this work has been done in the academic year 1973-74 during which I had the opportunity to work at the University of California at Los Angeles. This is an appropriate place to acknowledge my gratitude to the Department of Mathematics of UCLA and Prof. Coddington for providing this opportunity.

2. Subspaces. Let H be a Hilbertspace over the complex numbers $\mathbb{C}$, and $H^2 = H \oplus H$, considered as a Hilbertspace. The domain $D(T)$ and range $R(T)$ of a subspace T in H^2 are defined by

$$D(T) = \{f \in H \mid \{f,g\} \in T, \text{ some } g \in H\},$$

$$R(T) = \{g \in H \mid \{f,g\} \in T, \text{ some } f \in H\}.$$

For $f \in D(T)$ we let

$$T(f) = \{g \in H \mid \{f,g\} \in T\}.$$

The subspace T is an operator if $T(0) = \{0\}$ and then we write $T(f) = Tf$. We consider subspaces as linear relations with

$$\alpha T = \{\{f,\alpha g\} \in H^2 \mid \{f,g\} \in T\}, \quad \alpha \in \mathbb{C},$$

$$T + S = \{\{f,g+h\} \in H^2 \mid \{f,g\} \in T, \{f,h\} \in S\},$$

$$T^{-1} = \{\{g,f\} \in H^2 \mid \{f,g\} \in T\}.$$

The algebraic sum is defined by

$$T \dotplus S = \{\{f+h,g+k\} \in H^2 \mid \{f,g\} \in T, \{h,k\} \in S\}.$$

It is called direct if $T \cap S = \{\{0,0\}\}$ and orthogonal if $T \perp S$ in H^2. In the last case we write $T \dotplus S = T \oplus S$. If $T \subset S$ then the orthogonal complement of T in S is denoted by $S \ominus T$. If $S = H^2$ we write $H^2 \ominus T = T^{\perp}$.

In $H^2 \times H^2$ we define the semi-bilinear form $\langle\, ,\, \rangle$ by

$$\langle \{f,g\}, \{h,k\} \rangle = (g,h) - (f,k), \quad \{f,g\}, \{h,k\} \in H^2.$$

It is non-degenerate. For subsets A, B of H^2, $\langle A,B\rangle = 0$ means that $\langle\{f,g\},\{h,k\}\rangle=0$ for all $\{f,g\} \in A$ and all $\{h,k\} \in B$. The adjoint of a subspace T is the subspace T^* defined by

$$T^* = \{\{h,k\} \in H^2 \mid \langle T, \{\{h,k\}\} \rangle = 0\}.$$

It is easy to see that $T^{**} = T$, $S \subset T$ implies that $T^* \subset S^*$, etc. Two subspaces T_1, T_2 are said to be formally adjoint if $\langle T_1, T_2\rangle = 0$, i.e., $T_1 \subset T_2^*$ or $T_2 \subset T_1^*$. A subspace S is called symmetric if $\langle S,S\rangle = 0$ or, equivalently, $S \subset S^*$. A selfadjoint subspace H is one which satisfies $H = H^*$. We have the following result (cf. [2], [7]).

Theorem 2.1. Let S be a symmetric subspace in H^2 and let

$$M_S(\ell) = \{\{h,k\} \in S^* \mid k = \ell h\}, \quad \ell \in \mathbb{C}.$$

Then

(i) $\dim M_S(\ell)$ is constant for $\ell \in \mathbb{C}^+$ and for $\ell \in \mathbb{C}^-$, where $\mathbb{C}^{\pm} = \{\ell \in \mathbb{C} \mid \operatorname{Im} \ell \gtrless 0\}$,

(ii) $S^* = S \dotplus M_S(\ell) \dotplus M_S(\bar{\ell})$, $\ell \in \mathbb{C}^+$ (direct sum),

(iii) there exist selfadjoint extensions H of S in H^2, i.e., subspaces H in H^2 which satisfy $S \subset H = H^* \subset S^*$ if and only if $\dim M_S(\ell) = \dim M_S(\bar{\ell})$ $\ell \in \mathbb{C}^+$,

(iv) there always exist Hilbertspaces R containing H as a subspace and selfadjoint subspaces H in R^2 such that $S \subset H$.

For any subspace T in H^2 we may write $T = T_s \oplus T_\infty$, where

$$T_\infty = \{\{f,g\} \in T \mid f=0\}, \quad T_s = T \ominus T_\infty$$

The subspace T_s is called the operator part of T. It is a closed operator in H with $D(T_s) = D(T)$ dense in $T^*(0)^{\perp}$ and $R(T_s) \subset T(0)^{\perp}$.

The following theorem is due to R. Arens [1].

Theorem 2.2. If $H = H_s \oplus H_\infty$ is a selfadjoint subspace in H^2, then H_s is a densely defined selfadjoint operator in $H(0)^\perp$.

3. The basic linear ordinary differential operators. Let L be a system of n first order ordinary differential expressions on $\imath = (a,b) \subset \mathbb{R}$:

$$L = P_1 D + P_0 \ , \ D = d/dx,$$

where P_j is an nxn matrix whose entries belong to $C^j(\imath)$, $j = 0,1$, and $\det P_1(x) \neq 0, x \in \imath$. The system L is called regular if $\bar{\imath} = [a,b]$ is compact, the entries of P_j belong to $C^j(\bar{\imath})$, $j = 0,1$, and $\det P_1(x) \neq 0$, $x \in \bar{\imath}$. Otherwise L is called singular. The Lagrange adjoint of L is defined by

$$L^+ = -DP_1{}^* + P_0{}^* = Q_1 D + Q_0,$$

where $Q_1 = -P_1{}^*$, $Q_0 = P_0{}^* - (P_1{}')^*$. Thus if L is regular then so is L^+ and $L^{++} = L$. The system L is called formally symmetric if $L = L^+$.

Let $H = L^2(\imath)$, the Hilbertspace of nx1 matrix-valued functions on $\imath$ with innerproduct

$$(f,g) = \int_a^b g^* f \quad , \quad f,g \in L^2(\imath).$$

In H^2 we define the minimal operators T_0 and $T_0{}^+$ associated with L and L^+ by

$$T_0 = \{\{f,Lf\} \mid f \in C_0^1(\imath)\}^c,$$

$$T_0{}^+ = \{\{f^+,L^+f^+\} \mid f^+ \in C_0^1(\imath)\}^c,$$

where c denotes the closure in H^2. It is easy to see that $\langle T_0, T_0{}^+\rangle = 0$, i.e., that T_0 and $T_0{}^+$ are formally adjoint. Their adjoints T and T^+ are called the maximal operators and are given by

$$T = (T_0{}^+)^* = \{\{f,Lf\} \mid f \in H \cap AC_{loc}(\imath),\ Lf \in H\},$$

$$T^+ = (T_0)^* = \{\{f^+,L^+f^+\} \mid f^+ \in H \cap AC_{loc}(\imath),\ L^+f^+ \in H\}.$$

By Green's formula the semi-bilinear form on $H^2 \times H^2$ restricted to $T \times T^+$ deals with the behavior of the functions $f \in D(T)$, $f^+ \in D(T^+)$ at the end points of $\imath$:

$$(3.1) \quad \langle\{f,Lf\}, \{f^+,L^+f^+\}\rangle = (f^+(x))^* P_1(x) f(x) \Big|_{x\to a}^{x\to b}$$

$$= \lim_{x\to b}(f^+(x))^* P_1(x) f(x) - \lim_{x\to a} (f^+(x))^* P_1(x) f(x).$$

If L is regular then the functions $f \in D(T)$, $f^+ \in D(T^+)$ can be continuously extended to all of $\bar{\imath}$ and (3.1) reads

$$(3.2) \quad \langle\{f,Lf\}, \{f^+,L^+f^+\}\rangle = (f^+(b))^* P_1(b) f(b) - (f^+(a))^* P_1(a) f(a).$$

Suppose that the subspaces S and its adjoint S^* in H^2 satisfy

$$(3.3) \qquad T_0 \subset S \quad \text{and} \quad T_0{}^+ \subset S^*.$$

Then $S \subset T$, $S^* \subset T^+$, S and S^* are operators and it can be shown that there exist finite dimensional subspaces C and C^+ of T and T^+ such that

$$S = T \cap (C^+)^* \text{ and } S^* = T^+ \cap C^*.$$

It follows from (3.1) that S is a restriction of T and S* is a restriction of T^+ by means of a finite number of conditions at the two endpoints of ι. Conversely, if S is a restriction of T by a finite number of such two point-boundary conditions, then (3.3) holds. For example, let L be regular. If $S = T_o$ then we have using (3.2)

$$S = \{\{f, Lf\} \in T \mid f(a) = f(b) = 0\},$$

and if S = T then we have

$$S^* = \{\{f^+, L^+f^+\} \in T^+ \mid f^+(a) = f^+(b) = 0\} = T_o^+.$$

4. Integral-ordinary differential-boundary subspaces. In order to obtain subspaces with side conditions other than two point-boundary conditions we make T_o and T_o^+ smaller. We introduce two finite dimensional subspaces B, B^+ in H^2 and define the formally adjoint operators

$$S_o = T_o \cap (B^+)^* \quad \text{and} \quad S_o^+ = T_o^+ \cap B^*.$$

It can be shown that the adjoints of S_o, S_o^+ are given by

$$S_o^* = T^+ \dotplus B^+ \text{ and } (S_o^+)^* = T \dotplus B.$$

Without loss of generality we may and do assume that these algebraic sums are direct. We want to characterize all subspaces S and their adjoints S* that satisfy

$$(4.1) \quad \begin{cases} S_o \subset S, & \dim S \ominus S_o = d, \\ S_o^+ \subset S^*, & \dim S^* \ominus S_o^+ = d^+. \end{cases}$$

If (4.1) holds then it can be shown that we must have that

$$d + d^+ = \dim T \ominus T_o + \dim B + \dim B^+,$$

and that there exist subspaces C, C^+ of $T \dotplus B$ and $T^+ \dotplus B^+$ with $\dim C = d$, $\dim C^+ = d^+$ such that

$$S = (T \dotplus B) \cap (C^+)^* \text{ and } S^* = (T^+ \dotplus B^+) \cap C^*.$$

The subspaces C and C^+ represent the side conditions which now are a mixture of two point boundary and integral conditions. We want to make these conditions more explicit. In order to do so we shall make use of the following notations.

To indicate that a matrix A has p rows and q columns we write $A(p \times q)$. Thus $f \in H$ is of the form $f(n \times 1)$. The $p \times q$ zero matrix and the $p \times p$ identity matrix are denoted by 0_p^q and I_p. By $F(n \times p) \in H$, $\delta(n \times p) \in D(T)$ is linearly independent mod $D^p(T_o)$ or ${}^l\varphi(n \times p)$ is a basis for S(0) we mean that the p columns of these matrices have these properties. For $F(n \times p)$, $G(n \times q) \in H$ we define the "matrix innerproduct" (F,G) to be the $q \times p$ matrix whose i,j-th element is

$$(F,G)_{ij} = \int_a^b \sum_{k=1}^n \bar{G}_{ki} F_{kj}.$$

If $C(p \times r)$, $D(q \times s)$ are constant matrices, then

$$(FC,G) = (F,G)C \;,\; (F,GD) = D^* (F,G).$$

If $\{\delta, L\delta\}$ $(n \times p) \in T$, $\{\delta^+, L^+\delta^+\}$ $(n \times q) \in T^+$ then

$$<\{\delta,L\delta\} \;,\; \{\delta^+,L^+\delta^+\}> = (L\delta,\delta^+)-(\delta,L^+\delta^+)$$

is well defined. We denote by $(F:G)$ the matrix whose columns are obtained by placing the columns of G next to those of F in the order indicated.

We shall also make use of the following definitions:

$$B_2 = \{\{\sigma,\tau\} \in B \mid \sigma \in \mathcal{D}\,(T)\},\; B_1 = B \ominus B_2,$$

$$B_2^+ = \{\{\sigma^+,\tau^+\} \in B^+ \mid \sigma^+ \in \mathcal{D}\,(T^+)\},\; B_1^+ = B^+ \ominus B_2^+,$$

and

$$S_1 = T_o \cap (B_1^+)^* \;,\; S_1^+ = T_o^+ \cap B_1^*$$

It is not difficult to see that

$$S_o \subset S_1 \subset (S_1^+)^* = T \dotplus B_1 \subset (S_o^+)^* = T \dotplus B_1 \dotplus ((S_o^+)^*)_\infty \text{ (direct sums)},$$

and that a similar result holds starting with S_o^+. Moreover, it is easy to see that B_1, $S_1^* = T^+ \dotplus B_1^+$ are operators, $\mathcal{D}^o\,(S_1)$ is dense in H, $\dim (S_o^+)^*(0) = \dim B_2$, etc. The following result is a special case of a theorem that will be stated and proved elsewhere.

Theorem 4.1. Let S, S^* satisfy (4.1) and suppose that

$$(4.2)\quad \begin{cases} {}^1\varphi \quad (n \times s_1) & \text{is a basis for } S(0), \\ {}^1\varphi^+ \quad (n \times s_1^+) & \text{is a basis for } S^*(0), \end{cases}$$

$$(4.3)\quad \begin{cases} ({}^1\varphi : {}^2\varphi) \quad (n \times (s_1 + s_2)) & \text{is a basis for } (S_o^+)^*(0), \\ ({}^1\varphi^+ : {}^2\varphi^+) \quad (n \times (s_1^+ + s_2^+)) & \text{is a basis for } S_o^*(0). \end{cases}$$

Then there exist

$$(4.4)\quad \begin{cases} u = \{\gamma, L\gamma\} \in T \;,\; \alpha = \{{}^1\sigma, {}^1\tau\} \in B_1 \;,\; (n \times s_2^+), \\ u^+ = \{\gamma^+,L^+\gamma^+\} \in T^+ \;,\; \alpha^+ = \{{}^1\sigma^+,{}^1\tau^+\} \in B_1^+ \;,\; (n \times s_2), \end{cases}$$

$$(4.5)\quad \begin{cases} v = \{\delta, L\delta\} \in T \;,\; \beta = \{{}^2\sigma, {}^2\tau\} \in B_1 \;,\; (n \times (d-s_1 - s_2^+)) , \\ v^+ = \{\delta^+,L^+\delta^+\} \in T^+ \;,\; \beta^+ = \{{}^2\sigma^+,{}^2\tau^+\} \in B_1^+ \;,\; (n \times (d^+- s_1^+- s_2)), \end{cases}$$

and a constant matrix E $(s_2 \times s_2^+)$ such that

$$(4.6)\quad \begin{cases} \delta + {}^2\sigma & \text{is linearly independent mod } \mathcal{D}\,(S_1), \\ \delta^+ + {}^2\sigma^+ & \text{is linearly independent mod } \mathcal{D}\,(S_1^+), \end{cases}$$

$$(4.7)\quad <v + \beta,\; v^+ + \beta^+> = 0_{d^+-s_1^+-s_2}^{d-s_1-s_2^+} \;,$$

and if

$$(4.8)\quad \Psi = {}^2\varphi\,[E - \tfrac{1}{2}\langle u+\alpha, u^+ + \alpha^+\rangle],\qquad \Psi^+ = {}^2\varphi^+\,[E + \tfrac{1}{2}\langle u+\alpha, u^+ + \alpha^+\rangle]^*,$$

$$(4.9)\quad \zeta = -{}^2\varphi\,\langle v+\beta, u^+ + \alpha^+\rangle,\qquad \zeta^+ = {}^2\varphi^+\,\langle u+\alpha, v^+ + \beta^+\rangle^*,$$

then

$$(4.10)\quad S = \{\{h + \sigma, Lh + \tau + \varphi\} \mid h \in \mathcal{D}(T),\ \{\sigma,\tau\} \in B_1,\ \varphi \in (S_o^+)^*(0) \text{ with}$$

$$(h + \sigma, {}^1\varphi^+) = 0^1_{s_1^+},$$

$$(h + \sigma, \zeta^+) - \langle\{h + \sigma, Lh + \tau\}, v^+ + \beta^+\rangle = 0^1_{d^+ - s_1^+ - s_2},$$

$$\varphi = {}^1\varphi c + {}^2\varphi[(h + \sigma, \Psi^+) - \langle\{h + \sigma, Lh + \tau\}, u^+ + \alpha^+\rangle],$$

where $c(s_1 \times 1) \in \mathbb{C}^{s_1}$ is arbitrary$\}$,

$$(4.11)\quad S^* = \{\{h^+ + \sigma^+, L^+h^+ + \tau^+ + \varphi^+\} \mid h^+ \in \mathcal{D}(T^+), \{\sigma^+,\tau^+\} \in B_1^+, \varphi^+ \in S_o^*(0) \text{ with}$$

$$({}^1\varphi, h^+ + \sigma^+) = 0_1^{s_1},$$

$$(\zeta, h^+ + \sigma^+) + \langle v + \beta, \{h^+ + \sigma^+, L^+h^+ + \tau^+\}\rangle = 0_1^{d - s_1 - s_2^+},$$

$$\varphi^+ = {}^1\varphi^+ c^+ + {}^2\varphi^+[(\Psi, h^+ + \sigma^+) + \langle u + \alpha, \{h^+ + \sigma^+, L^+h^+ + \tau^+\}\rangle]^*,$$

where $c^+(s_1^+ \times 1) \in \mathbb{C}^{s_1^+}$ is arbitrary$\}$.

<u>Conversely if the bases in</u> (4.3) <u>and the matrix</u> $E\ (s_2 \times s_2^+)$ <u>are given, the elements of</u> (4.4) <u>and</u> (4.5) <u>exist satisfying</u> (4.6) <u>and</u> (4.7) <u>and</u> Ψ, Ψ^+, ζ and ζ^+ <u>are defined by</u> (4.8) <u>and</u> (4.9), <u>then</u> S <u>defined by</u> (4.10) <u>satisfies</u> (4.11), <u>and</u> (4.1) <u>and</u> (4.2) <u>hold</u>.

If L is regular then in Theorem 4.1 we may replace the elements mentioned in (4.4), (4.5) by constant matrices and fixed bases of B_1 and B_1^+, and express (4.6), (4.7), (4.8) and (4.9) in terms of these matrices and bases. The subspaces S, S* are then determined by these matrices, the values of the functions $h \in \mathcal{D}(T)$, $h^+ \in \mathcal{D}(T^+)$ at the endpoints of $\imath$ and certain integrals involving h, Lh, h^+ and L^+h^+.

<u>Example 4.2</u>. If L is regular and $B_1 = B_1^+ = \{\{0,0\}\}$ then S in (4.10) of Theorem 4.1 involves an expression and side conditions of the form (1.1), (1.2) and so does S* in (4.11). The parameter c occuring in (1.1) is the same parameter c that occurs in the description (4.10) of S. It's presence stipulates the subspace (non-operator) character of S.

<u>Example 4.3. (Interface conditions)</u>. Let L be arbitrary and let $c \in \imath$ be fixed. Let B be spanned by $\{\sigma^1,\tau^1\}$ $(n \times n)$ where $\sigma^1 = \tau^1 = 0_n^n$ on (a,c), $\sigma^1 \in C^1[c,b)$, $\sigma^1(c) = -P_1^{-1}(c)$, σ^1 vanishes in a neighborhood of b and $\tau^1 = L\sigma^1$ on $[c,b)$. Similarly, let B^+ be spanned by $\{\sigma^{+1},\tau^{+1}\}$ $(n \times n)$ where $\sigma^{+1} = \tau^{+1} = 0_n^n$ on (a,c), $\sigma^{+1} \in C^1[c,b)$, $\sigma^{+1}(c) = (P_1{}^*(c))^{-1}$, σ^{+1} vanishes in a neighborhood of b and $\tau^{+1} = L^+\sigma^{+1}$ on $[c,b)$. Then $T \cap B = T^+ \cap B^+ = B_2 = B_2^+ = \{\{0,0\}\}$ and hence $B_1 = B$, $B_1^+ = B^+$. Let S and S* satisfy (4.1) with $d+d^+ = \dim T \ominus T_o + 2n$. Then S and S* are operators and by Theorem 4.1 can be described as follows.

$$S = \{\{f,Lf\} \mid f \in H \cap AC_{loc}(\iota\setminus\{c\}),\ Lf \in H, \text{ and}$$
$$(\delta^+(x))^* P_1(x)\, f(x) \Big|_{x\to a}^{x\to b} + (\delta^+(c))^* P_1(c)\, f(c-o) -$$
$$[(\delta^+(c))^* P_1(c) + (D^+)^*]\, f(c+o) = 0_{d}^{1}{}_{+}\},$$

$$S^* = \{\{f^+, L^+f^+\} \mid f^+ \in H \cap AC_{loc}(\iota\setminus\{c\}),\ L^+f^+ \in H \text{ and}$$

$$\delta^*(x)\, P_1^*(x)\, f^+(x) \Big|_{x\to a}^{x\to b} + \delta^*(c)P_1^*(c)\, f^+(c-o) - [(\delta^*(c))P_1^*(c) - D^*]f^+(c+o)$$
$$= 0_d^1\}.$$

Here $\{\delta, L\delta\} \in T$, $\{\delta^+, L^+\delta^+\} \in T^+$ and the constant matrices $D(n \times d)$, $D^+(n \times d^+)$ are such that $\{\delta,L\delta\}+\{\sigma^1,\tau^1\}D$, $\{\delta^+,L^+\delta^+\} + \{\sigma^{+1},\tau^{+1}\}D^+$ satisfy (4.6) with ${}^2\sigma=\sigma^1 D$, ${}^2\sigma^+ = \sigma^{+1} D^+$ and

$$(\delta^+(x))^* P_1(x)\delta(x) \Big|_{x\to a}^{x\to b} - (D^+)^*\sigma(c) + (\delta^+(c))^* D + (D^+)^* P_1^{-1}(c)\, D = 0_{d^+}^{d},$$

which is (4.7) of Theorem 4.1.

5. Selfadjoint integral-ordinary differential-boundary subspaces. In this section we assume that L is formally symmetric and $B = B^+$. Then $S_o = S_o^+$ is a symmetric operator and we want to characterize the selfadjoint extensions H of S_o in H^2. As to the existence of such extensions we remark that it can be proved that

$$\dim M_{S_o}(\ell) = \dim M_{T_o}(\ell) + \dim B,\ \ell \in \mathbb{C}^{\pm}.$$

Hence as a consequence of Theorem 2.1 (iii) we have the following result (cf. [5]).

Theorem 5.1. There exist selfadjoint extensions H in H^2 of $S_o = T_o \cap B^*$ where T_o is the symmetric minimal operator associated with $L^o = L^{+o}$ in H if and only if T_o has selfadjoint extensions in H^2.

We now also assume that $\dim M_{S_o}(\ell) = \dim M_{S_o}(\bar{\ell}) = d$, say, $\ell \in \mathbb{C}^+$.

Using Theorem 4.1 we obtain the following characterization of the selfadjoint extension H of S_o in H^2 which is Theorem (3.3) of [5] applied to differential operators.

Theorem 5.2. Let H satisfy

(5.1) $S_o \subset H = H^* \subset S_o^*$,

and suppose that

(5.2) ${}^1\varphi$ $(n \times s_1)$ is a basis for $H(0)$,

(5.3) $({}^1\varphi : {}^2\varphi)$ $(n \times (s_1 + s_2))$ is a basis for $S_o^*(0)$.

Then there exist

(5.4) $u = \{\gamma, L\gamma\} \in T$, $\alpha = \{{}^1\sigma, {}^1\tau\} \in B_1$, $(n \times s_2)$,

(5.5) $v = \{\delta, L\delta\} \in T$, $\beta = \{{}^2\sigma, {}^2\tau\} \in B_1$, $(n \times (d-s_1-s_2))$, and a constant matrix $E = E^*$ $(s_2 \times s_2)$ such that

(5.6) $\delta + {}^2\sigma$ is linearly independent mod $D(S_1)$,

(5.7) $\langle v+\beta\ ,\ v+\beta\rangle = 0^{d-s_1-s_2}_{d-s_1-s_2}$,

and if

(5.8) $\Psi = {}^2\varphi[E - \frac{1}{2}\langle u+\alpha\ ,\ u+\alpha\rangle]$,

(5.9) $\zeta = -{}^2\varphi\ \langle v+\beta\ ,\ u+\alpha\rangle$,

then

(5.10) $H = \{\{h + \sigma,\ Lh + \tau + \varphi\}\ |\ h \in D(T)\ ,\ \{\sigma,\tau\} \in B_1,\ \varphi \in S_o^*(0)$ with

$(h + \sigma,\ {}^1\varphi) = 0^1_{s_1}$,

$(h + \sigma, \zeta) - \langle\{h + \sigma,\ Lh + \tau\},\ v+\beta\rangle = 0^1_{d-s_1-s_2}$,

$\varphi = {}^1\varphi c + {}^2\varphi[(h + \sigma, \Psi) - \langle\{h + \sigma,\ Lh + \tau\},\ u+\alpha\rangle]$,

where $c\ (s_1 \times 1) \in \mathbb{C}^{s_1}$ is arbitrary$\}$.

Moreover, if the bases in (5.2), (5.3) are orthonormal then the operator part H_s of H is given by

(5.11) $H_s(h + \sigma) = Lh + \tau - {}^1\varphi(Lh + \tau,\ {}^1\varphi) +$

$+\ {}^2\varphi[(h + \sigma, \Psi) - \langle\{h + \sigma,\ Lh + \tau\}, u+\alpha\rangle$.

Conversely, if (5.3) and the constant matrix $E = E^*(s_2 \times s_2)$ are given, the elements in (5.4), (5.5) exist satisfying (5.6), (5.7) and ψ, ζ are defined by (5.8), (5.9), then H defined by (5.10) satisfies (5.1) and (5.2). If the basis in (5.3) is orthonormal then H_s is given by (5.11).

Proof. Let H satisfy (5.1), (5.2) and let (5.3) hold. By Theorem 4.1 with ${}^1\varphi^+ = {}^1\varphi$, ${}^2\varphi^+ = {}^2\varphi$, $s_1^+ = s_1$, $s_2^+ = s_2$, $d^+ = d$, and $S = H = H^* = S^*$, there exist elements as in (4.4), (4.5) satisfying (4.6), (4.7) and a constant matrix $E(s_2 \times s_2)$ such that if $\Psi, \Psi^+, \zeta, \zeta^+$ are defined by (4.8), (4.9) then $H = H^*$ is described by (4.10) as well as by (4.11). Thus H satisfies (5.10). We shall show that $v+\beta$ satisfies (5.7), Ψ in (4.8) can be written as in (5.8) for some hermitian matrix E, and ζ in (4.9) can be written as in (5.9).

The operator S_1 is densely defined (and symmetric) in H and therefore there exist a $w(n \times (s_1 + s_2)) \in s_1$ such that

$$\langle\{0, ({}^1\varphi : {}^2\varphi)\}, w\rangle = I_{s_1 + s_2}.$$

We put

$$A = -\langle\{0, ({}^1\varphi : {}^2\varphi)\},\ v+\beta\rangle^*, \quad ((s_1 + s_2) \times (d-s_1-s_2)),$$

$$B = -\langle\{0, ({}^1\varphi : {}^2\varphi)\},\ u^+ + \alpha^+ - u - \alpha\rangle^*, \ ((s_1 + s_2) \times s_2).$$

Then by (4.10), (4.11) of Theorem 4.1 and the facts that $S = S^*$, $S_1 \subset S_1^*$ and $u+\alpha$, $u^+ + \alpha^+$, $v+\beta \in S_1^*$, we have that for all $\{h+\sigma, Lh+\tau+\varphi\} \in S^1 = H^1$

$$<\{h+\sigma, Lh+\tau+\varphi\}, v+\beta+wA+\{0,\zeta\}> = 0^1_{d-s_1-s_2},$$

$$<\{h+\sigma, Lh+\tau+\varphi\}, u^+ + \alpha^+ - u - \alpha + wB + \{0,\Psi^+\} - \{0,\Psi\}> = 0^1_{s_2},$$

and hence

(5.12) $\quad v+\beta + wA + \{0,\zeta\} \in H,$

(5.13) $\quad u^+ + \alpha^+ - u - \alpha + wB + \{0,\Psi^+\} - \{0,\Psi\} \in H.$

From (5.12) it follows that

$$<v+\beta, v+\beta> = <v+\beta+wA+\{0,\zeta\}, v+\beta+wA+\{0,\zeta\}> = 0^{d-s_1-s_2}_{d-s_1-s_2},$$

which is (5.7). From (5.13) and the description of φ in (5.10) we deduce that

$$\Psi^+ - \Psi = -{}^2\varphi <u^+ + \alpha^+ - u - \alpha + wB, u+\alpha> = -{}^2\varphi <u^+ + \alpha^+ - u - \alpha, u+\alpha>,$$

and from (4.8) it follows that

$$\Psi^+ - \Psi = {}^2\varphi\,[E^* - E - \tfrac{1}{2} <u^+ + \alpha^+, u+\alpha> + \tfrac{1}{2} <u+\alpha, u^+ + \alpha^+>].$$

Since the elements of ${}^2\varphi$ are linearly independent it follows that

(5.14) $$E - \tfrac{1}{2} <u+\alpha, u^+ + \alpha^+ - u - \alpha> = E^* + \tfrac{1}{2} <u^+ + \alpha^+ - u - \alpha, u+\alpha>$$

$$= [E - \tfrac{1}{2} <u+\alpha, u^+ + \alpha^+ - u - \alpha>]^*.$$

Now, Ψ in (4.8) can be written as

$$\Psi = {}^2\varphi\,[E - \tfrac{1}{2} <u+\alpha, u^+ + \alpha^+>] = {}^2\varphi[F - \tfrac{1}{2} <u+\alpha, u+\alpha>],$$

where $F = E - \tfrac{1}{2} <u+\alpha, u^+ + \alpha^+ - u - \alpha> = F^*$ by (5.14). This proves (5.8) where we have written E instead of F. From (5.12) and (5.13) it follows that $<v+\beta, u^+ + \alpha^+> = <v+\beta, u+\alpha>$, which shows that ζ in (4.9) can be written as in (5.9). The orthonormality of the bases in (5.2) and (5.3) and (5.10) clearly imply (5.11).

The second part of Theorem 5.2 is a trivial consequence of the second part of Theorem 4.1. This completes the proof.

Example 5.3. Consider Example 4.3, where we suppose that $L = L^+$. Then $P_1 = P_1^*$ and therefore we may and do assume that $B = B^+$.
Suppose also that T_o has selfadjoint extensions. Then all selfadjoint extensions H in H^2 of $S_o = T_o \cap B^*$ are operators described by

$$H = \{\{f, Lf\} \mid f \in H \cap AC_{loc}(\iota \setminus \{c\}),\ Lf \in H \text{ and}$$
$$\delta^*(x)P_1(x)\ (x) \Big|_{x\to a}^{x\to b} + \delta^*(c)P_1(c)f(c-o) - [\delta^*(c)P_1(c) + D^*]f(c+o)$$
$$= 0^1_d\},$$

where $d = \tfrac{1}{2}\dim T \ominus T_o + n$ and $\{\delta, L\delta\}(n \times d) \in T, D\ (n \times d)$ are such that $\delta + \sigma^1 D$ is linearly independent mod $D(S_1) = D(S_o)$ and

$$\delta^*(x)P_1(x)\delta(x)\Big|_{x\to a}^{x\to b} - D^*\delta(c) + \delta^*(c)\, D + D^*\, P_1^{-1}(c)\, D = 0_d^d \, .$$

Example 5.4. Let $L = L^+$ be regular. We define the symmetric operator $S_o \subset T_o$ by

$$D(S_o) = \{f \in D\,(T_o) \mid \int_a^b d\mu^*_j\, f = 0,\ j = 1, \ldots, k\},$$

where μ_j are $n \times 1$ matrix-valued functions of bounded variation on $\bar{\iota}$. It can be shown (cf. [5]) that there exist $\{\sigma,0\}$ $(n \times k) \in H$ such that

$$D(S_o) = \{f \in D(T_o) \mid (T_o f,\sigma) = 0_k^{\,1}\}.$$

Thus, if B is spanned by $\{\sigma,0\}$ then $S_o = T_o \cap B^*$. Since $\dim M_{T_o}(\ell) = \dim M_{T_o}(\bar{\ell}) = n$, $\ell \in ¢^{\pm}$, selfadjoint extensions H of S_o in H^2 exist and are characterized by Theorem 5.2.

6. Symmetric subspaces and generalized spectral families. This section is a continuation of section 2. Let S be a symmetric subspace in H^2 and let $H = H_s \oplus H_\infty$ be a selfadjoint extension of S in R^2, where R is a Hilbertspace containing H as a subspace (cf. Theorem 2.1(iv)). By Theorem 2.2 and the Spectral Theorem for selfadjoint operators we have that

$$H_s = \int_{-\infty}^{\infty} \lambda \, d\, E_s(\lambda),$$

where $E_s = \{E_s(\lambda) \mid \lambda \in \mathbb{R}\}$ is the unique suitably normalized spectral family of projections in $H(0)^\perp = R \ominus H(0)$ for H_s. The resolvent R_s of H_s can be written as

$$R_s(\ell) = \int_{-\infty}^{\infty} \frac{d\, E_s(\lambda)}{\lambda - \ell}, \quad \ell \in ¢^{\pm} .$$

For $\lambda \in \mathbb{R}$ let the linear operators $E(\lambda)$ on R be defined by

$$E(\lambda)\, f = \begin{cases} E_s(\lambda) f & f \in H(0)^\perp, \\ 0 & f \in H(0). \end{cases}$$

Then $E = \{E(\lambda) \mid \lambda \in \mathbb{R}\}$ is called the spectral family of projections in R for H. The selfadjoint subspace extension H of S in R^2 is called minimal if the set $\{E(\lambda)f \mid f \in H\} \cup H$ is fundamental in R.
The resolvent R_H of H defined by

$$R_H(\ell) = (H - \ell I)^{-1}, \quad \ell \in ¢^{\pm},$$

where I is the identity operator on R, is an operator valued function on $¢^{\pm}$. For $\ell \in ¢^{\pm}$ $R_H(\ell)$ is bounded on R and can be expressed in terms of E:

$$R_H(\ell) = \int_{-\infty}^{\infty} \frac{d\, E(\lambda)}{\lambda - \ell} . \tag{6.1}$$

Let P be the orthogonal projection of R onto H and let

$$R(\ell)f = PR_H(\ell)f \quad , f \in H,\ \ell \in \mathbb{C}^{\pm},$$

$$F(\lambda)f = PE(\lambda)f \quad , f \in H,\ \lambda \in \mathbb{R}.$$

Then R is called the generalized resolvent and F the generalized spectral family for S corresponding to H. We remark that all generalized spectral families for S can always be constructed using the above method by starting with a minimal selfadjoint extension of S.

For $f \in H$ it follows from (6.1) that

$$(R(\ell)f,f) = \int_{-\infty}^{\infty} \frac{d(F(\lambda)f,f)}{\lambda - \ell} \quad , \ell \in \mathbb{C}^{\pm},$$

and an inversion of this equality yields for $f \in H$

$$((F(\lambda) - F(\mu))f,f) = \lim_{\varepsilon \downarrow 0} \frac{1}{\pi} \int_{\mu}^{\lambda} \operatorname{Im}(R(\nu+i\varepsilon)f,f)d\nu,$$

where λ, μ are continuity points of F. For more details we refer to [4], [5] and [7]. In the next section we shall give a method of calculating all generalized spectral families for $S_o = T_o \cap B^*$ as considered in section 5.

7. Eigenfunction expansions. As in section 5 we assume that $L = L^+$ and $B = B^+$ and hence that $S_o = S_o^+$ is symmetric in H^2. We do not assume that $\dim M_{S_o}(\ell) = \dim M_{S_o}(\bar{\ell})$, $\ell \in \mathbb{C}^{\pm}$.

Let $H = H_s \oplus H_\infty$ be a fixed minimal selfadjoint extension of S_o in R^2 and $F = \{F(\lambda) \mid \lambda \in \mathbb{R}\}$ the corresponding generalized spectral family as described in the previous section. We shall indicate how one can obtain a suitable expression for F which immediately leads to eigen function expansions.

Let $c \in \imath$ be fixed. Let $p = \dim B$ and $\{\sigma,\tau\}$ $(n \times p)$ a basis for B. Let $s^1(x,\ell)$ $(n \times n)$ and $u(x,\ell)$ $(n \times p)$ be the unique solutions of

$$(L-\ell)\, s^1(\ell) = 0, \qquad s^1(c,\ell) = I_n,$$

$$(L-\ell)\, u(\ell) = \ell\sigma - \tau, \qquad u(c,\ell) = 0_n^p.$$

where $\ell \in \mathbb{C}$. Let $s^2(\ell) = u(\ell) + \sigma$ and $s(\ell) = (s^1(\ell) : s^2(\ell))$.

One can easily verify that for all $h \in H(0)^\perp = R \ominus H(0)$ and $\ell \in \mathbb{C}^{\pm}$

$$\{PR_s(\ell)h, \ell\, PR_s(\ell)h + Ph\} \in S_o^* = T \dotplus B.$$

Hence there exist a unique element $\{f,Lf\} \in T$ and a unique vector $a(p \times 1) \in \mathbb{C}^p$ such that

$$\{PR_s(\ell)h, \ell PR_s(\ell)h + Ph\} = \{f,Lf\} + \{\sigma,\tau\}\, a.$$

We define

$$\Gamma(PR_s(\ell)\, h) = \begin{pmatrix} f(c) \\ a \end{pmatrix} \ ((n+p) \times 1) \quad , \ell \in \mathbb{C}^{\pm}.$$

It can be proved that the linear map $h \mapsto \Gamma(PR_s(\ell)h)$ from $(H(0))^{\perp}$ into $\mathbb{C}^{n+p}$ is continuous. The Riesz Representation Theorem implies that there exist $G(\ell) = (G_1(\ell),\ldots, G_{n+p}(\ell))$, $G_j(\ell) \in (H(0))^{\perp}$, $j=1,\ldots, n+p$, such that

$$\Gamma(PR_s(\ell)\, h) = (h,G(\ell)), \quad h \in (H(0))^{\perp}, \ \ell \in \mathbb{C}^{\pm}.$$

We define

$$t(\lambda) = (H_s - \bar{\ell}_0)(E_s(\lambda) - E_s(0))\, G(\ell_0), \lambda \in \mathbb{R},$$

where $\ell_0 \in \mathbb{C}^{\pm}$ is fixed, and put

$$\rho(\lambda) = \Gamma(Pt(\lambda))$$
$$= ((H_s - \ell_0)t(\lambda),\ G(\ell_0)),\ \lambda \in \mathbb{R}.$$

<u>Theorem 7.1. The</u> $(n+p) \times (n+p)$ <u>matrix-valued function</u> ρ <u>on</u> $\mathbb{R}$ <u>is hermitian, non decreasing and continuous from the right, and for all</u> $h \in H$, $\alpha,\beta \in \mathbb{R}$

(7.1) $$(F(\beta) - F(\alpha))\, h = \int_{\alpha}^{\beta} s(\lambda)d_{\lambda}\ (h, \int_{0}^{\lambda} s(\mu)d\rho\ (\mu)).$$

<u>If</u> $h \in C_0(\iota)$ <u>then</u> (7.1) <u>can be written as</u>

$$(F(\beta) - F(\alpha))\, h = \int_{\alpha}^{\beta} s(\nu)\ d\rho\ (\nu)\ \hat{h}\ (\nu),$$

<u>where</u>

$$\hat{h}(\nu) = (h,\ s(\nu)) = \int_a^b s^*\ (x,\nu)\ h(x)\ dx.$$

The matrix ρ is called the <u>spectral matrix</u> (properly normalized ρ is unique). It can be shown that if H_s has a pure point spectrum then ρ consists of stepfunctions only.

Let $\hat{f}$, $\hat{g}$ be $(n+p) \times 1$ matrix-valued functions on $\mathbb{R}$ and define

$$(\hat{f},\hat{g}) = \int_{-\infty}^{\infty} \hat{g}^*\ d\rho\ \hat{f}.$$

Since ρ is non decreasing we have $(\hat{f},\hat{f}) \geq 0$ and we can define $\|\hat{f}\| = (\hat{f},\hat{f})^{\frac{1}{2}}$. Let $\hat{H}$ be the Hilbertspace defined by

$$\hat{H} = L^2\ (\rho) = \{\ \hat{f}\ |\ \|\hat{f}\| < \infty\ \}.$$

The eigenfunction expansion that follows from Theorem 7.1 takes the following form

<u>Theorem 7.2.</u> For $f \in H$

$$\hat{f}(\nu) = \int_a^b s^*(x,\nu)\ f(x)dx$$

<u>converges in norm in</u> $\hat{H}$, <u>and</u>

$$F(\infty)\ f = \int_{-\infty}^{\infty} s(\nu)\ d\rho\ (\nu)\ \hat{f}(\nu),$$

where this integral converges in norm in H. Moreover, $(F(\infty)f,g) = (\hat{f},\hat{g})$ for all $f,g \in H$, and the map $V: H \to \hat{H}$ given by $Vf = \hat{f}$ is a contraction, $\| Vf \| \leq \| f \|$. It is an isometry, $\| Vf \| = \| f \|$, for $f \in H \cap H(0)^{\perp} = H \ominus PH(0)$ and

$$f = \int_{-\infty}^{\infty} s(\nu)\, d\rho(\nu)\, \hat{f}(\nu), \quad f \in H \ominus PH(0).$$

The map V implies a splitting of H and $VH \subset \hat{H}$:

$$H = H_0 \oplus H_1 \oplus H_2, \qquad VH = VH_0 \oplus VH_2.$$

Here H_0 is the maximal subspace of H on which V is an isometry, i.e.,

$H_0 = \{f \in H \mid \| Vf \| = \| f \|\}$, $H_1 = \{f \in H \mid Vf = 0\}$ and $H_2 = H \ominus (H_1 \oplus H_2)$.

Once can prove that

$$H_0 = H \cap (H(0))^{\perp} = \{f \in H \mid F(\infty) f = f\},$$

$$H_1 = H \cap H(0) = \{f \in H \mid F(\infty) f = 0\}.$$

If $D(S_0)$ is dense in H or if H is an operator then $H = H_0$.
But there exist examples of subspaces S_0 and extension H^0 for which H_1 and H_2 are non trivial.

Theorem 7.3. We have $VH_0 = \hat{H}$ if and only if F is the spectral family for a selfadjoint subspace extension of S_0 in H^2 itself.

The proofs of the Theorem stated here can be found in [5].

References (including references to related work).

1. R. Arens, Operational calculus of linear relations, Pacific J. Math., 11 (1961), 9-23.
2. E.A. Coddington, Extension theory of formally normal and symmetric subspaces, Mem. Amer. Math.Soc. No. 134 (1973).
3. E.A. Coddington, Selfadjoint subspace extensions of nondensely defined symmetric operators, Advances in Math. 14(1974), 309-332.
4. E.A. Coddington, Selfadjoint problems for nondensely defined ordinary differential operators and their eigenfunction expansions, Advances in Math. 14(1974).
5. E.A. Coddington and A. Dijksma, Selfadjoint subspaces and eigenfunction expansions for ordinary differential subspaces, to appear in J.Diff.Equations.
6. A. Dijksma, and H.S.V. de Snoo, Eigenfunction expansions for nondensely defined differential operators, J.Diff.Equations 17(1975) 198-219.
7. A. Dijksma and H.S.V. de Snoo, Selfadjoint extensions of symmetric subspaces, Pacific J. Math. 54(1974) 71-100.
8. A.M. Krall, Differential-boundary operators, Trans. Amer. Math. Soc. 154 (1971) 429-458.
9. A M. Krall, The development of general differential and general differential-boundary systems, to be published.
10. H.J. Zimmerberg, Linear integro-differential-boundary-parameter problems, to appear in Ann. di Mat. Pura ed Appl.

Department of Mathematics
Rijksuniversiteit Groningen
Groningen, The Netherlands.

W. Eckhaus (ed.), New Developments in Differential Equations

SOME DEGENERATED DIFFERENTIAL OPERATORS ON VECTOR BUNDLES

by

R. Martini

1. Introduction

In this paper we shall deal with differential operators Θ , acting on a vector bundle E over a compact C^∞ differentiable manifold with boundary. The differential operators Θ are of the type

$$\Theta = \alpha\Phi + \alpha^{-\frac{1}{2}}\Psi + \Xi ,$$

where

i) α is a continuous real-valued function defined on M such that α is smooth and $\alpha > 0$ in the interior of the manifold M and such that α vanishes at the boundary of M.

ii) Φ is a strongly elliptic differential operator of the second order.

iii) Ψ is a differential operator of at most order one.

iv) Ξ is a differential operator of order zero.

Our aim is to prove that with an operators Θ of the type above there can be associated an infinitesimal generator of a strongly continuous semi-group, defined on certain weighted L^2-spaces of sections of M over E.

Let us summarize briefly the contents of this paper. In section 2 we give the notation of the mathematical concepts we shall use in our investigation about the differential operators of the type described above. In section 3 and 4 some auxiliary Hilbertable spaces are constructed. Section 5 and 6 contain our basic results and in section 7 and 8 we investigate relations between the class of differential operators considered by us and infinitesimal generators of semi-groups of operators. Finally, in section 9 some examples are given.

2. Notations

M will denote a paracompact C^∞ differentiable manifold of dimension n, possibly with boundary ∂M. So partitions of unity with respect to any open covering of M may be constructed.

$T(M)$ and $T^*(M)$ will denote the tangent space of M and the cotangent space of M respectively. $T'(M)$ is the subbundle of $T^*(M)$ of non zero cotangent vectors. All vector bundles over M are supposed to be complex vector bundles, unless mentioned otherwise explicitly, and are usually denoted by E, F, ...

Now let U be an open subset of M, then $\Gamma(U,E)$ is the set of all C^∞ sections of E over U. $\Gamma_o(U,E)$ is the subset of $\Gamma(U,E)$ consisting of all sections whose supports are compact in U and disjoint from the boundary ∂M.

A positive smooth measure on M is a Borel measure μ such that for each chart $c = (U,\Phi)$ for M the measure $\Phi(\mu)$ induced by Φ on the σ-algebra of the Borel sets of $\Phi(U)$ is absolutely continuous with respect to the Lebesgue measure and has a strictly positive C^∞ density σ. Thus

$$\mu(A) = \int_{\Phi(A)} \sigma dx$$

for each Borel set A of U.

As usual, $W^p(E)$ $(-\infty < p < \infty)$ denote the p-th order Sobolev space. In general, this is a Hilbertable space, but if M is provided with a fixed positive smooth measure we may identify $W^o(E)$ with the space of all measurable square-integrable sections of E over **M**. Hence $W^o(E)$ is a Hilbert space.

Let U be an open subset of M. Then with $U' \subset\subset U$ we denote that U' is a relatively compact open subset of U. It follows that the closure $\overline{U'}$ of U' in U is compact.

Let Ω be an open subset of the n-dimensional Euclidean space $\mathbb{R}^n$ and let u,v be two mappings of Ω into the r-dimensional unitary space C^r. Then $|s|$, (s,t) are defined by

$$|s|(x) = |s(x)|, \quad (s,t)(x) = (s(x),t(x)).$$

Here $||$ and $(\, , \,)$ denote the usual norm and inner product in C^r respectively.

By $\mathbb{R}^n_+$ we denote the half-space consisting of all $(x_1, \ldots, x_n) \in \mathbb{R}^n$ such that $x_n > 0$ and B^n_+ denotes the half-ball, defined by $B^n_+ = B^n \cap \mathbb{R}^n_+$, where B^n is the unit ball in $\mathbb{R}^n$. $\overline{\mathbb{R}^n_+}$ denotes the closure of $\mathbb{R}^n_+$ in $\mathbb{R}^n$.

Let V be a finite-dimensional vector space and let Ω be a non-empty open subset in some n-dimensional Euclidean space $\mathbb{R}^n$. Then $C^r(\Omega,V)$ $(r = 0, 1, \ldots; \infty)$ is the space of all r-times continuously differentiable functions, mapping Ω into V.

$C^r_o(\Omega,V)$ denotes the subset of $C^r(\Omega,V)$ consisting precisely of those elements of $C^r(\Omega,V)$ which have compact support in Ω.

Occasionally, we use the space (Ω,V) which as a set equals $C^\infty_o(\Omega,V)$ but which is provided with its natural inductive limit topology. Its topological dual, the space of V-valued distributions, is denoted by $\mathcal{D}(\Omega,V)'$. For a vector bundle E over a manifold M we have the corresponding spaces $\mathcal{D}(M,E)$ and $\mathcal{D}(M,E)'$.

Let E and F be two vector bundles with the same base space M. A linear differential operator Θ from E to F is by definition a linear mapping Θ : $\Gamma_o(M,E) \to \Gamma_o(M,F)$ which is local: i.e., for each open subset U of M and each section $s \in \Gamma_o(M,E)$ such that the restriction $S|U = 0$ we have Θ $s|U = 0$. According to a theorem of Peetre we know that for any chart $c = (U,\phi)$ of M such that E and F admit trivializations (π_E,ν_E) and (π_F,ν_F) the local expression $\tilde{\Theta}$ defined by

$$\tilde{\Theta}\,(\nu_E s^{-1}) = \nu_E(\,\Theta\, s)^{-1}$$

is a linear differential operator in the ordinary sense. Conversely, if Θ is a linear mapping of $\Gamma_0(M,E)$ into $\Gamma_0(M,F)$ such that for any chart and any trivializations the local expression $\tilde{\Theta}$ is a linear differential operator in the ordinary sense it follows that Θ is linear differential operator. Thus the linear differential operators are the only local linear operators.

With a differential operator Θ there is for each natural number k associated a k-th order symbol $\sigma_k(\,\Theta\,)$ such that for each $\xi_x \in T'(M)\sigma_k(\,\Theta\,)(\xi_x)$ is a linear map from the fiber E_x into F_x. The set of all differential operators Θ from E into F such that $\sigma_m(\,\Theta\,)$ vanishes for any $m > k$ is denoted by $\mathrm{Diff}_k(E,F)$. A differential operator Θ is said to be of order k if Θ belongs to $\mathrm{Diff}_k(E,F)$ but Θ does not belong to $\mathrm{Diff}_{k-1}(E,F)$.

By definition, a k-th order differential operator Θ is elliptic in $x \in M$ if for each $\xi_x \in T_x'(M)$ $\sigma_k(\Theta)$ (ξ_x) is a linear isomorphism of E_x into F_x. A k-th order differential operator from E into F is elliptic if it is elliptic at each point $x \in M$. In addition, suppose that $F = E$ and that E is provided with a Hermitian structure <,>. Then a k-th order differential operator Θ is said to be strongly elliptic in $x \in M$ if for each $\xi_x \in T_x'(M)$ and for each $e_x \in E_x$ with $e_x \neq 0$ we have

$$\mathrm{Re} < \sigma_k(\,\Theta\,)(\xi_x)e_x, e_x >> 0. \, .$$

A k-th order differential operator in strongly elliptic if it is strongly elliptic in each point. For more detailed information about differential operators on vector bundles see e.g. PALAIS [4], chp. IV or NARASIMHAN [3], chp. 3.

3. Some auxiliary spaces

Throughout this and the next section M will denote a fixed compact C^∞ differentiable manifold of dimension $n \geq 1$, possibly with boundary ∂M and E is a fixed C^∞ vector bundle over M of rank r and with projection π, $\pi : E \to M$. We suppose that M is equipped with a strictly positive smooth measure μ and that E is equipped with a Hermitian structure < , > . α is supposed to be a real-valued continuous function defined on M such that $\alpha(x) > 0$ when $x \notin \partial M$.

Now we have the following definitions.

Definition 1. $L^2(E)$ is defined to be the completion of the set of all sections $s \in \Gamma(M,E)$ under the norm

$$(3.1) \qquad ||s||_0 = \{\int_M < s,\, s > \, d\mu\}^{\frac{1}{2}}.$$

Definition 2. $H(E,\alpha)$ is the completion of the set of all sections $s \in \Gamma_o(M,E)$ under the norm

(3.2) $$||s||_H = \{ \int_M < s, s > \frac{d\mu}{\alpha}\}^{\frac{1}{2}}.$$

Remark. $L^2(E)$ is a Hilbert space with the inner product

(3.3) $$(s,t)_0 = \int_M < s, t > d\mu$$

and may be considered as the space of all sections of M over E which are square-integrable over M with respect to the measure μ. $H(E,\alpha)$ is a Hilbert space with the inner product

(3.4) $$(s,t)_H = \int_M < s, t > \frac{d\mu}{\alpha}$$

and may be considered as the space of all measurable sections of M over E which are square-integrable over M with respect to the measure $\frac{d\mu}{\alpha}$.

Definition. Let $c = (U,\Phi)$ be a chart for M such that there is a trivialization (π,ν) of E over U and let $\tilde{s} = \nu s\Phi^{-1}$ be the local expression of the section $s \in \Gamma_o(M,E)$ corresponding to the chart c and the trivialization (π,ν). Suppose $U' \subset\subset U$ and put $\tilde{\alpha} = \alpha\Phi^{-1}$. Then by

(3.5) $$p_{c,\nu,U'}(s) = \{ \int_{\Phi(U')} |\tilde{s}|^2 \frac{dx}{\tilde{\alpha}} + \sum_i \int_{\Phi(U')} |D_i\tilde{s}|^2 dx\}^{\frac{1}{2}}$$

we introduce a semi-norm on the linear space $\Gamma_o(M,E)$.

Now let τ be the coarsest locally convex topology on $\Gamma_o(M,E)$ such that for all charts $c = (U,\Phi)$, all $U' \subset\subset U$ and all trivializations (π,ν) the semi-norm $p_{c,\nu,U'}$ is continuous. Then by $V_0(E,\alpha)$ we denote the completion of $\Gamma_o(M,E)$ with respect to the locally convex topology τ.

Theorem 1. Let $\mathfrak{a} = \{c_k = (U_k,\Phi_k)\}$ be an atlas for M consisting of a finite number of charts such that for each k E admits a trivialization (π,ν_k) over U_k and let $\{U_k'\}$ be an open covering of M such that for each k $U_k' \subset\subset U_k$; denote by $\tilde{s}_k = \nu_k s\Phi_k^{-1}$ the local expression over U_k of the section $s \in \Gamma_o(M,E)$ and put $\tilde{\alpha}_k = \alpha\Phi_k^{-1}$.
Then an inner product for $\Gamma_o(M,E)$ such that the associated norm generates the locally convex topology τ is given by

(3.6) $$(s,t)_V = \sum_k \int_{\Phi_k(U_k')} (\tilde{s}_k,\tilde{t}_k)\frac{dx}{\tilde{\alpha}_k} + \sum_i \int_{\Phi_k(U_k')} (D_i\tilde{s}_k, D_i\tilde{t}_k)dx.$$

Proof. Obviously, (,) is an inner product.

Now suppose that $c = (U,\Phi)$, $\bar{c} = (U,\Phi)$ are two charts with the same domain for M and (π,ν), $(\pi,\bar{\nu})$ two trivializations of E over U. Let $U' \subset\subset U$, $\tilde{s} = \nu s \Phi^{-1}$, $\bar{s} = \bar{\nu} s \bar{\Phi}^{-1}$, $\tilde{\alpha} = \alpha\Phi^{-1}$ and $\bar{\alpha} = \alpha\bar{\Phi}^{-1}$. Let $\chi = \bar{\Phi}\,\bar{\Phi}^{1}$, $\eta = \nu\bar{\nu}^{-1}$ be the transition functions. Then it follows that $\tilde{s} = \eta\bar{s}\chi$. Hence by the chain rule we get

$$D\tilde{s}(x) = D\eta(\bar{s}(\chi(x))) \cdot D\bar{s}(\chi(x)) \cdot D\chi(x).$$

Therefore

$$\sum_i |D_i\tilde{s}(x)|^2 \le ||D\eta(\bar{s}(\chi(x)))||^2 \sum_i |D_i\bar{s}(\chi(x))|^2 ||D\chi(x)||^2$$

for each $x \in \Phi(U')$. Now $U' \subset\subset U$. Hence there is a constant k_1 such that for each $x \in \Phi(U')$ we have

$$\sum_i |D_i\tilde{s}(x)|^2 \le k_1 \sum_i |D_i\bar{s}(\chi(x))|^2.$$

Thus

$$\int_{\Phi(U')} \sum_i |D_i\tilde{s}(x)|^2 dx \le k_1 \int_{\Phi(U')} \sum_i |D_i\bar{s}(\chi(x))|^2 dx =$$

$$= k_1 \int_{\bar{\Phi}(U')} \sum_i |D_i\bar{s}(y)|^2 |J(y)| dy,$$

where J is the Jacobian of the transition function χ^{-1}.
Hence there exists a constant k_2 such that

$$\sum_i \int_{\Phi(U')} |D_i\tilde{s}(x)|^2 dx \le k_2 \sum_i \int_{\bar{\Phi}(U')} |D_i\bar{s}(y)|^2.$$

From $\tilde{s} = \eta\bar{s}\chi$, $U' \subset\subset U$ it follows also that there exists a constant k_3 such that for each $x \in U'$ we have

$$\int_{\Phi(U')} |\tilde{s}(x)|^2 \frac{dx}{\tilde{\alpha}(x)} = \int_{\Phi(U')} |\eta \cdot \bar{s}(\chi(x))|^2 \frac{dx}{\bar{\alpha}(\chi(x))} =$$

$$= \int_{\bar{\Phi}(U')} |\eta \cdot \bar{s}(y)|^2 |J(y)| \frac{dy}{\bar{\alpha}(y)} \le k_3 \int_{\bar{\Phi}(U')} |\bar{s}(y)|^2 \frac{dy}{\bar{\alpha}(y)}.$$

Now let $p_{c,\nu,U'}$ be a semi-norm. Then from the estimates made above we see that there exists a constant k_4 such that

$$p^2_{c,\nu,U'}(s) = \sum_k p^2_{c,\nu,U' \cap U'_k}(s) \le \sum_k p^2_{c_k,\nu_k,U' \cap U'_k}(s) \le$$

$$\le c_4 \sum_k p^2_{c_k,\nu_k,U'_k}(s) = (s,s)_V = ||s||^2_V.$$

Thus the norm $||\ ||_V$ generates the locally convex topology τ.

Since M is compact and locally compact there exists an atlas $\mathfrak{a}$ having the properties stated in theorem 1. Thus recalling that a Hilbertable space is a complete locally convex space such that the locally convex structure is generated by an inner product we have the following corollary of theorem 1.

Corollary. $V_o(E,\alpha)$ is a Hilbertable space.

Remark. Obviously, we have $V_o(E,\alpha) \subset H(E,\alpha) \subset L^2(E)$ with continuous inclusion maps.

Considering trivial bundles over open subsets of an Euclidean space the definitions in this section may be changed in an obvious way to trivial bundles over non-necessarily bounded open subsets of an Euclidean space $\mathbb{R}^n$. In this case, by the existence of a canonical global trivialization, $V_o(E,\alpha)$ is a Hilbert space.

If α is non-vanishing on M it follows that $H(E,\alpha)$ and $V_o(E,\alpha)$ are isomorphic to the Sobolev spaces $W^o(E)$ and $W^1_o(E)$ respectively.

4. The embedding J of $V_o(E,\alpha)$ into $H(E,\alpha)$

In this section we shall deal with the continuous linear embedding J of $V_o(E,\alpha)$ into $H(E,\alpha)$. The question whether this embedding is compact (completely continuous) is now our main interest. Considering first manifolds without boundary we have the following consequence of Rellich's theorem.

Theorem 2. Let M be a compact C^∞ manifold without boundary. Then the embedding J of $V_o(E,\alpha)$ into $H(E,\alpha)$ is compact.

Proof. M is supposed to be a compact manifold and α is a continuous positive function defined on M. Hence is bounded and bounded away from zero on M. Thus $V_o(E,\alpha)$ is the same as the Sobolev space $W^1_o(E)$ and $H(E,\alpha)$ is isomorphic to the Sobolev space $W^o(E)$.

Now from Rellich's embedding theorem (see e.g. PALAIS [4], ch. X, section 4) it follows that the embedding of $W^1_o(E)$ into $W^o(E)$ is compact. Hence the embedding of $V_o(E,\alpha)$ into $H(E,\alpha)$ is compact too.

For manifolds with boundary the situation is more complicated. In this case we have a result of which the proof is divided into a number of parts, the last one of which gives our result.

Lemma 1. Let α be a positive continuous function defined on $\mathbb{R}^n_+$, depending only on the last variable; i.e., there exists a positive continuous function $\bar{\alpha}$ defined on $\mathbb{R}_+$ such that $\alpha(x) = \bar{\alpha}(x_n)$ when $x = (x_1, \ldots, x_n) \in R^n_+$.

In addition, suppose that

$$\int_0^a \frac{x_n}{\bar{\alpha}(x_n)}\, dx_n < \infty$$

<u>for some $a > 0$. Then for all $s \in V_o(\mathbb{R}^n_+ \times C^r,\alpha)$ we have the estimate</u>

$$(4.1)\qquad \int_0^a \int_{\mathbb{R}^{n-1}} \frac{|s(x)|^2}{\alpha(x)}\, dx \le \int_0^a \frac{x_n}{\bar{\alpha}(x_n)}\, dx_n \cdot \int_{\mathbb{R}^n} |D_n s|^2 dx.$$

<u>Proof</u>. Suppose $s \in C_o^\infty(\mathbb{R}^n_+, C^r)$ and let

$$x = (x',x_n),\ x' \in \mathbb{R}^{n-1},\ x_n \in \mathbb{R}.$$

Then we have

$$\int_0^a \int_{\mathbb{R}^{n-1}} \frac{|s(x)|^2}{\alpha(x)}\, dx = \int_0^a \frac{1}{\bar{\alpha}(x_n)} \int_{\mathbb{R}^{n-1}} |s(x',x_n) - s(x',o)|^2 dx'dx_n =$$

$$= \int_0^a \frac{1}{\bar{\alpha}(x_n)} \int_{\mathbb{R}^{n-1}} \left|\int_0^{x_n} D_n s(x',u)du\right|^2 dx'dx_n.$$

Using the inequality of Cauchy-Schwarz we see that the last expression is not greater than

$$\int_0^a \frac{1}{\bar{\alpha}(x_n)} \int_{\mathbb{R}^{n-1}} x_n \cdot \int_0^{x_n} |D_n s(x',u)|^2 du\ dx'dx_n =$$

$$= \int_0^a \frac{x_n}{\bar{\alpha}(x_n)} \int_{\mathbb{R}^{n-1}} \int_0^{x_n} |D_n s(x',u)|^2 du\ dx'dx_n \le$$

$$\le \int_0^a \frac{x_n}{\bar{\alpha}(x_n)}\, dx_n \cdot \int_{\mathbb{R}^n} |D_n s(x)|^2 dx.$$

Since for fixed $a > 0$ both sides of inequality (4.1) are continuous as functions of $s \in V_o(\mathbb{R}^n_+ \times C^r,\alpha)$ and the set $C_o^\infty(\mathbb{R}^n_+, C^r)$ is dense in $V_o(\mathbb{R}^n_+ \times C^r,\alpha)$ we have established inequality (4.1) in case $s \in V_o(\mathbb{R}^n_+ \times C^r,\alpha)$. This completes the proof of lemma 1.

<u>Lemma 2</u>. <u>For each $\varepsilon > 0$ there exists a $\delta > 0$ such that for all $s \in W^1(\mathbb{R}^n \times C^r)$ and all $y \in \mathbb{R}^n$ with $|y| < \delta$.</u>

$$(4.2)\qquad \int_{\mathbb{R}^n} |s(x+y) - s(x)|dx \le \varepsilon\{ \int_{\mathbb{R}^n} |s(x)|^2 + \sum_i \int_{\mathbb{R}^n} |D_i s(x)|^2 dx\}.$$

<u>Proof</u>. Let $\varepsilon > 0$ and suppose that $s \in C_o^\infty(\mathbb{R}^n, C^r)$. Then it follows from Plancherel's theorem that

$$(4.3)\qquad \int_{\mathbb{R}^n} |s(x+y) - s(x)|^2 dx = \frac{1}{(2\pi)^n} \int_{\mathbb{R}^n} |e^{iy\cdot\xi} - 1|^2 |\hat{s}(\xi)|^2 d\xi,$$

where $\hat{s}$ denotes the Fourier transform of s defined by

$$\hat{s}(\xi) = \int_{\mathbb{R}^n} e^{-i\xi\cdot x}\, s(x)dx.$$

Evidently, the right hand side of (4.3) equals

$$(4.4) \qquad \frac{1}{(2\pi)^n} \int_{\mathbb{R}^n} \frac{|e^{i\xi\cdot y}-1|^2}{(1+|\xi|^2)} (1+|\xi|^2)|\hat{s}(\xi)|^2 d\xi$$

and it can easily be established that there exists a $\delta > 0$, depending only on ε, such that for all $\xi \in \mathbb{R}^n$ and all $y \in \mathbb{R}^n$ with $|y| < \delta$

$$\frac{|e^{i\xi\cdot y}-1|^2}{1+|\xi|^2} < \varepsilon.$$

Thus for all $y \in \mathbb{R}^n$ with $|y| < \delta$ it follows that (4.4) is dominated by

$$\frac{\varepsilon}{(2\pi)^n} \int_{\mathbb{R}^n} (1+|\xi|^2)|\hat{s}(\xi)|^2 d\xi = \varepsilon\{\int_{\mathbb{R}^n} |s(x)|^2 dx + \sum_i \int_{\mathbb{R}^n} |D_i s(x)|^2 dx\}.$$

For fixed $y \in \mathbb{R}^n$ both sides of inequality (4.2) are continuous as functions of $s \in W'(\mathbb{R}^n \times C^r)$ and $C_o^\infty(\mathbb{R}^n, C^r)$ is dense in $W^1(\mathbb{R}^n \times C^r)$. Hence for all $y \in \mathbb{R}^n$ with $|y| < \delta$ and all $s \in W^1(\mathbb{R}^n \times C^r)$ inequality (4.2) is valid. This completes the proof of lemma 2.

Lemma 3. *Let α be as in lemma 1. Denote by $\tilde{s}$ the canonical extension to $L^2(\mathbb{R}^n \times C^r) = L^2(\mathbb{R}^n, C^r)$ of an* $s \in L^2(B_+^n \times C^r) = L^2(B_+^n, C^r)$. *Then*

$$\lim_{y \to 0} \int_{\mathbb{R}^n} |(\frac{\tilde{s}}{\alpha^{\frac{1}{2}}})(x+y) - (\frac{\tilde{s}}{\alpha^{\frac{1}{2}}})(x)|^2 dx = 0$$

uniformly on the closed unit ball K of $V_o(B_+^n \times C^r, \alpha)$; *i.e., uniformly on the set K consisting of all* $s \in V_o(B_+^n \times C^r, \alpha)$ *such that*

$$\int_{B_+^n} |s|^2 \frac{dx}{\alpha} + \sum_i \int_{B_+^n} |D_i s|^2 dx \leq 1.$$

Proof. We may restrict ourselves to the case where the limit is taken over all $y = (y_1, \ldots, y_n) \in \mathbb{R}^n$ with $y_n \geq 0$. In this case we have

$$(4.5) \qquad \int_{\mathbb{R}^n} |(\frac{\tilde{s}}{\alpha^{\frac{1}{2}}})(x+y) - (\frac{\tilde{s}}{\alpha^{\frac{1}{2}}})(x)|^2 dx = \int_0^{y_n} \int_{\mathbb{R}^{n-1}} \frac{|\tilde{s}(x)|^2}{\alpha(x)} dx +$$

$$+ \int_0^\infty \int_{\mathbb{R}^{n-1}} |\frac{\tilde{s}(x+y)}{\alpha^{\frac{1}{2}}(x+y)} - \frac{\tilde{s}(x)}{\alpha^{\frac{1}{2}}(x)}|^2 dx.$$

Now for any $a > 0$ we obtain

$$(4.6)\qquad \int_0^a \int_{\mathbb{R}^{n-1}} \left|\frac{\tilde{s}(x+y)}{\alpha^{\frac{1}{2}}(x+y)} - \frac{\tilde{s}(x)}{\alpha^{\frac{1}{2}}(x)}\right|^2 dx \le 2\int_0^a \int_{\mathbb{R}^{n-1}} \frac{|\tilde{s}(x+y)|^2}{\alpha(x+y)}dx +$$

$$+ 2\int_0^a \int_{\mathbb{R}^{n-1}} \frac{|\tilde{s}(x)|^2}{\alpha(x)}dx = 2\int_{y_n}^{a+y_n} \int_{\mathbb{R}^{n-1}} \frac{|\tilde{s}(x)|^2}{\alpha(x)}dx +$$

$$+ 2\int_0^a \int_{\mathbb{R}^{n-1}} \frac{|\tilde{s}(x)|^2}{\alpha(x)}dx \le 4\int_0^{a+y_n} \int_{\mathbb{R}^{n-1}} \frac{|\tilde{s}(x)|^2}{\alpha(x)}dx.$$

Let $\varepsilon > 0$ be given. By lemma 1 we may fix an $a > 0$ and a δ_1 such that for all $y \in \mathbb{R}^n$ with $y_n \ge 0$ and $|y| < \delta$,

$$(4.7)\qquad \int_0^{a+y_n} \int_{\mathbb{R}^{n-1}} \frac{|\tilde{s}(x)|^2}{\alpha(x)}dx < \frac{\varepsilon}{10} .$$

Thus for all $y \in \mathbb{R}^n$ with $y_n \ge 0$ and $|y| < \delta_1$ the right hand side of (4.5) is not greater than

$$(4.8)\qquad \frac{\varepsilon}{2} + \int_a^\infty \int_{\mathbb{R}^{n-1}} \left|\frac{\tilde{s}(x+y)}{\alpha^{\frac{1}{2}}(x+y)} - \frac{\tilde{s}(x)}{\alpha^{\frac{1}{2}}(x)}\right|^2 dx.$$

Next it follows that

$$(4.9)\qquad \int_a^\infty \int_{\mathbb{R}^{n-1}} \left|\frac{\tilde{s}(x+y)}{\alpha^{\frac{1}{2}}(x+y)} - \frac{\tilde{s}(x)}{\alpha^{\frac{1}{2}}(x)}\right|^2 dx =$$

$$= \int_a^\infty \int_{\mathbb{R}^{n-1}} \left|\frac{\tilde{s}(x+y)}{\alpha^{\frac{1}{2}}(x+y)} - \frac{s(x+y)}{\alpha^{\frac{1}{2}}(x)} + \frac{\tilde{s}(x+y)}{\alpha^{\frac{1}{2}}(x)} - \frac{\tilde{s}(x)}{\alpha^{\frac{1}{2}}(x)}\right|^2 dx \le$$

$$\le \int_a^\infty \int_{\mathbb{R}^{n-1}} |\alpha^{-\frac{1}{2}}(x+y) - \alpha^{-\frac{1}{2}}(x)|^2 |\tilde{s}(x+y)|^2 dx +$$

$$+ \int_a^\infty \int_{\mathbb{R}^{n-1}} \frac{|\tilde{s}(x+y) - \tilde{s}(x)|^2}{\alpha(x)} dx.$$

Using the fact that $\alpha^{-\frac{1}{2}}$ is uniformly continuous and bounded on the set $\{x = (x_1, \ldots, x_n) \in \mathbb{R}^n \,|\, x_n > a, |x| \le 1+\delta_1\}$ there exists a constant c and a δ_2 with $0 < \delta_2 < \delta_1$ such that for all $y \in \mathbb{R}^n$ with $|y| < \delta_2$ and $y_n \ge 0$ the last expression is dominated by

$$\tfrac{1}{4}\varepsilon \int_a^\infty \int_{\mathbb{R}^{n-1}} |\tilde{s}(x+y)|^2 dx + c\int_a^\infty \int_{\mathbb{R}^{n-1}} |\tilde{s}(x+y) - \tilde{s}(x)|^2 dx.$$

Evidently this expression is not greater than

$$(4.10)\qquad \tfrac{1}{4}\varepsilon \int_0^\infty \int_{\mathbb{R}^{n-1}} |\tilde{s}(x)|^2 dx + c\int_0^\infty \int_{\mathbb{R}^{n-1}} |\tilde{s}(x+y) - \tilde{s}(x)|^2 dx.$$

Now by lemma 2 there exists a δ_3 with $0 < \delta_3 < \delta_2$ such that for all $y \in \mathbb{R}^n$ with $y_n \geq 0$ and $|y| < \delta_3$ the second term of (4.10) is smaller than $\frac{1}{4}\varepsilon$. Thus for all $y \in \mathbb{R}^n$ with $y_n \geq 0$ and $|y| < \delta_3$ we see that expression (4.8) is estimated by ε. This completes the proof of lemma 3.

In the next part of our proof we need a characterization of relatively compact sets in the space $L^2(\mathbb{R}^n \times C^r) = L^2(\mathbb{R}^n, C^r)$. This characterization is contained in the following theorem, due to Fréchet-Kolmogorov.

Theorem 3. *A subset K of $L^p(\mathbb{R}^n, C^r)(1 \leq p < \infty)$ is relatively compact in $L^p(\mathbb{R}^n \times C^r)$ if and only if the following three conditions are satisfied:*

(i) $$\sup_{s \in K} \{ \int_{\mathbb{R}^n} |s(x)|^p dx\}^{\frac{1}{p}} < \infty;$$

i.e., the subset K is uniformly bounded in $L^p(\mathbb{R}^n, C^r)$;

(ii) $$\lim_{y \to 0} \int_{\mathbb{R}^n} |s(x+y) - s(x)|^p dx = 0$$

uniformly in $s \in K$;

(iii) $$\lim_{A \to \infty} \int_{|x| \geq A} |s(x)|^p dx = 0$$ *uniformly in $s \in K$.*

For a proof of this theorem see e.g. DUNFORD-SCHWARTZ [1], p. 301.

We continue our discussion with the following lemma.

Lemma 4. *Let α be as in lemma 1. Then the continuous and linear embedding of $V_o(B_+^n \times C^r)$ into $H(B_+^n \times C^r)$ is compact.*

Proof. As in the proof of lemma 3 we denote by $\tilde{s}$ the canonical extension of an $s \in L^2(B_+^n, C^r)$ to $L^2(\mathbb{R}^n, C^r)$ and we let K be the closed unit ball of $V_o(B_+^n \times C^r, \alpha)$.

Applying the theorem of Fréchet-Kolmogorov we see that it is sufficient to prove that the following conditions are satisfied;

(i) $$\sup_{s \in K} \{ \int_{\mathbb{R}^n} |(\frac{\tilde{s}}{\alpha^{\frac{1}{2}}})(x)|^2 dx\}^{\frac{1}{2}} < \infty;$$

(ii) $$\lim_{y \to 0} \int_{\mathbb{R}^n} |(\frac{\tilde{s}}{\alpha^{\frac{1}{2}}})(x+y) - (\frac{\tilde{s}}{\alpha^{\frac{1}{2}}})(x)|^2 dx = 0$$

uniformly in $s \in K$;

(iii) $$\lim_{A \to \infty} \int_{|x| > A} |(\frac{\tilde{s}}{\alpha^{\frac{1}{2}}})(x)|^2 dx$$ uniformly in $s \in K$.

Evidently, condition (i) and (iii) are satisfied. So we only have to verify condition (ii). But this follows immediately from lemma 3. This completes the proof of lemma 4.

We may now state our result

Theorem 4. Let M be a compact C^∞ differentiable manifold with boundary ∂M and let d be the metric on M induced by a Riemannian structure g on M. Denote by $\rho(x)$ the distance from the point $x \in M$ to the boundary ∂M of M. In addition, suppose that the real-valued function α is of the form $\alpha = \rho^p (0 \le p < 2)$. Then the continuous linear embedding of $V_0(E,\alpha)$ into $H(E,\alpha)$ is compact.

Proof. Let $c_i = (U_i, \Phi_i)(i = 1, 2, \ldots, \ell)$ be charts for M such that

$$\Phi_i(U_i) = \overline{\mathbb{R}^n_+}(i = 1, 2, \ldots, \ell),\ \partial M = \bigcup_{i=1}^{\ell} \Phi_i^{-1}(B^{n-1}(\tfrac{1}{2})),$$

where

$$B^{n-1}(\tfrac{1}{2}) = \{x = (x_1, \ldots, x_n) \in \mathbb{R}^n \mid x_n = 0,\ |x| < \tfrac{1}{2}\}.$$

Choose additional charts $c_i = (U_i, \Phi_i)(i = \ell+1, \ldots, m)$ such that

$$\Phi_i(U_i) = \mathbb{R}^n (i = \ell+1, \ldots, m),\ M = (\bigcup_{i=1}^{\ell} \Phi_i^{-1}(\overline{B}^n_+ (\tfrac{1}{2}))) \cup$$
$$\cup (\bigcup_{i=\ell+1}^{m} \Phi_i^{-1}(B^n(\tfrac{1}{2}))),$$

where

$$B^n(\tfrac{1}{2}) = \{x \in \mathbb{R}^n \mid |x| < \tfrac{1}{2}\},\ \overline{B}^n_+(\tfrac{1}{2}) = \{x \in B^n(\tfrac{1}{2}) \mid x_n \ge 0\}.$$

Let $\{\lambda_i\}(i = 1, 2, \ldots, m)$ be a C^∞ partition of unity for M with

$$\operatorname{supp}(\lambda_i) \subset \Phi_i^{-1}(\overline{B}^1_+)(i = 1, 2, \ldots, \ell),\ \operatorname{supp}(\lambda_i) \subset \Phi_i^{-1}(B^n)$$
$$(i = \ell+1, \ldots, m)$$

and let $\{\mu_i\}$ $(i = 1, 2, \ldots, m)$ be C^∞ real-valued functions defined on M such that

$$\operatorname{supp}(\mu_i) \subset \Phi_i^{-1}(\overline{B}^n_+)(i = 1, 2, \ldots, \ell),\ \operatorname{supp}(\mu_i) \subset \Phi_i^{-1}(B^n)$$
$$(i = \ell+1, \ldots, m)$$

and such that μ_i equals one on $\operatorname{supp}(\lambda_i)(i = 1, 2, \ldots, m)$.
Finally, let the charts $c_i(i = 1, 2, \ldots, m)$ be chosen such that the vector bundle admits for any $i = 1, 2, \ldots, m$ a trivialization (π, ν_i) over U_i.

Now we define the following mappings:

(i) $$\hat{P} : \Gamma_o(M,E) \to \bigoplus_{i=1}^{m} \Gamma_o(X_n, C^r),$$

where

$$X_i = B_+^n \ (i = 1, 2, \ldots, \ell), \ X_i = B^n \ (i = \ell+1, \ldots, m),$$

and where the mapping $\hat{P}$ is given by

$$\hat{P}(s) = (\nu_1\mu_1 s\phi_1^{-1}, \ldots, \nu_m\mu_m s\phi_m^{-1})$$

(ii) $$\hat{Q} : \bigoplus_{i=1}^{m} \Gamma_o(X_i, C^r) \to \Gamma_o(M,E),$$

given by

$$\hat{Q}(t_1, \ldots, t_m) = \sum_{i=1}^{m} \lambda_i \nu_i^{-1} t_i \phi_i.$$

It follows that $\hat{Q}$ is a cross section for $\hat{P}$. Indeed, we have

$$\hat{Q}\hat{P}(s) = \sum_{i=1}^{m} \lambda_i \nu_i^{-1} \nu_i \mu_i s \phi_i^{-1} \phi_i = \sum_{i=1}^{m} \lambda_i \mu_i s = \sum_{i=1}^{m} \lambda_i s = s,$$

since $\{\lambda_i\}$ is a partition of unity and μ_i equals one on supp (λ_i) $(i = 1, 2, \ldots, m)$.

It can easily be seen that $\hat{P}$ is a continuous mapping when $\Gamma_o(M,E)$ and $\bigoplus_{i=1}^{m}\Gamma_o(X_n, C^r)$ are equipped with the relative topology of $V(E,\alpha)$ and $\bigoplus_{i=1}^{m} V(E_i, \tilde{\alpha}_i)$ respectively, where

$$E_i = B_+^n \times C^r \ (i = 1, 2, \ldots, \ell), \ E_i = B^n \times C^r (i = \ell+1, \ldots, m)$$

and $\tilde{\alpha}_i = \alpha\phi_i$ $(i = 1, 2, \ldots, m)$.

Also $\hat{Q}$ is a continuous mapping when $\Gamma_o(M,E)$ and $\bigoplus_{i=1}^{m}\Gamma_o(X_i, C^r)$ are equipped with the relative topology of $H(E,\alpha)$ and $\bigoplus_{i=1}^{m} H(E_i, \tilde{\alpha}_i)$ respectively. Hence $\hat{P}$ and $\hat{Q}$ extend to unique continuous linear mappings

$$P : V(E,\alpha) \to \bigoplus_{i=1}^{m} V(E_i, \tilde{\alpha}_i)$$

and

$$Q : \bigoplus_{i=1}^{m} H(E_i, \tilde{\alpha}_i) \to H(E,\alpha)$$

respectively. Moreover, the property that $\hat{Q}$ is a cross section of $\hat{P}$ is preserved. Hence we have the commutative diagram

$$\begin{array}{ccc} V(E,\alpha) & \xrightarrow{\quad P \quad} & \bigoplus_{i=1}^{m} V(E_i,\tilde{\alpha}_i) \\ \Big\downarrow J & & \Big\downarrow \tilde{J} \\ H(E,\alpha) & \xleftarrow{\quad Q \quad} & \bigoplus_{i=1}^{m} H(E_i,\tilde{\alpha}_i) \end{array}$$

where J and $\tilde{J}$ are continuous inclusion mappings.

Since the property being a compact linear mapping is preserved by composition with continuous linear mappings and since the direct sum of compact mappings is compact again it is sufficient to show that for any $i = 1, 2, \ldots, m$ the mapping

$$\tilde{J}_i : V(E_i,\tilde{\alpha}_i) \to H(E_i,\tilde{\alpha}_i)$$

is compact. Recalling that $\bar{U}_i \cap \partial M$ ($i = \ell+1, \ldots, m$) is empty, α is strictly positive on the interior $\mathring{M}$ of M and μ is a strictly positive smooth measure on M it follows that $\alpha_i (i = \ell+1, \ldots, m)$ is bounded and bounded away from zero. Hence for $i = \ell+1, \ldots, m$ the spaces $H(E_i,\tilde{\alpha}_i)$ and $V(E_i,\tilde{\alpha}_i)$ are isomorphic to the Sobolev spaces $W_0^0(B^n \times C^r)$ and $W_0^1(B^n \times C^r)$ respectively.

Now by Rellich's theorem it follows that the inclusion mapping of $W^1(B^n(1) \times C^r)$ into $W^0(B^n \times C^r)$ is compact. Thus the inclusion mapping of $V(E_i,\tilde{\alpha}_i)$ into $H(E_i,\tilde{\alpha}_i)$ is compact too. Hence what remains to be proved is the fact that for $i = 1, 2, \ldots, \ell$ the inclusion mapping of $V(B^n \times C^r,\tilde{\alpha}_i)$ into $H(B_+^n \times C^r,\tilde{\alpha}_i)$ is compact.

Let the integer $i(i \le \ell)$ be fixed and let δ be the distance of $\phi_i^{-1}(\overline{B_+^n})$ and the complement U_i^c of U_i.
Evidently, $\delta > 0$. Denote by D the set of all points $x \in M$, which have a distance to $\phi_i^{-1}(\overline{B_+^n})$ not greater than $\frac{1}{2}\delta$. Denote by S the set of all points $x \in \phi_i^{-1}(\overline{B_+^n})$ with distance $\rho(x)$ to the boundary ∂M not greater than $\frac{1}{4}\delta$. Define $\tilde{D} = \phi_i(D)$, $\tilde{S} = \phi_i(\tilde{S})$, $\tilde{g}$ the Riemannian structure on $\overline{\mathbb{R}_+^n}$ induced by ϕ_i, $\tilde{d}$ the metric on $\overline{\mathbb{R}_+^n}$ associated with $\tilde{g}$, $\tilde{\rho}(y)$ the distance with respect to the metric $\tilde{d}$ of a point $y \in R_+^n$ to the boundary of $\mathbb{R}_+^n$.

Now by the special choice of D and S it follows that $\tilde{\rho}(y) = \rho(\phi^{-1}(y))$ for all points $y \in \tilde{S}$.

We assert that there exist constants $A > 0$ and $A' > 0$ such that $\tilde{\rho}(y) \le Ay_n$ and $y_n \le A'\tilde{\rho}(y)$ for all $y = (y_1, \ldots, y_n) \in \tilde{S}$. This fact can be proved as follows.
Let I be the canonical mapping of the tangent space $T(\overline{\mathbb{R}_+^n})$ with Riemannian structure $\tilde{g}$ onto the tangent space $T(\mathbb{R}_+^n)$ with Euclidean structure e. Moreover, for any $x \in \mathbb{R}_+^n$ let A_x be the norm of the restricyion I_x of I to $T_x(\mathbb{R}_+^n)$. Then A_x is a continuous positive function of $x \in \mathbb{R}_+^n$ and by the compactness of $\tilde{D}$ it follows that

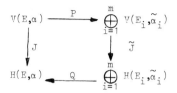

$A = \sup\{A_x | x \in \tilde{D}\}$ is finite. Denote by $\ell(c)$, $\ell'(c)$ the lengths of a C^1 curve in $\tilde{D}$ with respect to the structures $\tilde{g}$ and e respectively. Evidently we have $\ell(c) \leq \leq A\ell'(c)$. Hence for all $y \in \tilde{S}$, taking least upper bounds, it follows that $\tilde{\rho}(y) \leq \leq y_n$. Changing g and e in the reasoning made above we see that there exists also a constant $A' > 0$ such that for all $y \in \tilde{S}$ $y_n \in A'\tilde{\rho}(y)$.

Now it follows that

$$\tilde{\alpha}_i(y) = \alpha(\phi_i^{-1}(y)) = \{\rho(\phi_i^{-1}(y))\}^p = \{\tilde{\rho}(y)\}^p \leq (Ay_n)^p = A^p y_n^p$$

and

$$y_n^p \leq \{A'\tilde{\rho}(y)\}^p = (A')^p \tilde{\alpha}_i(y)$$

for all $y \in \tilde{S}$. Thus $V_0(B_+^n \times C^r, \tilde{\alpha}_i)$ and $H(B_+^n \times C^r, \tilde{\alpha}_i)$ are isomorphic to $V_0(B_+^n \times C^r, \bar{\alpha})$ and $H(B_+^n \times C^r, \bar{\alpha})$ respectively, where $\bar{\alpha}(y) = y_n^p$ for each $y = (y_1, \ldots, y_n) \in R_+^n$. Hence what remains to be proved is the fact that the embedding of $V_0(B_+^n \times C^r, \bar{\alpha})$ into $H(B_+^n \times C^r, \bar{\alpha})$ is compact. However, by assumption

$$\int_0^a y_n \frac{dy_n}{\bar{\alpha}} < \infty.$$

Thus by applying lemma 4 we complete the proof of this theorem.

5. The sesqui-linear form B_λ

In this and the following sections M will denote a fixed compact C^∞ differentiable manifold of dimension $n \geq 2$ with boundary ∂M and E will be a fixed C^∞ vector bundle over M with projection $\pi: E \to M$. M is supposed to be equipped with a strictly positive smooth measure μ and E is supposed to be provided with a Hermitian structure $h = \langle\ ,\ \rangle$. α denotes a real-valued continuous function defined on M such that $\alpha(x) > 0$ when $x \notin \partial M$, $\alpha(x) = 0$ when $x \in \partial M$ and such that its restriction to the interior $M \setminus \partial M$ is C^∞.

In this section our aim is to introduce a family of continuous sesqui-linear forms $(B_\lambda)(\lambda \in C)$ on the product space $V_0(E,\alpha) \times V_0(E,\alpha)$. These sesqui-linear forms are connected with a given degenerated differential operator θ, acting on the vector bundle E. Given a fixed norm for $V(E,\alpha)$ we shall prove that the sesqui-linear form B_λ is coercive for all complex numbers λ with real part Re λ sufficiently large.

Now let Θ be a differential operator, acting on the vector bundle E, of the form

$$(5.1) \qquad \Theta = \alpha\Phi + {}^{-\frac{1}{2}}\Psi + \Xi,$$

where

a. Φ is a strongly elliptic differential operator of the second order.

b. Ψ is a differential operator of at most order one.

c. Ξ is a differential operator of order zero or vanishes identically.

Thus Θ is a local linear mapping of $\Gamma_o(M,E)$ into $\Gamma_o(M,E)$. With the differential operator Θ we associate a family of differential operators $(\Theta_\lambda)(\lambda \in C)$ defined by

$$(5.2) \qquad \Theta_\lambda = \lambda - \Theta$$

and with the help of the family of differential operators $(\Theta_\lambda)(\lambda \in C)$ we introduce a family of sesquilinear forms $(\hat{B}_\lambda)(\lambda \in C)$ on the space $\hat{V}(E,\alpha) \times \hat{V}(E,\alpha)$, defined by

$$(5.3) \qquad \hat{B}_\lambda(s,t) = (\Theta_\lambda s,t)_H.$$

The first property of the family of sesqui-linear forms $(\hat{B}_\lambda)(\lambda \in C)$ we shall prove is stated in the following lemma

Lemma 5. The sesqui-linear form $\hat{B}_\lambda$ defined on $\hat{V}(E,\alpha) \times \hat{V}(E,\alpha)$ is continuous for any complex number λ.

Proof. Let $c = (U,\phi)$ be a chart for M such that E admits a trivialization (π,ν) over U and let $U' \subset\subset U$. Then for all $s,t \in \Gamma_o(M,E)$ it follows that

$$(5.4) \qquad \int_{U'} < \Theta_\lambda s,t > \frac{d\mu}{\alpha} = \lambda \int_{U'} < s,t > \frac{d\mu}{\alpha} - \int_{U'} < \Phi s,t > d\mu - $$
$$+ \int_{U'} < \Psi s,t > \frac{d\mu}{\alpha^{\frac{1}{2}}} - \int_{U'} < \Xi s,t > \frac{d\mu}{\alpha}$$

and we have the following equalities and estimates:

$$(5.5) \qquad |\lambda \int_{U'} < s,t > \frac{d\mu}{\alpha}| \le |\lambda| \{\int_{U'} < s,s > \frac{d\mu}{\alpha}\}^{\frac{1}{2}} \cdot \{\int_{U'} < t,t > \frac{d\mu}{\alpha}\}^{\frac{1}{2}} =$$
$$= |\lambda| \{\int_{U'} < \nu^{-1}\tilde{s}\phi, \nu^{-1}\tilde{s}\phi > \frac{d\mu}{\tilde{\alpha}\phi}\}^{\frac{1}{2}} \{\int_{U'} < \nu^{-1}\tilde{t}\phi, \nu^{-1}\tilde{t}\phi > \frac{d\mu}{\tilde{\alpha}\phi}\}^{\frac{1}{2}} =$$
$$= |\lambda| \{ \int_{\phi(U')} < \nu^{-1}\tilde{s}, \nu^{-1}\tilde{s} > \frac{\sigma dx}{\tilde{\alpha}}\}^{\frac{1}{2}} \cdot \{ \int_{\phi(U')} < \nu^{-1}\tilde{t}, \nu^{-1}\tilde{t} > \frac{\sigma dx}{\tilde{\alpha}}\}^{\frac{1}{2}}.$$

Here $\tilde{s}$ and $\tilde{t}$ denote the local expressions of the sections s and t respectively and σ denotes the density of the measure $\phi(\mu)$ with respect to the Lebesque measure on $\phi(U)$.

By the relatively compactness of U' in U and strictly positivity of σ there exists a constant $k_5 > 0$ such that (5.5) is dominated by

$$(5.6) \qquad k_5 |\lambda| \{ \int_{\phi(U')} |\tilde{s}|^2 \frac{dx}{\tilde{\alpha}}\}^{\frac{1}{2}} \{ \int_{\phi(U')} |\tilde{t}|^2 \frac{dx}{\tilde{\alpha}}\} =$$

$$= k_5|\lambda|p_{c,\nu,U'}(s) \cdot p_{c,\nu,U'}(t).$$

Using matrix notation we have for all $s,t \in \Gamma_o(M,F)$ the evaluations

$$(5.7) \qquad \int_{U'} < \Phi s,t > d\mu = \int_{U'} < \nu^{-1}(\widetilde{\Phi s})\phi, \nu^{-1}\tilde{t}\phi > d\mu =$$

$$= \int_{\phi(U')} < \nu^{-1}(\widetilde{\Phi s}), \nu^{-1}\tilde{t} > \sigma dx = \int_{\phi(U')} \bar{\tilde{t}}'h(\widetilde{\Phi s})dx,$$

here the prime denotes matrix transposition and h is a certain Hermitian matrix function.

If the local expression of Φ with respect to the chart c and the trivialization (π,ν) is given by

$$(5.8) \qquad \widetilde{\Phi s} = \sum_{i,j} a^{ij}D_iD_j\tilde{s} + \sum_j b^jD_j\tilde{s} + c\tilde{s},$$

where $a^{ij},b^j,c \in C^\infty(\phi(U),L(C^r,C^r))(i = 1, 2, \ldots, n)$ then (5.7) equals

$$(5.9) \qquad \int_{\phi(U')} \bar{\tilde{t}}'h(\sum_{i,j} a^{ij}D_iD_j\tilde{s} + \sum_j b^jD_j\tilde{s} + c\tilde{s})dx.$$

By integration by parts we see that the last expression equals

$$- \sum_{i,j} \int_{\phi(U')} D_i(\bar{\tilde{t}}'ha^{ij})D_j\tilde{s}dx + \sum_j \int_{\phi(U')} \bar{\tilde{t}}'hb^jD_j\tilde{s}dx +$$

$$+ \int_{\phi(U')} \bar{\tilde{t}}'hcsdx = \sum_{i,j} \int_{\phi(U')} (\overline{D_i\tilde{t}})'ha^{ij}D_j\tilde{s}dx +$$

$$+ \sum_j \int_{\phi(U')} \bar{\tilde{t}}'(hb^j - \sum_i D_i(ha^{ij}))D_j\tilde{s}dx + \int_{\phi(U')} \bar{\tilde{t}}'hc\tilde{s}.$$

Now h,a^{ij},b^jc are C^∞ en $\phi(U)$ and $\tilde{\alpha}^{\frac{1}{2}}$ is continuous on $\phi(U)$, in addition $U'\subset\subset U$. Hence there exist constants $k_6 > 0$, $k_7 > 0$ and $k_8 > 0$ such that

$$(5.10) \qquad \left| \int_{\phi(U')} < \nu^{-1}(\widetilde{\Phi s}), \nu^{-1}\tilde{t} > \sigma dx \right| \le k_6 \sum_{i,j} \int_{\phi(U')} |D_i\tilde{t}||D_j\tilde{s}|dx +$$

$$+ k_7 \sum_j \int_{\phi(U')} \left|\frac{\tilde{t}}{\tilde{\alpha}^{\frac{1}{2}}}\right||D_j\tilde{s}|dx + k_8 \int_{\phi(U')} \left|\frac{\tilde{t}}{\tilde{\alpha}^{\frac{1}{2}}} \frac{\tilde{s}}{\tilde{\alpha}^{\frac{1}{2}}}\right|dx \le$$

$$\le k_6 \sum_{i,j} \int_{\phi(U')} |D_i\tilde{t}|^2dx\}^{\frac{1}{2}} \cdot \{ \int_{\phi(U')} |D_j\tilde{s}|^2dx\}^{\frac{1}{2}} +$$

$$+ k_7 \sum_j \{ \int_{\phi(U')} \left|\frac{\tilde{t}}{\tilde{\alpha}}\right|^2dx\}^{\frac{1}{2}} \cdot \{ \int_{\phi(U')} |D_j\tilde{s}|^2dx\}^{\frac{1}{2}} +$$

$$+ k_8\{ \int_{\phi(U')} \frac{|\tilde{t}|^2}{\tilde{\alpha}} dx\}^{\frac{1}{2}} \cdot \{ \int_{\phi(U')} \frac{|\tilde{s}|^2}{\tilde{\alpha}} dx\}^{\frac{1}{2}} =$$

$$= k_6 \sum_i\{\int_{\phi(U')} |D_i\tilde{t}|^2dx\}^{\frac{1}{2}} \sum_j\{\int_{\phi(U')} |D_j\tilde{s}|^2dx\}^{\frac{1}{2}} +$$

$$+ k_7\{\int_{\phi(U')} |\tilde{t}|^2 \frac{dx}{\tilde{\alpha}}\}^{\frac{1}{2}} \cdot \sum_j\{\int_{\phi(U')} |D_j\tilde{s}|^2dx\}^{\frac{1}{2}} +$$

$$+ k_8\{\int_{\phi(U')} \frac{|\tilde{t}|^2}{\tilde{\alpha}} dx\}^{\frac{1}{2}} \cdot \{\int_{\phi(U')} \frac{|\tilde{s}|^2}{\tilde{\alpha}} dx\}^{\frac{1}{2}} \le$$

$$\le nk_6\{\sum_i \int_{\phi(U')} |D_i\tilde{t}|^2dx\}^{\frac{1}{2}} \cdot \{\sum_j \int_{\phi(U')} |D_j\tilde{s}|^2dx\}^{\frac{1}{2}} +$$

$$+ n^{\frac{1}{2}}k_7\{\int_{\phi(U')} |\tilde{t}|^2 \frac{dx}{\tilde{\alpha}}\}^{\frac{1}{2}} \cdot \{\sum_j \int_{\phi(U')} |D_j\tilde{s}|^2dx\}^{\frac{1}{2}} +$$

$$+ k_8\{\int_{\phi(U')} |\tilde{t}|^2 \frac{dx}{\tilde{\alpha}}\}^{\frac{1}{2}} \cdot \{\int_{\phi(U')} |\tilde{s}|^2 \frac{dx}{\tilde{\alpha}}\}^{\frac{1}{2}} \le$$

$$\le nk_6\, p_{c,\nu,U'}(t) \cdot p_{c,\nu,U'}(s) + n^{\frac{1}{2}}k_7\, p_{c,\nu,U'}(t)p_{c,\nu,U'}(s) +$$

$$+ k_8\, p_{c,\nu,U'}(t)p_{c,\nu,U'}(s) = (nk_6 + n^{\frac{1}{2}}k_7 + k_8) \cdot p_{c,\nu,U'}(t) \cdot p_{c,\nu,U'}(s).$$

Similarly, we obtain for all $s,t \in \Gamma_o(M,E)$ the evaluations

$$(5.11) \qquad \int_{U'} < \widetilde{\Psi s},t > \frac{d\mu}{\alpha^{\frac{1}{2}}} = \int_{U'} < \nu^{-1}(\widetilde{\Psi s})\phi ,\nu^{-1}\tilde{t}\phi > \frac{d\mu}{\tilde{\alpha}^{\frac{1}{2}}\phi} =$$

$$= \int_{\phi(U')} < \nu^{-1}(\tilde{\Psi}s),\nu^{-1}\tilde{t} > \sigma\frac{dx}{\tilde{\alpha}^{\frac{1}{2}}} = \int_{\phi(U')} \bar{\tilde{t}}'h(\tilde{\Psi}s)\frac{dx}{\tilde{\alpha}^{\frac{1}{2}}} .$$

If the local expression of the differential operator Ψ with respect to the chart c and the trivialization (π,ν) is given by

$$(5.12) \qquad \widetilde{\Psi s} = \sum_i d^iD_i\tilde{s} + e\tilde{s},$$

where $d^i,e \in C^\infty(\phi(U), L(C^r,C^r))$ $(i = 1, 2, \ldots, n)$ then the last expression equals

$$(5.13) \qquad \int_{\phi(U')} \bar{\tilde{t}}'h(\sum_i d^iD_i\tilde{s} + e\tilde{s})\frac{dx}{\tilde{\alpha}^{\frac{1}{2}}} = \sum_i \int_{\phi(U')} (\frac{\bar{\tilde{t}}'}{\tilde{\alpha}^{\frac{1}{2}}})hd^iD_i\tilde{s}\, dx +$$

$$+ \int_{\phi(U')} \frac{\bar{\tilde{t}}'}{\tilde{\alpha}^{\frac{1}{2}}} \tilde{\alpha}^{\frac{1}{2}}e \frac{\tilde{s}}{\tilde{\alpha}^{\frac{1}{2}}} dx.$$

Thus by the continuity of $h, d^i, e, \tilde{\alpha}^{\frac{1}{2}}$ on $\phi(U)$ and since $U' \subset\subset U$ it follows that there exist constants $k_9 > 0$ and $k_{10} > 0$ such that

$$(5.14) \quad \left|\int_{U'} \langle \Psi s, t \rangle \frac{d\mu}{\tilde{\alpha}^{\frac{1}{2}}}\right| \le k_9 \sum_i \int_{\phi(U')} \left|\frac{\tilde{t}}{\tilde{\alpha}^{\frac{1}{2}}}\right| |D_i\tilde{s}|\, dx +$$

$$+ k_{10} \int_{\phi(U')} \left|\frac{\tilde{t}}{\tilde{\alpha}^{\frac{1}{2}}}\right| \left|\frac{\tilde{s}}{\tilde{\alpha}^{\frac{1}{2}}}\right| dx \le$$

$$\le k_9 \sum_i \left\{\int_{\phi(U')} \frac{|\tilde{t}|^2}{\tilde{\alpha}}\, dx\right\}^{\frac{1}{2}} \cdot \left\{\int_{\phi(U')} |D_i\tilde{s}|^2 dx\right\}^{\frac{1}{2}} +$$

$$+ k_{10} \left\{\int_{\phi(U')} \frac{|\tilde{t}|^2}{\tilde{\alpha}}\, dx\right\}^{\frac{1}{2}} \cdot \left\{\int_{\phi(U')} \frac{|\tilde{s}|^2}{\tilde{\alpha}}\, dx\right\}^{\frac{1}{2}} \le$$

$$\le n^{\frac{1}{2}} k_9 \left\{\int_{\phi(U')} |\tilde{t}|^2 \frac{dx}{\tilde{\alpha}}\right\}^{\frac{1}{2}} \left\{\sum_i \int_{\phi(U')} |D_i\tilde{s}|^2 dx\right\}^{\frac{1}{2}} +$$

$$+ k_{10} \left\{\int_{\phi(U')} |\tilde{t}|^2 \frac{dx}{\tilde{\alpha}}\right\}^{\frac{1}{2}} \cdot \left\{\int_{\phi(U')} |\tilde{s}|^2 \frac{dx}{\tilde{\alpha}} \le\right.$$

$$\le (n^{\frac{1}{2}} k_9 + k_{10}) p_{c,\nu,U'}(s) p_{c,\nu,U'}(t).$$

Estimating the last term of the right hand side of (5.4) we obtain for some constant $k_{11} > 0$

$$(5.15) \quad \left|\int_{U'} \langle s, \Xi t \rangle \frac{d\mu}{\alpha}\right| = \left|\int_{\phi(U')} \langle \nu^{-1}s, \nu^{-1}(\widetilde{\Xi t}) \rangle \sigma \frac{dx}{\tilde{\alpha}}\right| =$$

$$= \left|\int_{\phi(U')} \bar{\tilde{s}}' h f \tilde{t} \frac{dx}{\tilde{\alpha}}\right| \le k_{11} \int_{\phi(U')} |\tilde{s}| |\tilde{t}| \frac{dx}{\tilde{\alpha}} \le$$

$$\le k_{11} \left\{\int_{\phi(U')} |\tilde{s}|^2 \frac{dx}{\tilde{\alpha}}\right\}^{\frac{1}{2}} \cdot \left\{\int_{\phi(U')} |\tilde{t}|^2 \frac{dx}{\tilde{\alpha}}\right\}^{\frac{1}{2}} \le$$

$$\le k_{11} p_{c,\nu,U'}(s) p_{c,\nu,U'}(t),$$

where $\widetilde{\Xi t} = f\tilde{t}$ and $f \in C^\infty(\phi(U), L(C^r, C^r))$.

Now choose an atlas $\mathfrak{a} = \{c_i = (U_i, \phi_i)\}$ $(i = 1, 2, \ldots, m)$ for M such that for any $i = 1, 2, \ldots, m$ E admits a trivialization (π, ν_i) over U_i and let $\{U_i'\}$ $(i = 1, 2, \ldots, m)$ be a covering of M such that $U_i' \subset\subset U_i$. Then it follows from the estimates made above that there exist constants $\{K_i\}$ $(i = 1, 2, \ldots, m)$ such that for any complex number λ

$$(5.16) \qquad |\hat{B}_\lambda(s,t)| = |\int_M < \Theta_\lambda s,t > \frac{d\mu}{\sigma}| \le \sum_i |\int_{U_i'} < \Theta_\lambda s,t > \frac{d\mu}{\sigma}| \le$$

$$\le \sum_i K_i p_{c_i,\nu_i,U_i'}(s) p_{c_i,\nu_i,U_i'}(t) \le$$

$$\le \max_{i=1,2,\ldots,m} K_i \sum_i p_{c_i,\nu_i,U_i'}(s) \cdot \sum_i p_{c_i,\nu_i,U_i'}(t).$$

Thus $\hat{B}_\lambda$ is a continuous sesqui-linear form on $\hat{V}(E,\sigma) \times \hat{V}(E,\sigma)$. This completes the proof of lemma 5.

Lemma 5 implies that for any complex number λ. $\hat{B}_\lambda$ has a unique continuous extension B_λ to $V_0(E,\sigma) \times V_0(E,\sigma)$.

The next property we want to prove about the family $(\hat{B}_\lambda)(\lambda \in \mathbb{C})$ is the following

Lemma 6. Let $|| \; ||_V$ be a fixed norm for $V_0(E,\sigma)$. Then the sesqui-linear form B_λ defined on $V_0(E,\alpha) \times V_0(E,\sigma)$ is coercive for all complex numbers λ with real part $\operatorname{Re} \lambda$ sufficiently large; i.e., there exist constants $K > 0$ and $\lambda_1 > 0$ such that

$$(5.17) \qquad \operatorname{Re} B_\lambda(s,s) \ge K||s||_V^2$$

for all λ with $\operatorname{Re} \lambda \ge \lambda_1$ and all $s \in V_0(E,\sigma)$.

Proof. Let $\mathfrak{A} = \{c_k = (U_k,\phi_k)\}(k = 1, 2, \ldots, m)$ be a fixed atlas for M such that for any $k = 1, 2, \ldots, m$ U_k admits a trivialization of E over U_k. In addition, let be given a modified C^∞ partition of unity with respect to the atlas $\mathfrak{A}$ such that $\sum_k \omega_k^2(x) = 1$ for each $x \in M$. Denote by σ_k the density of the measure $\phi_k(\mu)$ with respect to the Lebesgue measure on $\phi_k(U)$. Then dropping for a moment the index k we have the following equalities and estimates for an $s \in \Gamma_0(M,E)$

$$(5.18) \qquad -\int_M \omega^2 < \Phi s,s > d\mu = -\int_{\phi(U)} \tilde{\omega}^2 < \nu^{-1}(\widetilde{\Phi s}),\nu^{-1}\tilde{s} > \sigma d\dot{x} =$$

$$= -\int_{\phi(U)} \omega^2 \bar{\tilde{s}}' h(\widetilde{\Phi s}) dx =$$

$$= -\int_{\phi(U)} \tilde{\omega}^2 \bar{\tilde{s}}' h(\sum_{i,j} a^{ij} D_i D_j \tilde{s} + \sum_j b^j D_j \tilde{s} + c\tilde{s}) dx =$$

$$= - \sum_{i,j} \int_{\phi(U)} \tilde{\omega}^2 \bar{\tilde{s}}' h a^{ij} D_i D_j \tilde{s} - \sum_{j} \int_{\phi(U)} \tilde{\omega}^2 \bar{\tilde{s}}' h b^j D_j \tilde{s} \, dx +$$

$$- \int_{\phi(U)} \tilde{\omega}^2 \bar{\tilde{s}}' h c \tilde{s} \, dx =$$

$$= \sum_{i,j} \int_{\phi(U)} \overline{\widetilde{\omega s}}' h a^{ij} (D_i D_j \widetilde{\omega s} - (D_i D_j \tilde{\omega}) \tilde{s} - (D_j \tilde{\omega})(D_i \tilde{s}) +$$

$$- (D_i \tilde{\omega})(D_j \tilde{s}))dx - \sum_{j} \int_{\phi(U)} \tilde{\omega}^2 \bar{\tilde{s}}' h b^j D_j \tilde{s} \, dx +$$

$$- \int_{\phi(U)} \tilde{\omega}^2 \bar{\tilde{s}}' h c \tilde{s} \, dx = \sum_{i,j} \int_{\phi(U)} (\overline{D_i \widetilde{\omega s}})' h a^{ij} D_j \widetilde{\omega s} \, dx +$$

$$+ \sum_{j} \int_{\phi(U)} \overline{\widetilde{\omega s}}' (\sum_i D_i h a^{ij}) D_j \widetilde{\omega s} \, dx +$$

$$+ \sum_{j} \int_{\phi(U)} \overline{\widetilde{\omega s}}' h((\sum_i (a^{ij} + a^{ji}) D_i \tilde{\omega}) - \tilde{\omega} b^j) D_j \tilde{s} \, dx +$$

$$+ \int_{\phi(U)} \overline{\widetilde{\omega s}}' h((\sum_{i,j} a^{ij} D_i D_j \tilde{\omega}) - \tilde{\omega} c) \tilde{s} \, dx =$$

$$\sum_{i,j} \int_{\phi(U)} (\overline{D_i \widetilde{\omega s}})' h a^{ij} D_j \widetilde{\omega s} \, dx +$$

$$+ \sum_{j} \int_{\phi(U)} \overline{\widetilde{\omega s}}' ((\sum_i \tilde{\omega} D_i (h a^{ij}) + h(a^{ij} + a^{ji})(D_j \tilde{\omega})) - \tilde{\omega} h b^j) D_j \tilde{s} \, dx +$$

$$+ \int_{\phi(U)} \overline{\widetilde{\omega s}}' ((\sum_{i,j} (D_i h a^{ij})(D_j \tilde{\omega}) + h a^{ij} D_i D_j \tilde{\omega}) - \tilde{\omega} h c) \tilde{s} \, dx.$$

Since Φ is a strongly elliptic differential operator of the second order we may apply Gårding's inequality, formulated for systems, and then we obtain the following. There exist constants $k_{12} > 0$ and k_{13} such that

$$\text{Re} \sum_{i,j} \int_{\phi(U)} \overline{D_i \widetilde{\omega s}}' h a^{ij} D_j \widetilde{\omega s} dx \geq k_{12} (\int_{\phi(U)} |\widetilde{\omega s}|^2 dx + \tag{5.19}$$

$$+ \sum_{i} \int_{\phi(U)} |D_i \widetilde{\omega s}|^2 dx) - k_{13} \int_{\phi(U)} |\widetilde{\omega s}|^2 dx =$$

$$= k_{12} \sum_i \int_{\phi(U)} |D_i \tilde{\omega} s|^2 dx + (k_{12} - k_{13}) \int_{\phi(U)} |\tilde{\omega} s|^2 dx.$$

Using the inequality

$$2|x||y| \le \varepsilon|x|^2 + \frac{1}{\varepsilon}|y|^2, \tag{5.20}$$

which is valid for any $\varepsilon > 0$, we see that the expression (5.19) is not smaller than

$$\tfrac{1}{2}k_{12} \sum_i \int_{\phi(U)} |\tilde{\omega} D_i \tilde{s}|^2 dx + 3k_{12} \sum_i \int_{\phi(U)} |(D_i \tilde{\omega})\tilde{s}|^2 dx + \tag{5.21}$$

$$+ (k_{12} - k_{13}) \int_{\phi(U)} |\tilde{\omega} s|^2 dx.$$

Now $h, a^{ij}, b^j, \tilde{\omega}$ are C^∞ on $\phi(U)$ and $\tilde{\omega}$ is supported in $\phi(U)$. Thus there exists a constant k_{14} such that

$$\Big|\sum_j \int_{\phi(U)} \tilde{\omega}\bar{\tilde{s}}'((\sum_i \tilde{\omega} D_i h a^{ij} + h(a^{ij} + a^{ji})(D_j \tilde{\omega}) + \tag{5.22}$$

$$- \tilde{\omega} h b^j) D_j \tilde{s}\, dx\Big| \le k_{14} \sum_j \int_{\phi(U)} |\tilde{s}|^2 |\tilde{\omega} D_j \tilde{s}|\, dx.$$

Using the inequality (5.20) again we see that for every $\varepsilon > 0$ the last expression is estimated by

$$\frac{\varepsilon}{2} k_{14} \sum_j \int_{\phi(U)} |\tilde{\omega} D_j \tilde{s}|^2 dx + \frac{n}{2\varepsilon} k_{14} \int_{\phi(U)} |\tilde{s}|^2 dx. \tag{5.23}$$

Since $a^{ij}, h, c, \tilde{\omega}$ are C^∞ on $\phi(U)$ and $\tilde{\omega}$ is supported in $\phi(U)$ there exists a constant k_{15} such that

$$\Big| \int_{\phi(U)} \tilde{\omega}\bar{\tilde{s}}'((\sum_{i,j} (D_i h a^{ij})(D_j \tilde{\omega}) + h a^{ij} D_i D_j \tilde{\omega}) - \tilde{\omega} h c)\tilde{s} dx\Big| \le \tag{5.24}$$

$$\le k_{15} \int_{\phi(U)} |\tilde{s}|^2 dx.$$

Using similar arguments we see that there exists a constant k_{16} such that the expression (5.21) is greater than

(5.25) $$\tfrac{1}{2}k_{12} \sum_i \int_{\phi(U)} |\tilde{\omega} D_i \tilde{s}|^2 \, dx - k_{16} \int_{\phi(U)} |\tilde{s}|^2 \, dx.$$

Combining (5.18) - (5.25), (5.23) with a suitably chosen ε, and remembering that $\tilde{\alpha}$ is bounded on $\phi(U)$ we see that there exist constants $k_{17} > 0$ and k_{18} such that

(5.26) $$\mathrm{Re} - \int_M \omega^2 < \Phi s, s > d\mu \geq k_{17} \sum_i \int_{\phi(U)} |\tilde{\omega} D_i \tilde{s}|^2 dx +$$

$$- k_{18} \int_{\phi(U)} |\tilde{s}|^2 \frac{dx}{\tilde{\alpha}}.$$

For the part of the sesqui-linear form B_λ which contains the first order differential operator Ψ we obtain

(5.27) $$\int_M \omega^2 < \Psi s, s > \frac{d\mu}{\alpha^{\frac{1}{2}}} = \int_{\phi(U)} \tilde{\omega}^2 < \nu^{-1}(\tilde{\Psi s}), \nu^{-1}\tilde{s} > \sigma \frac{dx}{\alpha^{\frac{1}{2}}} =$$

$$= \int_{\phi(U)} \tilde{\omega}^2 \bar{\tilde{s}}' h(\tilde{\Psi s}) \frac{dx}{\tilde{\alpha}^{\frac{1}{2}}} = \int_{\phi(U)} \tilde{\omega}^2 \bar{\tilde{s}}' h(\sum_j d^j D_j \tilde{s} + e\tilde{s}) \frac{dx}{\tilde{\alpha}^{\frac{1}{2}}} =$$

$$= \sum_j \int_{\phi(U)} \tilde{\omega}^2 \bar{\tilde{s}}' h d^j D_j \tilde{s} \frac{dx}{\tilde{\alpha}^{\frac{1}{2}}} + \int_{\phi(U)} \tilde{\omega}^2 \bar{\tilde{s}}' h e \tilde{s} \frac{dx}{\tilde{\alpha}^{\frac{1}{2}}}.$$

Since $\tilde{\omega}$,h,e are C^∞ on $\phi(U)$ and $\tilde{\alpha}$ is bounded on $\phi(U)$ there exists a constant k_{19} such that

(5.28) $$\left|\sum_j \int_{\phi(U)} \tilde{\omega}^2 \bar{\tilde{s}}' h d^j D_j \tilde{s} \frac{dx}{\tilde{\alpha}^{\frac{1}{2}}}\right| \leq k_{19} \sum_j \int_{\phi(U)} \left|\frac{\tilde{s}}{\tilde{\alpha}^{\frac{1}{2}}}\right| |\tilde{\omega} D_j \tilde{s}| dx.$$

Using the inequality (5.20) again we see that for any $\varepsilon > 0$ the last expression is estimated by

(5.29) $$\frac{\varepsilon}{2} k_{19} \sum_j \int_{\phi(U)} |\tilde{\omega} D_j \tilde{s}|^2 dx + \frac{n}{2\varepsilon} k_{19} \int_{\phi(U)} |\tilde{s}|^2 \frac{dx}{\tilde{\alpha}}.$$

ω,h,e are C^∞ on $\phi(U)$ and $\tilde{\alpha}$ is bounded on $\phi(U)$. Thus there exists a constant k_{20} such that

(5.30) $$|\int_M \omega^2 < \Psi s,s > \frac{d\mu}{\alpha^{\frac{1}{2}}}| \le \tfrac{1}{2}k_{19} \sum_j \int_{\phi(U)} |\tilde{\omega} D_j \tilde{s}|^2 dx + k_{20} \int_{\phi(U)} |\tilde{s}|^2 \frac{dx}{\tilde{\alpha}} .$$

For the part of the sesqui-linear form B_λ which contains the differential operator Ξ we have: There exists a constant k_{21} such that .

(5.31) $$|\int_M \omega^2 < \Xi s,s > \frac{d\mu}{\alpha}| = |\int_{\phi(U)} \tilde{\omega}^2 < \nu^{-1}(\widetilde{\Xi s}),\nu^{-1}\tilde{s} > \sigma\frac{dx}{\tilde{\alpha}} =$$

$$= |\int_{\phi(U)} \tilde{\omega}^2 \bar{\tilde{s}}' h(\widetilde{\Xi s})\frac{dx}{\tilde{\alpha}}| = |\int_{\phi(U)} \tilde{\omega}^2 \bar{\tilde{s}}' hf\tilde{s} \frac{dx}{\tilde{\alpha}}| \le$$

$$\le k_{21} \int_{\phi(U)} |\tilde{s}|^2 \frac{dx}{\tilde{\alpha}} .$$

Now from (5.26), (5.30) and (5.31) it follows that there exist constants $A_k > 0$ $(k = 1, 2, \dots, m)$ and B_k $(k = 1, 2, \dots, m)$ such that

(5.32) $$\mathrm{Re} \int_M \omega_k^2 < \Theta_\lambda s,s > \frac{d\mu}{\alpha} \ge \mathrm{Re}\, \lambda \int_M \omega_k^2 < s,s > \frac{d\mu}{\alpha} +$$

$$+ A_k \sum_j \int_{\phi(U_k)} |\tilde{\omega}_k D_j \tilde{s}_k|^2 dx - B_k \int_{\phi(U_k)} |\tilde{s}|^2 \frac{dx}{\tilde{\alpha}_k} .$$

Hence we obtain

(5.33) $$\mathrm{Re}\, B_\lambda(s,s) = \mathrm{Re} \int_M < \Theta_\lambda s,s > \frac{d\mu}{\alpha} \ge \mathrm{Re}\, \lambda \int_M < s,s > \frac{d\mu}{\alpha} +$$

$$+ \sum_k A_k \sum_j \int_{\phi(U_k)} |\tilde{\omega}_k D_j \tilde{s}_k|^2 dx - \sum_k B_k \int_{\phi(U_k)} |\tilde{s}|^2 \frac{dx}{\tilde{\alpha}_k} .$$

Choose a number C such that the sets $\{U'_k\}$ $(k = 1, 2, \dots, m)$ defined by

$$U'_k = \{x \in M | \omega_k(x) > C\}$$

form a covering of M. Then there exist constants $k > 0$ and L such that

(5.34) $$\mathrm{Re}\, B_\lambda(s,s) \ge (\mathrm{Re}\, \lambda - L) \int_M < s,s > \frac{d\mu}{\alpha} +$$

$$+ K\{\sum_{k} \int_{\phi(U_k')} |\tilde{s}|^2 \frac{dx}{\tilde{\alpha}_k} + \sum_{j} \int_{\phi(U_k')} |D_j\tilde{s}_k|^2 dx\}.$$

Hence there exist constants $\lambda_1 > 0$ and $K > 0$ such that for each $s \in \Gamma_0(M,F)$ and all complex numbers λ with $\mathrm{Re}\,\lambda \geq \lambda_1$ we have

(5.35) $\quad \mathrm{Re}\, B_\lambda(s,s) \geq K||s||_V^2$.

By the continuity of B_λ and the $V_0(E,\alpha)$-norm it follows that (5.35) holds for all complex numbers λ with $\mathrm{Re}\,\lambda \geq \lambda_1$ and all $s \in V_0(E,\alpha)$. This completes the proof of lemma 6.

Summarizing lemma 5 and lemma 6 and using the theorem of Lax-Milgram we have established the following theorem.

Theorem 5. Let $\hat{B}_\lambda$ be the sesqui-linear form (5.3) defined on $\hat{V}(E,\alpha) \times \hat{V}(E,\alpha)$, where $\hat{V}(E,\alpha)$ is the space $\Gamma_0(\mathbf{M},\mathbf{E})$ provided with the relative topology of $V(E,\alpha)$. Then $\hat{B}_\lambda$ is a continuous sesqui-linear form on $\hat{V}(E,\alpha) \times \hat{V}(E,\alpha)$ and the unique extension B_λ of $\hat{B}_\lambda$ to $V_0(E,\alpha) \times V_0(E,\alpha)$ is with respect to any fixed norm $||\ ||_V$ for $V_0(E,\alpha)$ coercive for Re λ sufficiently large; i.e., there exist constants $K > 0$ and $\lambda_1 > 0$, depending on $||\ ||_V$ such that

$$\mathrm{Re}\, B_\lambda(s,s) \geq K||s||_V$$

for all complex numbers λ with $\mathrm{Re}\,\lambda \geq \lambda_1$ and all $s \in V_0(E,\alpha)$.

Moreover, for any fixed norm $||\ ||_V$ for $V_o(E,\alpha)$ there exist a family of continuous linear maps $(\mathcal{L}_\lambda)(\lambda \in C), \mathcal{L}_\lambda : V_0(E,\alpha) \to V_0(E,\alpha)$ such that

(5.36) $\quad B_\lambda(s,t) = (\mathcal{L}_\lambda s,t)_V$

for all $s,t \in V_0(E,\alpha)$ and such that $\mathcal{L}_\lambda$ is a linear isomorphism for $\mathrm{Re}\,\lambda \geq \lambda_1$.

6. Bijections associated with the differential operator Θ.

In this section the space $V_0(E,\alpha)$ is provided with a fixed norm $||\ ||_V$ and related inner product $(\ ,\ )_V$.

Consider the diagram

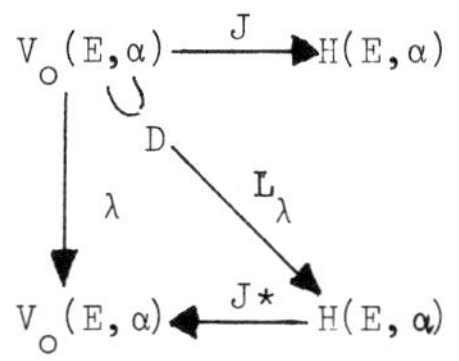

where J is the inclusion map of $V_0(E,\alpha)$ into $H(E,\alpha)$ and J^* its adjoint defined by

(6.1) $$(Js,t)_H = (s,J^*t)_V$$

for all $s \in V_0(E,\alpha)$ and all $t \in H(E,\alpha)$.

Since $\Gamma_0(M,E) \subset V_0(E,\alpha) \subset H(E,\alpha)$ and $\Gamma_0(M,E)$ is dense in $H(E,\alpha)$ it follows that J has dense range. This implies that J^* is injective. Moreover, it can easily be seen that the range $R(J^*)$ of J^* consists of all $s \in V_0(E,\alpha)$ such that the map $u \to (s,u)_V$ is continuous on $V_0(E,\alpha)$ with respect to the relative topology of $H(E,\alpha)$. Now from (5.36) and (6.1) it follows that for all complex numbers λ

(6.2) $$B_\lambda(s,t) = ((J^*)^{-1}\circ\mathcal{L}_\lambda s,t)_H$$

if $s,t \in V_0(E,\alpha)$ and $\mathcal{L}_\lambda s \in R(J^*)$. Hence (6.2) holds for the set D_λ of all $s \in V_0(E,\alpha)$ such that the map $u \to B_\lambda(s,u)$ is continuous when $V_0(E,\alpha)$ is equipped with the relative topology of $H(E,\alpha)$ and all $t \in V_0(E,\alpha)$.

Now it is a consequence of the definition of the sesqui-linear form B_λ that the set D_λ is independent of λ. So we may drop the index λ.

Define the linear map $L_\lambda = (J^*)^{-1}\circ\mathcal{L}_\lambda$ of D into $H(E,\alpha)$. From theorem 5 we know that $\mathcal{L}_\lambda$ is a linear isomorphism for all complex numbers λ with $\operatorname{Re}\lambda \geq \lambda_1$. Hence L_λ has for $\operatorname{Re}\lambda \geq \lambda_1$ as inverse the continuous linear map $G_\lambda = \mathcal{L}_\lambda^{-1}\circ J^*$ with domain $H(E,\alpha)$ and range D.

Summarizing the above discussion we have the following lemma

Lemma 7. *Let J^* be the adjoint of the inclusion map J of $V_0(E,\alpha)$ into $H(E,\alpha)$ and let D be the set consisting of all $s \in V_0(E,\alpha)$ such that the map $u \to B_\lambda(s,u)$ is continuous when $V_0(E,\alpha)$ is provided with the relative topology of $H(E,\alpha)$. Then the linear map $L_\lambda = (J^*)^{-1}\circ\mathcal{L}_\lambda$ of D into $H(E,\alpha)$ is for $\operatorname{Re}\lambda \geq \lambda_1$ a bijection of D onto $H(E,\alpha)$. Here $\mathcal{L}_\lambda$ and λ_1 are given in theorem 5. Moreover, for $\operatorname{Re}\lambda \geq \lambda_1$ the inverse G_λ of L_λ is continuous.*

We want to elucidate the structure of the operators $\{L_\lambda\}$ $(\lambda \in C)$ and to give a more concrete description of the set D. This we shall do in the following lemma

Lemma 8. *Let $L^2(E)$ be identified anti-linearly with a subset of the topological dual $\mathcal{D}(M,E)'$ of $\mathcal{D}(M,E)$ by the identification $s \to (t \to (t,s)_0)$. Extend the domain Θ and Θ_λ $(\lambda \in C)$ to $\mathcal{D}(M,E)'$ by the definitions*

(6.3) $$(\Theta s)(t) = s({}^t\Theta_\lambda t) \quad (\Theta_\lambda s)(t) = s({}^t\Theta_\lambda t)$$

for each $s \in \mathcal{D}(M,E)'$ and each $t \in \mathcal{D}(M,E)$. Here ${}^t\Theta$ and ${}^t\Theta_\lambda$ denote the formal adjoint in $L^2(E)$ of the operator Θ and Θ_λ respectively. Then

$$D = \{s \in V_o(E,\alpha) \mid \Theta s \in H(E,\alpha)\}$$

and $L_\lambda s = \Theta_\lambda s$ for each $s \in D$ and each complex number λ.

Proof. Fix a complex number λ. Recalling the definition of $\hat{B}_\lambda$ in (5.3) it follows that for all $s,u \in \Gamma_o(M,E)$

$$(6.4) \qquad \hat{B}_\lambda(s,u) = (\Theta_\lambda s,u)_H = (s,{}^t\Theta_\lambda(\tfrac{u}{\alpha}))_o.$$

Therefore since B_λ is the continuous extension of $\hat{B}_\lambda$ to $V_o(E,\alpha) \times V_o(E,\alpha)$ we have for all $s \in V_o(E,\alpha), u \in \Gamma_o(M,E)$

$$(6.5) \qquad B_\lambda(s,u) = (s,{}^t\Theta_\lambda(\tfrac{u}{\alpha}))_o = \overline{({}^t\Theta_\lambda(\tfrac{u}{\alpha}),s)_o} = \overline{s({}^t\Theta_\lambda(\tfrac{u}{\alpha}))} =$$

$$= \overline{(\Theta_\lambda s)(\tfrac{u}{\alpha})} .$$

From (6.2) we obtain that for all $s \in D$ and all $u \in \Gamma_o(E,\alpha)$

$$(6.6) \qquad B_\lambda(s,u) = (L_\lambda s,u)_H = \overline{(u,L_\lambda s)_H} = \overline{(L_\lambda s)(\tfrac{u}{\alpha})}.$$

Thus for all $s \in D$ and all $u \in \Gamma_o(M,E)$ we have

$$(\Theta_\lambda s)(\tfrac{u}{\alpha}) = (L_\lambda s)(\tfrac{u}{\alpha}).$$

Remembering that L_λ maps into $H(E,\alpha)$ this implies that for all $s \in D$ we have $L_\lambda s = \Theta_\lambda s$ and that $s = \lambda s - \Theta_\lambda s$ belongs to $H(E,\alpha)$.

Conversely, suppose that $s \in V_o(E,\alpha)$ such that $\Theta s \in H(E,\alpha)$. Then also $\Theta_\lambda s \in H(E,\alpha)$ and from equality (6.5) we obtain for all $u \in \Gamma_o(M,E)$

$$(6.7) \qquad B_\lambda(s,u) = \overline{(\Theta_\lambda s)(\tfrac{u}{\alpha})} = \overline{(\tfrac{u}{\alpha},\Theta_\lambda s)_o} = (\Theta_\lambda s,\tfrac{u}{\alpha})_o = (\Theta_\lambda s,u)_H.$$

Since $\Gamma_o(M,E)$ is dense in $V_o(E,\alpha)$ we may extend the equality between both ends of (6.7) to: For all $u \in V_o(E,\alpha)$ we have

$$B_\lambda(s,u) = (\Theta_\lambda s,u)_H.$$

Hence the map $u \to B_\lambda(s,u)$ is continuous on $V_o(E,\alpha)$ when $V_o(E,\alpha)$ is provided with the relative topology of $H(E,\alpha)$. This implies $s \in D_\lambda = D$, which completes the proof of lemma 8.

Combining lemma 7 and lemma 8 we obtain the following theorem

Theorem 6. Let $L^2(E)$ be identified conjugate linearly with a subset of the topolo-

gical dual $\mathcal{D}(M,E)'$ *of* $\mathcal{D}(M,E)$ *by the identification* $s \to (t \to (t,s)_o)$. *Extend the domain of* Θ *and* Θ_λ $(\lambda \in C)$ *to* $\mathcal{D}(M,E)'$ *by the definitions*

$$(\Theta s)(t) = s({}^t\Theta_\lambda t) \quad (\Theta_\lambda s)(t) = s({}^t\Theta_\lambda t)$$

for each $s \in \mathcal{D}(M,E)'$ *and each* $t \in \mathcal{D}(M,E)$. *Here* ${}^t\Theta$ *and* ${}^t\Theta_\lambda$ *denote the formal adjoint in* $L^2(E)$ *of the operator* Θ *and* Θ_λ *respectively. Then the restriction of* Θ_λ *to the set*

$$D = \{s \in V_o(E,\alpha) \mid \Theta s \in H(E,\alpha)\}$$

is a bijection of D onto $H(E,\alpha)$ *for all complex numbers* λ *with* $\mathrm{Re}\,\lambda$ *sufficiently large, say for* $\mathrm{Re}\,\lambda \geq \lambda_1$. *Moreover, for* $\mathrm{Re}\,\lambda \geq \lambda_1$ *the inverse* G_λ *of* $\Theta_\lambda|D$ *is continuous.*

7. Relations with semi-groups of operators

In this section we shall connect the results of the preceding sections with the analytic theory of semi-groups of operators. Our connection will be made by means of the theorem of Hille-Yosida, formulated in the version we need; for Banach spaces. See YOSIDA [5], p. 248-249.

Theorem 7. *Let X be a Banach space and let A be a linear operator with domain D(A) X and range R(A) X. Then A is an infinitesimal generator of a uniquely determined strongly continuous semi-group T of class* (C_o) *satisfying* $||T(t)|| \leq \exp(\beta t)$ *for all* $t \in [0,\infty)$, *where* $\beta \geq 0$ *if and only if the following two conditions are fulfilled.*

i) *D(A) is dense in X.*

ii) *The resolvent* $R(\lambda;A) = (\lambda I - A)^{-1}$ *exists for all real* λ *sufficiently large and satisfies*

$$||(\lambda I - A)^{-1}|| \leq (\lambda - \beta)^{-1}.$$

We shall prove the following statement

Theorem 8. *Let* Θ *be the differential operator defined in (5.1) and let L be the operator induced by* Θ *on D (see theorem 6). Then L is the infinitesimal generator of a strongly continuous semi-group of class* (C_o) *on the Hilbert space* $H(E,\alpha)$.

Proof. From the results of the preceding section it follows that the domain $D(L) = D$ is dense in $H(E,\alpha)$ and that $\lambda - L$ is a bijection of D onto $H(E,\alpha)$ for all real numbers λ with $\lambda \geq \lambda_1$. Hence it is sufficient to prove that there exists a constant β such that

$$||(\lambda - L)u||_H \geq (\lambda - \beta)||u||_H$$

for all $u \in D$ and all $\lambda \geq \lambda_1$. For when this is the case all the conditions in the theorem of Hille-Yosida are fulfilled.

Now from the estimates of the preceding section, especially (5.34) it follows that there exists a $\beta \geq 0$ such that

$$|B_\lambda(s,s)| \geq (\lambda - \beta)||s||_H^2 \tag{7.1}$$

for all $s \in V_o(E,\alpha)$ and all $\lambda \geq 0$.

From the remarks around (6.2) we obtain

$$B_\lambda(s,s) = (L_\lambda s,s)_H = ((\lambda - L)s,s)_H \tag{7.2}$$

when $s \in D$. Hence, when $s \in D$ and $\lambda \geq \lambda_1$ it follows from (7.1) and (7.2) that

$$|(L_\lambda s,s)_H| \geq (\lambda - \beta)||s||_H^2. \tag{7.3}$$

Then by Schwarz' inequality we obtain that for all $s \in D$

$$||L_\lambda s||_H||s||_H \geq (\lambda - \beta)||s||_H^2, \tag{7.4}$$

which completes the proof.

In the case of line bundles we have associating with the differential operator Θ an operator in the Hilbert space $L^2(E)$ instead of in the Hilbert space $H(E,\alpha)$ the following result

Theorem 9. Let E be a line bundle and let Θ be the differential operator defined in (5.1) and let L be the operator induced by Θ on D (see theorem 6). Moreover, let

$$d. \quad \alpha^{-\frac{1}{2}}d\alpha$$

be bouded on $M \setminus \partial M$. Here $d\alpha$ denotes the differential of α. Then we assert

i) L is closable in the Hilbert space $L^2(E)$.

ii) The smallest closed extension $\bar{L}$ of L in $L^2(E)$ is an infinitesimal generator of a strongly continuous semi-group of class (C_o) on $L^2(E)$.

Proof i. We have to prove (see e.g. YOSIDA [5], p. 77-78) that for each sequence (f_n) $(n = 1, 2, \ldots)$ in the domain $D(L)$ of L with $\lim_{n\to\infty} f_n = 0$ and $\lim_{n\to\infty} Lf_n = g$ it follows that $g = 0$.

Therefore suppose that $f_n \in D$ $(n = 1, 2, \ldots)$, $\lim_{n\to\infty} f_n = 0$ and $\lim_{n\to\infty} Lf_n = g \in L^2(E)$ in the sense of convergence in $L^2(E)$. Then for each $u \in \Gamma_0(M,E)$ we have

$$(u, Lf_n)_0 = ({}^tLu, f_n)_o, \tag{7.5}$$

where ${}^tL$ is the formal adjoint in $L^2(E)$ of L. By taking limits at both sides of equation (7.5) we obtain $(u,g)_0 = 0$ for each $u \in \Gamma_0(M,F)$. Hence $g = 0$.

Proof ii. Analogous to the proof of theorem 8 it suffices to show that there exists a constant β' such that

$$||(\lambda - L)s||_o \geq (\lambda - \beta')||s||_o \tag{7.6}$$

for all $s \in D$ and all $\lambda \geq \lambda_1$. For in this case it follows from the facts, $H(E,\sigma)$ is dense in $L^2(E)$, $\lambda - \bar{L} = \overline{\lambda - L}$ and the continuity of the norm $|| \; ||_0$, that for any $\lambda \geq \lambda_1$ $\lambda - \bar{L}$ is a bijection of $D(\bar{L})$ onto $L^2(E)$ satisfying the inequality

$$||(\lambda - \bar{L})s||_0 \geq (\lambda - \beta')||s||_0$$

for all $s \in D(\bar{L})$. Now the part of the proof which remains to be proved proceeds in practically the same direction as that of lemma 6 in section 5. So we use the same notation as in that proof. Then we have the following equality for any $s \in \Gamma_0(M,E)$ (compare 5.18)

$$\begin{aligned} -\int_M \omega^2 < \alpha\Phi s, s > d\mu = {} & \sum_{i,j} \int_{\phi(U)} \overline{\tilde{\alpha} D_i \tilde{\omega s}}\, h a^{ij} D_j \tilde{\omega s}\, dx \; + \\ & + \sum_j \int_{\phi(U)} \bar{\tilde{\omega s}}((\sum_i \tilde{\omega} D_i (h\tilde{\alpha} a^{ij}) + h\tilde{\alpha}(a^{ij} + a^{ji})(D_j\tilde{\omega})) - \tilde{\omega} h \tilde{\alpha} b^j) D_j \tilde{s}\, dx \; + \\ & + \int_{\phi(U)} \bar{\tilde{\omega s}}((\sum_{i,j} (D_i h\tilde{\alpha} a^{ij})(D_j \tilde{\omega}) + h\tilde{\alpha} a^{ij} D_i D_j \tilde{\omega}) - \tilde{\omega} h \tilde{\alpha} c) \tilde{s}\, dx. \end{aligned} \tag{7.7}$$

Now Φ is a uniformly strongly elliptic differential operator of the second order acting on a line bundle. Hence there exists a constant $k_{22} > 0$ such that for the first term of the right hand side of (7.7) we have

$$\operatorname{Re} \sum_{i,j} \int_{\phi(U)} \overline{\tilde{\alpha} D_i \tilde{\omega s}}\, h a^{ij} D_j \tilde{\omega s}\, dx \geq k_{22} \sum_i \int_{\phi(U)} |D_i \tilde{\omega s}|^2 \tilde{\alpha}\, dx. \tag{7.8}$$

Using inequality (5.19) we see that the right hand side of (7.8) is not smaller than

$$(7.9)\qquad \tfrac{1}{2}k_{22}\sum_i \int_{\phi(U)} |\tilde{\omega}D_i\tilde{s}|^2\tilde{\alpha}\,dx + 3k_{22}\sum_i \int_{\phi(U)} |(D_i\tilde{\omega})\tilde{s}|^2 dx.$$

By the assumption d. it follows that there exists a constant k_{23} such that the second term of the right hand side of (7.7) is estimated by

$$(7.10)\qquad k_{23}\sum_j \int_{\phi(U)} |\tilde{s}||\tilde{\alpha}^{\frac{1}{2}}\tilde{\omega}D_j\tilde{s}|dx$$

and using inequality (5.20) again it follows that for any $\varepsilon > 0$ (7.10) is dominated by

$$(7.11)\qquad \frac{\varepsilon}{2}k_{23}\sum_j \int_{\phi(U)} \tilde{\alpha}\tilde{\omega}^2|D_j\tilde{s}|^2 dx + \frac{n}{2\varepsilon}k_{23}\int_{\phi(U)} |\tilde{s}|^2 dx.$$

Also there exists a constant k_{24} such that the third term of (7.7) is estimated by

$$(7.12)\qquad k_{24}\int_{\phi(U)} |\tilde{s}|^2\, dx.$$

Combining (7.7)-(7.12), (7.10) with a suitably chosen ε, yields

$$(7.13)\qquad \mathrm{Re} - \int_M \omega^2 < \alpha\Phi s,s > d\mu \geq k_{25}\sum_i \int_{\phi(U)} |\tilde{\omega}D_i\tilde{s}|^2\tilde{\alpha}dx +$$

$$- k_{26}\int_{\phi(U)} |\tilde{s}|^2\, dx$$

for any $s \in \Gamma_o(M,E)$. Here k_{25} and k_{26} are constants such that $k_{25} > 0$.

Similar to equality (5.27) we have for any $s \in \Gamma_o(M,E)$

$$(7.14)\qquad \int_M \omega^2 < \alpha^{\frac{1}{2}}\Psi s,s > d\mu =$$

$$= \sum_j \int_{\phi(U)} \tilde{\omega}^2\bar{\tilde{s}}h\tilde{\alpha}^{\frac{1}{2}}d^jD_j\tilde{s}\, dx + \int_{\phi(U)} \tilde{\omega}^2\bar{\tilde{s}}h\tilde{\alpha}^{\frac{1}{2}}\tilde{e}\tilde{s}dx$$

and similar to (5.30) we see that there exist constants k_{26} and k_{27} such that for any $s \in \Gamma_0(M,E)$

$$(7.15) \qquad \left|\int_M \omega^2 < \alpha^{\frac{1}{2}}\Psi s,s > d\mu\right| \le k_{26} \sum_j \int_{\phi(U)} |\tilde{\omega} D_j \tilde{s}|^2 \tilde{\alpha} dx + k_{27} \int_{\phi(U)} |\tilde{s}|^2 dx.$$

Obviously, there exists a constant k_{28} such that for any $s \in \Gamma_0(M,E)$ we have

$$(7.16) \qquad \left|\int_M \omega^2 < \Xi s,s > d\mu\right| \le k_{28} \int_{\phi(U)} |\tilde{s}|^2 dx.$$

Now from (7.13), (7.15) and (7.16) it follows that there exist constants $E_k > 0$ ($k = 1, 2, \ldots, m$) and F_k ($k = 1, 2, \ldots, m$) such that

$$(7.17) \qquad \operatorname{Re} \int_M \omega_k^2 < \Theta_\lambda s,s > d\mu \ge \operatorname{Re} \lambda \int_M \omega_k^2 < s,s > d\mu +$$

$$+ A_k \sum_j \int_{\phi(U_k)} |\tilde{\omega}_k D_j \tilde{s}_k|^2 \tilde{\alpha} dx - B_k \int_{\phi(U_k)} |\tilde{s}|^2 dx$$

for any $s \in \Gamma_0(M,E)$. This yields

$$(7.18) \qquad \operatorname{Re} (\Theta_\lambda s,s)_0 = \operatorname{Re} \int_M < \Theta_\lambda s,s > d\mu \ge$$

$$\ge \operatorname{Re} \lambda ||s||_0^2 + \sum_k E_k \sum_j \int_{\phi(U_k)} |\tilde{\omega}_k D_j \tilde{s}_k|^2 \tilde{\alpha} dx - \sum_k F_k \int_{\phi(U_k)} |\tilde{s}|^2 dx.$$

Now let C be a number such that the sets $\{U_k'\}$ ($k = 1, 2, \ldots, m$) defined by

$$U_k' = \{x \in M | \omega_k(x) > C\}$$

form a covering of M. Then from (7.18) it follows that there exists a constant β' such that

$$(7.19) \qquad \operatorname{Re} (L_\lambda s,s)_0 = \operatorname{Re} (\Theta_\lambda s,s)_0 \ge (\lambda - \beta')||s||_0^2$$

for any $s \in \Gamma_0(M,E)$ and $\lambda \ge 0$. Because

$$(L_\lambda s,s)_0 = B_\lambda(s, \alpha s)$$

for all $U \in D$ and since elements of D can be approximated with respect to the topology of $V_o(E,\alpha)$ by functions belonging to $\Gamma_o(M,E)$ we see by taking limits at both sides of (7.19) that

$$(7.20) \qquad \mathrm{Re}\,(L_\lambda s,s) \geq (\lambda - \beta')||s||_o^2$$

for all $U \in D$ and all $\lambda \geq 0$. This completes the proof of theorem 9.

8. A method for the construction of semi-groups.

We recall that in section 4 we have investigated the embedding of $V_o(E,\alpha)$ into $H(E,\alpha)$. The question whether this embedding is compact was our main interest and a positive result concerning this point was stated in theorems 2 and 4.

In order to give a method for the construction of semi-groups we suppose that this indeed is the case. Thus we suppose the embedding J of $V_o(E,\alpha)$ into $H(E,\alpha)$ to be compact. In section 5 we proved that the operator $L_\lambda = \Theta_\lambda$ defines a bijection of the set $D = \{s \in V_o(E,\alpha) | \Theta s \in H(E,\alpha)\}$ onto $H(E,\alpha)$ for $\mathrm{Re}\,\lambda$ sufficiently large, let say for $\mathrm{Re}\,\lambda \geq \lambda_1$. By G_λ we have denoted the continuous inverse of L_λ for $\mathrm{Re}\,\lambda \geq \lambda_1$.

Now consider the following diagram

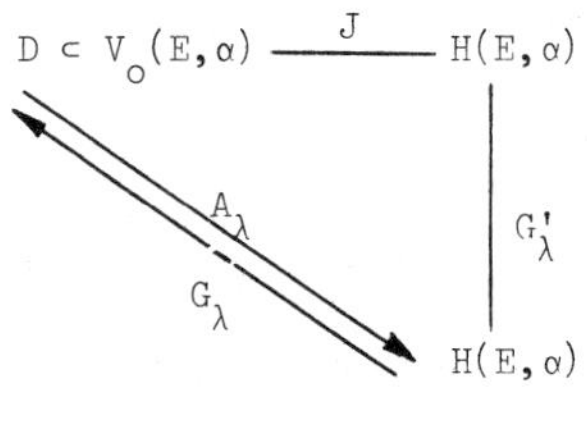

It follows that the operator $G'_\lambda = J \circ G_\lambda$, $G'_\lambda : H(E,\alpha) \to H(E,\alpha)$, is compact as a composition of a compact linear and a continuous linear operator. In addition if the operator Θ is symmetric on D in $H(E,\alpha)$ with respect to some norm $|| \; ||_H$; i.e.,

$$(8.1) \qquad (\Theta s,t)_H = (s,\Theta t)_H$$

for all $s,t \in D$, it follows that for any real number λ with $\lambda \geq \lambda_1$ the operator G'_λ is selfadjoint. Thus in this case it is possible to apply the spectral theory for compact selfadjoint operators to G'_λ. Then G'_λ has the representation

$$(8.2) \qquad G'_\lambda s = \sum_k \mu_k^\lambda \sum_{\ell \in I_k} (s,s_{k\ell})_H s_{k\ell}$$

for any $s \in H(E,\alpha)$, where $\{\mu_k^\lambda\}$ is the set of eigenvalues of G_λ' and where $\{s_{k\ell}\}$ $(\ell \in I_k)$ is an orthonormal set of eigenvectors belonging to the eigenvalue μ_k^λ.

Now G_λ' coincides with the resolvent $R(\lambda;L)$ and between the semi-group $\{T(t)\}$ $(t \geq 0)$ corresponding to the infinitesimal generator L and the resolvent $R(\lambda;L)$ there exists the relation (see (2.5) in chapter I)

$$(8.3) \qquad R(\lambda;L)f = \int_0^\infty e^{-\lambda t} T(t)f\, dt.$$

Hence

$$(8.4) \qquad G_\lambda f = \int_0^\infty e^{-\lambda t} T(t)f\, dt.$$

So by inversion of (8.4) we obtain a representation for the semi-group $\{T(t)\}$ $(t \geq 0)$ corresponding to the infinitesimal generator L.

9. Examples.

In this section we shall give some examples to illustrate the results of the preceding sections.

Example 1. Let M be the unit-ball B^n in the n-dimensional Euclidean space $\mathbb{R}^n$ and let E be the trivial complex bundle $B^n \times C^r$. Define the differential operator $\Theta : \Gamma_0(M,E) \to \Gamma_0(M,E)$ by

$$(\Theta s)(x) = (x,(1 - |x|^2)^p \Delta \hat{s}(x))$$

if $s(x) = (x,\hat{s}(x))$ and where Δ denotes the Laplacean $\sum_i D_i^2$ and where $|x|^2 = \sum_i x_i^2$. Then Θ satisfies the assumptions of theorem 8 in case $p > 0$. Moreover, take $\alpha(x) = (1 - |x|^2)^p$ and let $\rho(x)$ be the distance of a point $x \in M$ to the boundary of M. Then since for each $x \in M$ we have

$$\rho^p(x) \leq \alpha(x) \leq 2^p \rho^p(x)$$

it follows from theorem 4 that the embedding of $V_0(E,\alpha)$ into $H(E,\alpha)$ is compact provided that $p < 2$.

Also using Green's formula it can easily be seen that the extension of Θ defined in theorem 6 is symmetric on D. Thus in case $0 < p < 2$ the construction method of section 8 can be applied.

Example 2. Let M be the cylinder $S^1 \times I$ embedded canonically in the 3-dimensional Euclidean space $\mathbb{R}^3$. Here S^1 is the 1-dimensional unit-sphere and $I = [0,1]$. Let E be the trivial complex vector bundle over M.

Define the linear differential operator $\Theta : \Gamma_0(M,E) \to \Gamma_0(M,E)$ by

$$(\Theta s)(x) = (x, \alpha(x)\Delta\hat{s}(x))$$

if $s(x) = (x,\hat{s}(x))$ and where in cylinder-coordinates $\alpha(x) = x_3^p(1 - x_3)^q$ and where Δ is the Laplacean given in the same coordinates by

$$\frac{\partial^2}{\partial\phi^2} + \frac{\partial^2}{\partial x_3^2} .$$

Then Θ satisfies the conditions of theorem 7 in case $p > 0$ and $q > 0$ and the conditions of theorem 8 in case $p \geq 2$ and $q \geq 2$. Evidently, we can define a similar differential operator on the cylinder $S^n \times I$ embedded in the (n+2)-dimensional Euclidean space $\mathbb{R}^{n+2}$. Of course, S^n denotes the n-dimensional unit-sphere.

Example 3. Let M be an oriented n-dimensional Riemannian compact manifold with boundary ∂M and let

$$E = \Lambda(T^*(M))_C = \bigoplus_{i=1}^{n} \Lambda^i(T^*(M))_C = \bigoplus_{i=1}^{n} \Lambda^i(T^*(M)) \otimes C$$

be the vector bundle of complex-valued differential forms on M.

As usual, $d : \Gamma(M,E) \to \Gamma(M,E)$ denotes exterior derivation.

Let $*$ be the real linear automorphism of E such that if $e_1, e_2, \ldots, e_n$ is an oriented orthonormal basis for $T_x^*(M)$ and σ a permutation of the integers $1, 2, \ldots, n$ it follows that

$$*e_1 \wedge e_2 \wedge \ldots \wedge e_n = 1;$$

$$*1 = e_1 \wedge e_2 \wedge \ldots \wedge e_n;$$

$$*e_{\sigma(1)} \wedge e_{\sigma(2)} \wedge \ldots \wedge e_{\sigma(p)} = e_{\sigma(p+1)} \wedge \ldots \wedge e_{\sigma(n)}$$

if σ is even and

$$*e_{\sigma(1)} \wedge e_{\sigma(2)} \wedge \ldots \wedge e_{\sigma(p)} = -e_{\sigma(p+1)} \wedge \ldots \wedge e_{\sigma(n)}$$

if σ is odd. Obviously, $*$ maps $\Lambda^p(T^*(M))_C$ into $\Lambda^{n-p}(T^*(M))_C$ and $** = \bar{w}$, where $\bar{w}$ is the automorphism of E which on $\Lambda^p(T^*(M))_C$ is multiplication by $(-1)^{p(n-p)}$.
It follows that the operation $*$ induces a conjugate linear real automorphism of $\Gamma(M,E)$, which also will be denoted by $*$.

The Riemannian structure for M defines a strictly positive smooth measure μ on M. If the notation for the Riemannian structure is $(\ ,\ )_M$, then this measure has with respect to the chart $c = (U,\phi)$ a local density σ which is given by

$$\sigma(\phi(x)) = [\det g_{ij}(x)]^{\frac{1}{2}},$$

where

$$g_{ij}(x) = (d_x\phi_i, d_x\phi_j)_M.$$

It also induces a Hermitian structure $<,>_E$ for E. This we shall describe now. Let $e_1, e_2, \ldots, e_n$ be an oriented orthonormal basis for $T_x^*(M)$. Then the Hermitian structure $<,>_E$ is defined so that the set

$$\{e_{i_1} \wedge e_{i_2} \wedge \ldots \wedge e_{i_p}\} \ (1 \le p \le n,\ 1 \le i_1 < i_2 < \ldots < i_p \le n)$$

form an orthonormal basis for $\Lambda(T_x^*(M))$.

Let $\Gamma_o(M,E)$ be provided with the inner product, given by

$$(s,t)_o = \int_M < s,t > d\mu.$$

Then, if with respect to this inner product the formal adjoint of the exterior derivation d is denoted by ${}^t d$, it follows that

$${}^t d = \bar{w} * d * w,$$

where w is the automorphism of E which is multiplication by $(-1)^p$ on $\Lambda^p(T^*(M))_C$ (see PALAIS [4], chp. IV, p. 76-77).

Now let α be a continuous real-valued function defined on M such that the restriction of α to the interior $M \setminus \partial M$ is C^∞ and positive and such that $\alpha(x) = 0$ when $x \in \partial M$. Then for the differential operator

$$\alpha d : \Gamma_o(M,E) \to \Gamma_o(M,E)$$

it follows that its formal adjoint satisfies

$${}^t(\alpha d) = \alpha\bar{w} * d * w + \bar{w} * (d\alpha) \wedge \ * w.$$

Hence some calculations yield

$$\alpha d\,{}^t(\alpha d) = \alpha^2 d\bar{w} * d * w + \alpha\{(d\alpha) \wedge \bar{w} * d * w + d\bar{w} * (d\alpha) \wedge \ * w\}$$

and

$$({}^t(\alpha d))\alpha d = \alpha^2\bar{w} * d * wd + 2\alpha\{\bar{w} * (d\alpha) \wedge \ * wd\}.$$

So if the Laplacean Δ is defined by

$$\Delta = - \{d({}^{t}d) + ({}^{t}d)d\}$$

it follows from the equality

$$\Delta = - \{d\bar{w} \star d \star w + \dot{\bar{w}} \star d \star w\}$$

that the differential operator $\Theta : \Gamma_o(M,E) \to \Gamma_o(M,E)$ defined by

$$\Theta = - \{\alpha d\,{}^{t}(\alpha d) + ({}^{t}(\alpha d))\alpha d\}$$

equals

$$\Theta = \alpha^2\Delta + \alpha\Psi,$$

where Ψ is the differential operator of the first order given by

$$\Psi = - \{(d\alpha) \wedge \bar{w} \star d \star w + d\bar{w} \star (d\alpha) \wedge \star w + 2\bar{w} \star (d\alpha) \wedge \star wd\}$$

Since Δ is strongly elliptic the discussion above shows that the theory developed in the preceding section can be applied to the differential operator Θ just defined.

REFERENCES

[1] DUNFORD, N. and J.T. SCHWARTZ, Linear Operators, Vol. 1, Interscience Publishers (1958).

[2] MARTINI, R.,Differential Operators degenerating at the Boundary as Infinitesimal Generators of Semi-groups, Thesis, Delft (1975).

[3] NARASIMHAN, R., Analysis on Real and Complex Manifolds, North-Holland Publ. Comp. (1968).

[4] PALAIS, R., Seminar on the Atiyah-Singer Index Theorem, Ann. of Math., Study 57, Princeton (1965).

[4] YOSIDA, K., Functional Analysis, Grundlehren d. Math. Wiss., Springer-Verlag (1968).